5G Mobile Terminal Multi-Antenna Technology

In response to the continuous innovation in the field driven by the rapid development of mobile communication, this book systematically explores the technologies and applications of mobile terminal antennas.

It presents the fundamental theories, design methodologies, and physical implementations of mobile terminal antennas across a wide spectrum of configurations. This book thoroughly examines single-, dual-, and multi-antenna configurations for 5G mobile terminals, as well as single-, dual-, and multi-band designs. It also covers fixed single-beam and reconfigurable multi-beam technologies. The content is enhanced by numerous practical design examples featuring broadband miniaturized antennas, sub-6 GHz antennas, and millimeter-wave phased array antennas. These examples provide the theoretical foundations and technical references that are essential for designing advanced 5G mobile terminal antennas.

This book will be an invaluable resource for R&D engineers, university students, and researchers in electronic and communication engineering specializing in mobile terminal antennas. Professionals in the mobile terminal antenna industry will find it to be an authoritative training manual and reference guide.

Changjiang Deng is an Associate Professor and a Ph.D. Supervisor at the School of Integrated Circuits and Electronics, Beijing Institute of Technology. He is a Senior Member of IEEE, a Senior Member of the Chinese Institute of Electronics, and a committee member of the Antenna and RF Technology Society of the China Institute of Communications. His research interests include mobile phone antennas, base station antennas, intelligent reflecting surfaces (IRS), and millimeter-wave/terahertz antenna technologies.

Zhenghe Feng is a Professor in the Department of Electronic Engineering at Tsinghua University. He previously served as the Director of the Department of Electronic Engineering at Tsinghua University, Deputy Director of the State Key Laboratory of Microwave and Digital Communications, Executive Director of the Chinese Institute of Electronics, and Chairman of its Microwave Society. He is a Fellow of both the Chinese Institute of Electronics and IEEE. His research focuses on wireless communication, RF and microwave technology, electromagnetic field theory and numerical methods, smart antennas, and spatiotemporal signal processing.

5G Mobile Terminal Multi-Antenna Technology

Changjiang Deng and Zhenghe Feng

CRC Press
Taylor & Francis Group
Boca Raton London New York

CRC Press is an imprint of the
Taylor & Francis Group, an **informa** business

人民邮电出版社
POSTS & TELECOM PRESS

First edition published 2026
by CRC Press
2385 NW Executive Center Drive, Suite 320, Boca Raton FL 33431

and by CRC Press
4 Park Square, Milton Park, Abingdon, Oxon, OX14 4RN

CRC Press is an imprint of Taylor & Francis Group, LLC

Translated by DeepL

ISBN: 978-1-041-25797-4 (hbk)
ISBN: 978-1-041-29540-2 (pbk)
ISBN: 978-1-003-77003-9 (ebk)

DOI: 10.1201/9781003770039

Typeset in Minion
by codeMantra

Contents

Preface to the Chinese Edition

Mobile terminals represented by cell phones are the main carriers in the mobile Internet era. From 1G to 5G Mobile Terminals, the shape of antennas of mobile terminals has changed dramatically, from external to built-in, from thick to thin, and from single antenna to multi-antenna, and the ground clearance size used for antenna design decreases with the increase in the screen-to-body ratio. In addition, the function of antennas is getting more and more powerful, and wide-band miniaturized antennas, sub-6 GHz antennas, and millimeter-wave phased array antennas have become the research hotspots one after another. The thin, light, and miniaturized trend of 5G Mobile Terminals equipment puts forward higher requirements on antenna size, precision, and other aspects, which brings a great challenge to the antenna design of mobile terminals and pushes the antenna technology to continuously innovate and break through.

5G Mobile Terminals is a complex antenna microsystem. The number of antennas laid out inside the terminal is more than 20; how to lay out, how to realize miniaturized design, how to realize decoupling, and how to regulate the beam are the problems that need to be solved in the antenna design process. Pattern analysis provides a window to visualize the working principle of antennas, and multi-antenna synergy is an important method to improve antenna performance. Taking patterns as an entry point, this book analyzes in detail the patterns of single-antenna, dual-antenna, multiple-antenna, and millimeter-wave phased-array antennas and illustrates the application of multi-antenna synergy in bandwidth expansion, MIMO antenna decoupling, and array beam scanning through specific design examples.

Chapter 1 of this book introduces the development of mobile terminal antennas; presents the evolution of antennas in terms of changes in antenna structure, frequency bands, and number; and analyzes the typical working mode of multi-antenna systems by taking commercial cell phone products as examples. Chapter 2 describes the basic theory of mobile terminal antennas and introduces basic antenna parameters, antenna patterns, typical mobile terminal antenna forms, MIMO antenna decoupling principle, and phased-array antenna beam scanning principle, which provides theoretical support for the subsequent design. Chapter 3 introduces the content of single antenna design for 5G Mobile Terminals, specifically describes the Smith Chart diagram and impedance matching method, and introduces a variety of single antenna design examples. Chapter 4 introduces the content of dual antenna design for 5G Mobile Terminals, and for the decoupling demand of sub-3 GHz band, the dual antenna design examples of 900 and 1800 MHz bands based on the feature module are introduced. Chapter 5 introduces the content of multi-antenna design for 5G Mobile Terminals, and for the decoupling demand of sub-6 GHz band, design examples of highly isolated antenna pairs based on odd/even mode theory, antenna pairs covering 3300–6000 MHz band, a multi-antenna module with four cells aggregation, and antenna pairs covering 3.5/4.9 GHz dual bands are introduced. Chapter 6 introduces the millimeter-wave phased array antenna design for 5G Mobile Terminals; proposes the design of small ground clearance, single-layer dielectric substrate for dual-polarized end-fire phased array antennas in response to beam scanning requirements; explores the potentials of dual-polarized patch phased array antennas, dual-polarized dielectric rod phased array antennas, and series-fed beam scanning antennas for end-fire antenna scanning; and analyzes sub-6 GHz antenna and millimeter-wave antenna integration designs.

Most of the design examples in this book come from the authors' many years of research accumulation, and the strong support from many domestic and foreign colleagues during the writing process is hereby expressed with sincere thanks! Special thanks to Prof. Sean Victor Hum of the University of Toronto, Canada, and Prof. Kamal Sarabandi of the University of Michigan, USA, for their guidance and help during the writing of this book. We are grateful to graduate students Liu Di and Zhu Yuqing for their hard work in organizing and proofreading this book. We thank Guo Jia, editor of the People's Posts and Telecommunications Publishing House, for her guidance and assistance during the publication of this book.

This book has been written with valuable advice from antenna engineers from Huawei, Honor, Xiaomi, OPPO, vivo, and other leading companies in the industry, but there is still room for exploration to capture the deeper perspectives of the industrial world, and some of the technical descriptions are still some distance away from practical engineering applications. Due to the authors' limited level and experience, this book inevitably has deficiencies, and we urge experts and readers to criticize and correct us.

Changjiang Deng

CHAPTER 1

Overview of Mobile Terminal Antennas

THIS CHAPTER FIRST INTRODUCES the development history of mobile communication from 1G to 5G; then focuses on the evolution process of mobile terminal antenna, including the changes of antenna structure, antenna band, and antenna number; gives a typical multi-antenna layout scheme for 5G Mobile Terminals; and systematically summarizes the typical working mode of a multi-antenna system in cell phones. Finally, it introduces several cell phone antenna product examples.

1.1 DEVELOPMENT OF MOBILE COMMUNICATIONS

As a precious resource independent of spatial and temporal dimensions, electromagnetic spectrum resources have realized extensive development and application in the last 100 years. Maxwell's system of Maxwell's equations, derived by Maxwell in 1864, laid the foundation of electromagnetic field theory. Hertz first experimentally confirmed the existence of electromagnetic waves in 1888, and Marconi's transatlantic communication in 1901 opened up a new era of communication [1,2]. Since then, radio technology has evolved rapidly and has given rise to a variety of applications such as communication, navigation, remote sensing, and so on. These applications have greatly changed the way people communicate and communicate, enhanced their ability to cognize and transform nature, and changed the production and lifestyle of the whole society. In the process of building an information society, radio technology will still provide an important boost to the national economy and social development [3].

In general, there are three major uses of electromagnetic waves: first, as sensors for signal detection, such as radar; second, as information carriers for transmitting data, such as wireless communication; and third, as energy carriers for transmitting energy, such as wireless charging. Among the many applications, wireless communication is undoubtedly one of the brightest pearls. At present, wireless communication has derived many proprietary application scenarios, such as mobile communication and satellite communication.

DOI: 10.1201/9781003770039-1

Different application scenarios have different requirements and technical characteristics. For example, in a satellite communication system, one side of the communication is the satellite in orbit, and the other side is the ground end-user. Although the application scenarios are different, they can all be simplified into a point-to-point communication model. For different application scenarios, the transmitter and receiver devices need to be purposefully designed, and the quality of the wireless channel between the transceiver and the receiver needs to be evaluated.

1.1.1 Characteristics of Mobile Communications

Mobile communication is the main application realization of wireless communication. A typical application scenario of mobile communication is shown in Figure 1.1. The mobile communication system includes three parts: base station, mobile terminal, and wireless channel. Each base station is responsible for the signal coverage of a certain area, and adjacent base stations realize seamless service through frequency multiplexing. Mobile terminals include portable devices such as cell phones and tablet computers, among which cell phones are most widely used. When a mobile terminal communicates with a base station, a propagation channel of electromagnetic waves, i.e., a wireless channel, is established between the two. The characteristics of these three parts are summarized as follows.

Base Station: The location is immovable and is used to provide stable wireless signal coverage. According to the different coverage distances, it can be categorized into macro base stations, micro base stations, repeaters, and indoor distribution systems. Since the location of the base station is fixed, the requirements for cost and power consumption can be appropriately relaxed, focusing on the realization of high performance. A base station can serve multiple end-users at the same time through time-division multiplexing, frequency-division multiplexing, or code-division multiplexing.

Mobile Terminal: The position can be moved arbitrarily for accessing the mobile communication network, emphasizing mobility and portability, with strict restrictions on cost and power consumption, and strong constraints on size and weight. Within the coverage area of the base station, the mobile terminal establishes a stable connection with a single base station. At the edge of the base station coverage area, the mobile terminal can receive

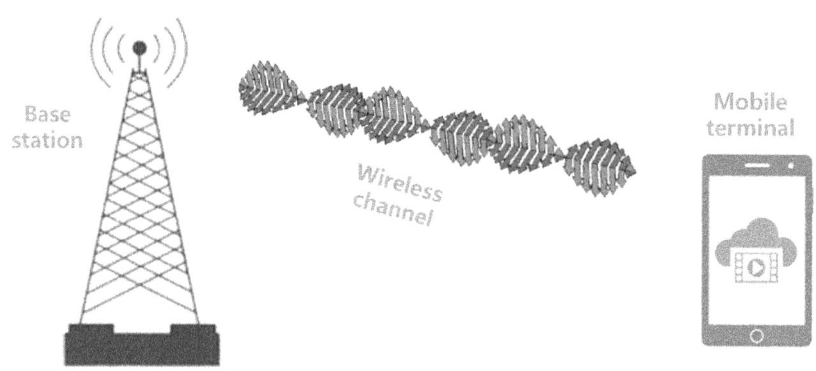

FIGURE 1.1 Typical application scenarios for mobile communications.

TABLE 1.1 Equipment Characteristics of Mobile Communication System

Equipment Type	Mobility	Power Consumption	Size	Performance
Base station	None	High	Unconstrained	High
Mobile terminal	High	Low	Restricted	Low

signals sent by multiple base stations at the same time, and generally realizes cross-area service switching through soft switching.

Wireless Channel: Because the position of the mobile terminal can be moved arbitrarily, the wireless channel of the mobile communication system is more complicated. Electromagnetic wave propagation is generally divided into line-of-sight propagation and non-line-of-sight propagation. Line-of-sight propagation channel is relatively simple, similar to the electromagnetic wave propagation channel in free space, and only needs to consider atmospheric attenuation. Non-line-of-sight propagation channels include reflection paths, refraction paths, diffraction paths, etc. Electromagnetic waves traveling through multiple paths to reach the receiver will produce a signal superposition or offset. Wireless channels are generally characterized in terms of time-varying and frequency selectivity.

Since the channel propagation environment is generally time-varying and difficult to predict, electromagnetic waves in a wireless channel are usually physical quantities that cannot be regulated. Reconfigurable intelligent hypersurface technology provides a novel paradigm to actively change the channel, which has great potential to be applied to 6G in the future, but how to fully utilize its potential still needs to be further investigated. In the vast majority of cases, the physical quantities that can be regulated are the electromagnetic waves on the base station side and the mobile terminal side. In order to efficiently generate and receive electromagnetic waves, while taking into account the characteristics of mobile communications, mobile communication systems put forward different requirements for base stations and mobile terminals, and the specific equipment characteristics are shown in Table 1.1. In order to improve the portability of the mobile terminal, the mobile terminal antenna performance is chosen to be sacrificed, while the space of the base station is more abundant, and the base station antenna performance can be chosen to be upgraded to make up for the loss caused by the decline in the antenna performance of the mobile terminal, thus ensuring that a stable wireless channel can be established between the base station and the mobile terminal.

1.1.2 Standard Division of Mobile Communication

The mission of mobile communication is to provide people with good electromagnetic signal coverage on a global scale. In order to enable mobile end-users all over the world to have sensorless access to mobile communication networks, a unified and common global standard is indispensable. This standard specifies parameters such as signaling system, operating frequency, modulation method, and so on. As long as this standard is complied with, terminals of different brands can access the same mobile network. This work is currently organized and coordinated by the International Telecommunication Union (ITU).

Since the 1980s, when mobile communications became commercially available on a large scale, mobile communications have undergone five generations of standards

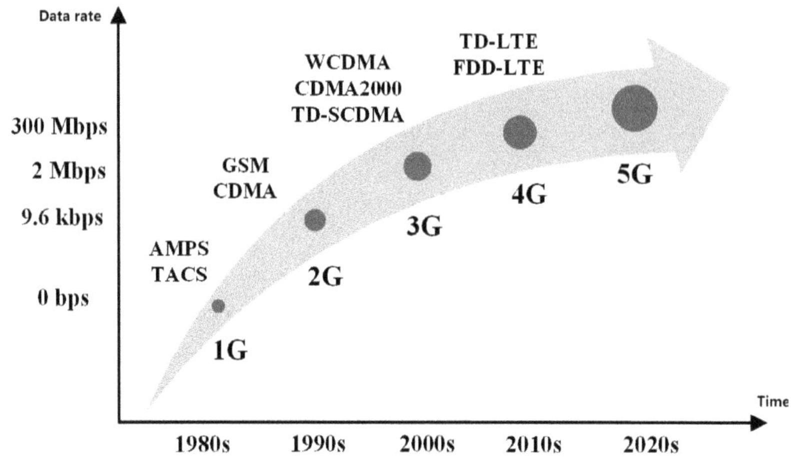

FIGURE 1.2 Evolution of mobile communication standards. (Note: AMPS is Advanced Mobile Phone System; TACS is Total Access Communication System; GSM is Global System for Mobile Communications; CDMA is Code-Division Multiple Access; WCDMA is Wideband CDMA, Code-Division Multiple Access; TD-SCDMA is Time Division-Synchronous Code Division Multiple Access; TD-LTE is Time-Division Long Term Evolution; FDD-LTE is Frequency-Division Duplex Long Term Evolution.)

evolution. Each generation of standards spans about 10 years, and all have major performance improvements over the previous generation. Figure 1.2 illustrates the evolution of mobile communication standards.

1G: This generation of mobile communication mainly solves the problem of mobile communication from scratch, typically represented by analog cellular mobile communication systems. The main standards are Advanced Mobile Phone System (AMPS) of the United States and Total Access Communication System (TACS) of Europe, which have not formed a worldwide standard. The key technology is the cellular network proposed by Bell Labs, which realizes frequency multiplexing by dividing the whole network into cellular sub-networks and greatly improves the system capacity. In 2008, the United States closed down AMPS, which marked the complete withdrawal of 1G from the world stage.

2G: Aiming at the problems of low spectrum utilization and limited service types exposed by 1G, 2G adopts digital communication technology to replace analog communication technology, and the main standards are GSM introduced in Europe and Interim Standard 95 (CDMA-based) introduced in the US. 2G mainly provides voice services, with both SMS and MMS sending and receiving functions, and for the first time has the capability of low-speed data transmission. For the first time, it has the ability of low-speed data transmission, and it is the world's first mobile communication standard that realizes global interconnection. Even after more than 30 years of development, 2G is still one of the main communication standards in many countries and has shown lasting vitality.

3G: After the realization of voice services, the new demand is for Internet services with faster data transmission rates, such as online browsing of pictures and text. 3G's main standards include WCDMA proposed by Europe, CDMA2000 proposed by the US, and

TD-SCDMA proposed by China. 3G realizes the convergence of mobile communications and the Internet, and it has the function of wireless access to the Internet as well as the ability to organize networks more flexibly. However, 3G is only a transitional standard in many countries and has not been popularized on a large scale due to the lack of uniformity in the standards of different countries and the limited increase in the transmission rate of data, which is unable to meet the demand for high-speed data transmission of multimedia.

4G: In order to provide high-speed data transmission services and to end the chaotic situation of multiple systems of the same generation competing with each other, 4G has only one LTE (Long Term Evolution) standard, which is used to unify the development of various interfaces. Compared with 3G, 4G has considerable improvements in many aspects. 4G's core network has fully shifted from the circuit domain to the IP (Internet Protocol) domain, with flatter management, and the utilization of wireless spectrum has been greatly improved to meet the demand for high-speed data transmission. 4G network can not only transmit pictures but also transmit video in real time, which is the mainstream mobile communication network in today's world and the cornerstone of the rapid development of mobile Internet.

5G: In order to meet the diversified needs of high-speed, low-latency, and massive connections, the development direction and scene of 5G are more diversified, which not only realizes high-speed data transmission between people but also meets the efficient communication between people and things, and between things and things. At present, 5G has realized large-scale commercialization all over the world, and in the future, it will make breakthroughs in many fields such as the Internet of Things, the Internet of cars, intelligent medical care, intelligent factories, and virtual reality, which is the standard that is currently being promoted.

1.2 EVOLUTION OF MOBILE TERMINAL ANTENNA

In the mobile communication system, the antenna is the core device responsible for electromagnetic wave sending and receiving, and it is the bridge connecting the wired circuit and the radio wave; its performance directly affects the performance of the whole wireless communication system. According to the widely adopted academic definition, an antenna is a structure with an interconversion region between guided and free space waves [4]. This definition suggests that an antenna is a conversion device. The parameters of antenna can be subdivided into two categories: one is the circuit characteristic parameters, including input impedance, VSWR, bandwidth, etc., and the other is the radiation characteristic parameters, including gain, polarization, and beam width. Different demands and application scenarios have different requirements for antenna parameters, such as mobile communication systems focusing on signal spatial coverage, satellite communication systems focusing on wireless link reliability, and radar systems being very sensitive to the antenna's beam pointing and side-lobe. These requirements and application scenarios promote the diversified development of antenna forms, and also promote the continuous improvement and innovation of antenna technology.

Antennas in mobile communication systems can be divided into two categories: base station antennas and mobile terminal antennas. From 1G to 5G, the forms and functions of these two types of antennas have undergone great changes to better meet the development needs of mobile communications. It should be noted that since this book does not deal with the design of base station antennas, the relevant content will not be expanded. Mobile terminals include a variety of forms, such as portable devices, embedded devices, and wearable devices. In this book, if there is no special explanation, mobile terminals refer to cell phones and mobile terminal antennas refer to cell phone antennas. This section will present the antenna evolution of the cell phone, a typical mobile terminal, from the three dimensions of antenna structure, frequency band, and number.

1.2.1 Changes in Antenna Structure

Cell phones are one of the most frequently used electronic devices in daily life. From 1G to 5G, the development of cell phone antennas can be roughly divided into three stages, which are external antenna stage, internal antenna stage, and conformal antenna stage. Figure 1.3 shows the changes in the appearance and size of cell phones over the past 40 years, from which the evolution of cell phone antennas can be understood. From the size point of view, the cell phone has experienced the process of change from large to small, and then from small to large, and the cell phone screen shows the trend of from nothing to more and more large screen ratio, and even the development of a full screen; from the antenna structure, the cell phone antenna develops from external to built-in, and the thickness changes from thick to thin. In addition, for the antenna, the screen can be regarded as a metal floor, and the antenna usually needs to be far away from the metal floor in order to effectively radiate electromagnetic waves. The larger the screen ratio, the smaller the ground clearance (non-metallic area inside the cell phone) available for antenna design, and the greater the challenge of antenna design.

External antennas were mainly used in the 1G era and the early 2G era. At that time, a cellular phone was mainly used as a practical communication tool, and there was no special requirement for its shape and size, which was typically represented by the mobile phone (commonly known as "Big Brother") in the 1990s. There are many types of external antennas, including monopole antennas, sleeve antennas, and helical antennas [5]. Motorola introduced the world's first commercial cell phone, the DynaTAC 8000X, in 1983, which

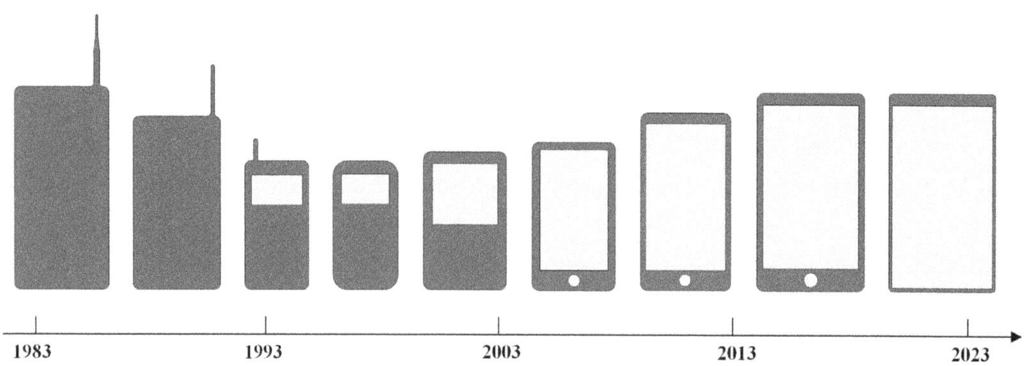

FIGURE 1.3 Changes in the appearance and size of cell phones in the last decades.

had an external monopole antenna in the form of an antenna of one-quarter wavelength (the operating wavelength).

From the mid to late 2G era, people wanted to remove the external antenna to improve the portability and reliability of cellular phones, and cellular phone antennas gradually evolved into built-in antennas. Changing the cell phone antenna from external to built-in not only makes the entire device more aesthetically pleasing but also reduces the probability of damage to the antenna due to accidental dropping or bumping of the cell phone and improves the reliability of the structure. However, reducing the size of the antenna and tucking it inside the cell phone will result in a decrease in antenna performance. But this cost is affordable for the entire mobile communication system—from the 2G era onwards, the construction of mobile communication networks worldwide has become more and more perfect, the number of base stations is more and more densely distributed, and the performance of base station antennas has become stronger and stronger, and these measures are sufficient to compensate for the loss caused by the decline in the antenna performance of mobile terminals.

There are more multi-antenna choices in the form of built-in antennas, including Printed Monopole Antenna, IFA (Inverted-F Antenna), PIFA (Planar Inverted-F Antenna), Slot Antenna, Loop Antenna, and so on. Nokia introduced the popular feature phone 2100 in 2003, and the phone antenna took the form of a PIFA, with the antenna placed on top of the phone. The appearance and size of the antenna can no longer be seen because it is covered by a shell.

As people's requirements for the appearance of cell phones are getting higher and higher, the thickness of cell phones is getting thinner and thinner, and the screen is getting bigger and bigger, resulting in the ground clearance that can be used for antenna design getting smaller and smaller, and the cell phone antenna is developing in the direction of co-conformity with the airframe. Unified planning of antenna design and cell phone structure has become a trend, with the metal frame of the cell phone and the metal back cover serving as part of the antenna in a unified integrated design. New materials and new technologies make the cell phone antenna have a higher degree of design freedom. The iPhone 4 smartphone launched by Apple uses a metal frame as the cell phone antenna, which not only has high mechanical strength but also has a beautiful appearance. It is worth mentioning that this is the first time that the metal frame was used for antenna design, leading the design trend of cell phone antenna; since then, the metal frame has become the main radiator of cell phone antenna.

1.2.2 Changes in Antenna Frequency Band

As people's needs are upgraded from low-rate voice communication to high-rate video communication, the transmission rate of mobile communication is getting higher and higher. According to Shannon's formula, increasing the spectrum bandwidth is an important way to improve the channel capacity and transmission rate. From 1G to 5G, each generation of standards divides a new mobile communication spectrum to meet the development needs of mobile communication. In order to facilitate the differentiation of each frequency band, the 3GPP (3rd Generation Partnership Project) organization, under the guidance of the ITU, divides the frequency of mobile communication into two frequency bands, FR1 (450–6000 MHz) and FR2 (24.25–52.6 GHz), i.e., the familiar sub-6 GHz band and millimeter-wave band. Based on different services or operators, each band can be subdivided into a number of narrow bands, and the specific major frequency bands are divided as shown in Tables 1.2 and 1.3.

In Table 1.2, the uplink and downlink of Frequency-Division Duplex (FDD) are differentiated by frequency, and the uplink and downlink of Time-Division Duplex (TDD) are differentiated by time slots. Prior to the establishment of the 3GPP organization, European countries and the United States divided the frequency bands individually. In the 4G era, the 3GPP organization unified the naming of communication bands and called them LTE bands, including LTE700, LTE2300, and so on. However, due to the inconsistency of the LTE frequency bands divided by each country, the 3GPP organization cannot define LTE

TABLE 1.2 Main Frequency Band Divisions of FR1

Frequency Band no.	Uplink Band (Transmit, MHz)	Downlink Band (receive, MHz)	Duplex Mode
N1	1920–1980	2110–2170	FDD
N2	1850–1910	1930–1990	FDD
N3	1710–1785	1805–1880	FDD
N5	824–849	869–894	FDD
N7	2500–2570	2620–2690	FDD
N8	880–915	925–960	FDD
N20	832–862	791–821	FDD
N28	703–748	758–803	FDD
N38	2570–2620	2570–2620	TDD
N41	2496–2690	2496–2690	TDD
N50	1432–1517	1432–1517	TDD
N51	1427–1432	1427–1432	TDD
N66	1710–1780	2110–2200	FDD
N70	1695–1710	1995–2020	FDD
N77	3300–4200	3300–4200	TDD
N78	3300–3800	3300–3800	TDD
N79	4400–5000	4400–5000	TDD
N80	1710–1785	SUL	SUL
N81	880–915	SUL	SUL
N82	832–862	SUL	SUL

Note: SUL is Supplementary Uplink.

TABLE 1.3 Main Frequency Band Divisions for FR2

Frequency Band no.	Uplink Band (Transmit, GHz)	Downlink Frequency Band (Receive, GHz)	Duplex Mode
N257	26.5–29.5	26.5–29.5	TDD
N258	24.25–27.5	24.25–27.5	TDD
N260	37–40	37–40	TDD

frequency bands in the same way as the GSM frequency bands, which are more regular, so it can only define a list of LTE-supported frequency bands for the reference of operators and communication equipment manufacturers in each country. In the 5G era, the 3GPP organization unified the use of N to identify each frequency band, and domestic operators are generally accustomed to using B to identify the frequency band, such as China Mobile supporting the B41 band.

For cell phone antenna design, it is not necessary to differentiate between uplink and downlink and between transmit and receive, but it is necessary to determine how much bandwidth and which frequency bands the cell phone antenna needs to cover. The FR1 and FR2 bands can be roughly categorized into four frequency bands, namely, 698–960 MHz (low band), 1710–2690 MHz (mid-high band (MHB)), 3300–6000 MHz (high band), and 24.25–29.5 GHz (millimeter-wave band). The design goal of a cell phone antenna is to cover some or all of these bands in as small a ground clearance as possible and to make the radiation efficiency as high as possible.

1.2.3 Changes in the Number of Antennas

From 1G to 5G, although the total size of the antenna in the whole cell phone is decreasing, the number of antennas within the cell phone is rapidly increasing, driven by the demand in the following three areas.

The first aspect is the increase in frequency bands. 2G era frequency bands include 824–960 MHz and 1710–2170 MHz, which can be covered by a single antenna. The 5G standard, on the basis of compatibility with the 2G standard, adds 3300–6000 MHz frequency band, and it is difficult for a single antenna to cover such a wide frequency band. In order to solve this problem, multiple antennas can be used to cover different frequency bands according to the actual application scenarios.

The second aspect is the application of new technologies. Considering that the wireless spectrum is an expensive and limited resource, the way to increase the transmission rate by simply relying on increasing the frequency band is subject to more and more constraints. Under the condition that the frequency band is fixed and unchanged, some new technologies have been introduced into mobile communications since the 2G era, and typical technologies include Multiple-Input Multiple-Output (MIMO) technology and phased array technology. Using these techniques, it is possible to arrange multiple antennas at the same time at the base station and the mobile terminal, improving the spectrum utilization by increasing the complexity of the spatial dimension.

The third aspect is the integration of multiple functions. With the changing needs of users, cell phones have moved from the era of feature phones into the era of smartphones, becoming an intelligent, integrated, multifunctional portable platform. The cell phone not only has the function of communication but also has the functions of taking pictures, playing games, navigation, and so on. Different functions may require different types of antennas, and their corresponding operating frequency bands are not the same, which further increases the number of antennas.

1.3 MULTI-ANTENNA LAYOUT WITHIN A CELL PHONE

A cell phone can be regarded as a microsystem that integrates modules with multiple purposes. Taking the use of wireless transmission as an example, according to the different functions, the modules that require the use of antennas include mobile cellular communication modules, Wi-Fi (Wireless Fidelity) modules, Bluetooth modules, satellite navigation modules, Near Field Communication (NFC) modules, wireless charging modules, etc. The following is a brief introduction to these modules. These modules are briefly described below.

The mobile cellular communication module is the most complex module in the antenna design of a cell phone and is used to establish communication with the base station. From 1G to 5G, the antennas of mobile cellular communication modules need to support wider and wider frequency bands, and the number of multi-antenna is also increasing. The number of antennas used for cellular communication in a cell phone is 6–14.

Wi-Fi modules are used for indoor LANs to establish communication with wireless routers. The Wi-Fi standard has now evolved to the 7th generation, IEEE 802.11be, and the operating frequency bands are industrial bands that do not require licensing, mainly including the 2.4 GHz band and the 5 GHz band. The frequency range of 2.4 GHz band is 2.4–2.497 GHz, and the frequency range of 5 GHz band is 5.15–5.85 GHz. Due to the higher operating frequency and shorter communication distance, the number of antennas of Wi-Fi module is less.

Bluetooth modules are used for low cost and low power consumption for close-range wireless connectivity. Bluetooth technology has now evolved to version 5.4. The operating frequency of Bluetooth modules used in different portable devices varies. The operating frequency band of Bluetooth modules in cell phones is the 2.4 GHz band. Bluetooth modules in this band not only share the same frequency band with Wi-Fi modules, but also often share an antenna to save antenna space.

Satellite navigation module is used to receive satellite signals to realize real-time positioning of the user's position. The mainstream standards of this module are the GPS (Global Positioning System) of the United States [working in the 1575.42 MHz band (L1) and 1176.45 MHz band (L5)], the Galileo system of Europe (working in the 1575.42 MHz band), the GLONASS system of Russia (working in the 1602 MHz band), and the BeiDou satellite navigation system of China (working in the 1561 MHz band). The antenna design of satellite navigation modules is relatively simple because they have a small bandwidth and generally only need to have reception capability. With the rise of direct satellite connection applications for cell phones, more frequency bands will be used for satellite communications.

NFC modules are used for ultra-close-range non-contact communication. Initially standardized by companies such as Nokia and Philips, NFC modules operate in the 13.56 MHz band. Because the wavelength of this band is very long, the NFC antenna inside the cell phone belongs to the electric small size antenna, the radiation ability is very weak, and it can realize electronic payment, identity authentication, data exchange, anti-counterfeiting, and other functions through magnetic field induction.

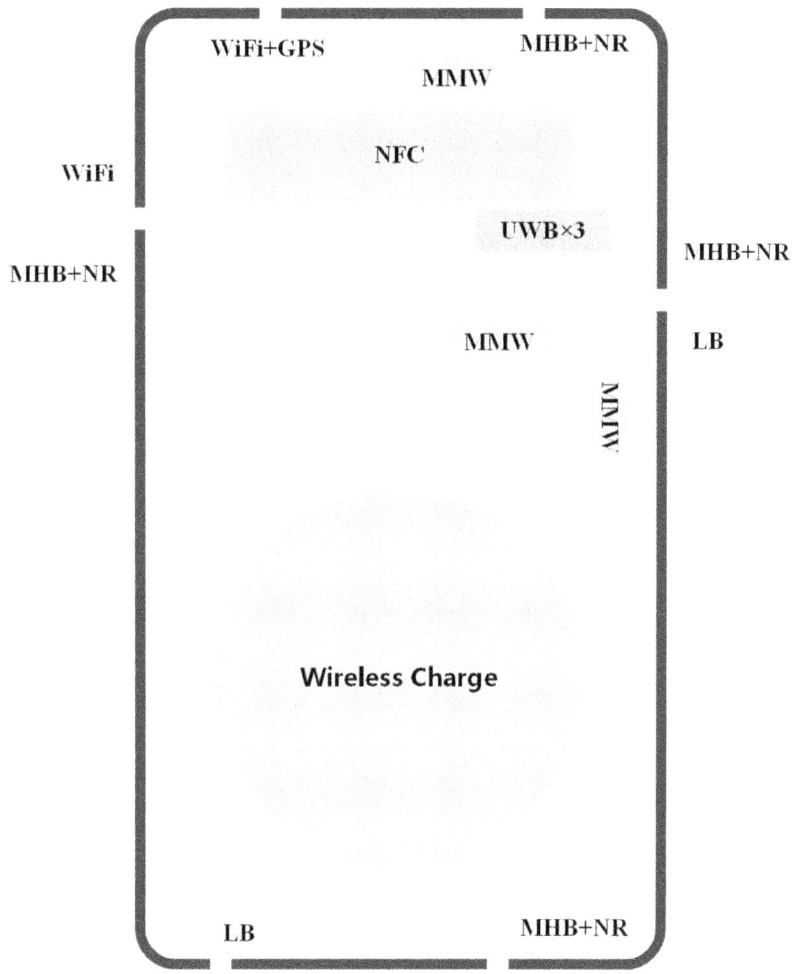

FIGURE 1.4 Typical layout scheme of multi-antenna within a cellular phone.

Wireless charging module is used to realize the wireless charging function of a cell phone. The mainstream standards of this module are Qi standard, PMA (Physical Medium Attachment) standard, and so on. Among them, the Qi standard is based on the principle of electromagnetic induction, with an electromagnetic wave frequency of 100–205 kHz, which is suitable for short-distance wireless charging, characterized by strong compatibility and high transmission efficiency. Most of the wireless charging products on the market now use this standard, which is also the most familiar wireless charging standard to the public.

Combining these modules together constitutes a multi-antenna system. A typical layout scheme for a multi-antenna within a cell phone is shown in Figure 1.4. The multiple antennas are distributed in various locations inside the cell phone platform. In this typical layout scheme, the multi-antenna used for mobile communication is mainly distributed on the metal bezel, and the energy is effectively radiated by setting breakpoints. Among them, the number of low-band antennas is 2, which are placed at the bottom and the side wall of the cell phone; the number of medium, high-band, and new airport antennas is 4, which are placed on the four edges of the metal bezel; the number of Wi-Fi antennas is 2, which are placed at the upper left corner of the cell phone; the number of GPS antennas is 1, which is placed at the top of the cell phone; and the number of millimeter-wave antennas is 3, which is placed at the inner side of the cell phone's back cover; and the number of ultra-wideband antennas is 3, which are placed on the inside of the back cover of the cell phone. NFC antennas and wireless charging coils are also placed on the inside of the back cover of the cell phone. As can be seen, this typical layout scheme contains at least 17 antennas. Currently, mobile communication has evolved to the 5G era, in which the working frequency of mobile communication is higher, and the function is more extensive. The mobile cellular communication module inside the cell phone needs to deploy multiple antennas in order to realize high-speed wireless data transmission and multi-functional applications.

Among these functional modules, the antenna of the Wi-Fi module and the antenna of the Bluetooth module can be shared, and the satellite navigation module operates in an extremely narrow frequency band, which can be covered by a single small-sized antenna. Therefore, the design difficulty of the cell phone antenna is to cover the mobile cellular communication band. Based on this, this book focuses on antenna design in mobile cellular communication modules.

1.4 TYPICAL OPERATION OF MULTI-ANTENNA SYSTEMS

Multiple antenna modules within a cellular phone have independent functions and are able to work without interfering with each other. For multi-antenna systems, the relative position and ground clearance size of each type of antenna within the cell phone is basically fixed. Considering that the antenna design schemes for Wi-Fi module, Bluetooth module, satellite navigation module, NFC module, and wireless charging module are more mature, this section focuses on the typical working mode of multi-antenna system used for 4G/5G mobile cellular communication module.

For the multi-antenna of the mobile cellular communication module in a cell phone, they can be divided into three groups of antennas that work independently depending on the frequency band: the first group of antennas covers 698–960 MHz (low band) and 1710–2690 MHz (MHB), the second group of antennas covers 3300–6000 MHz (high band), and the third group of antennas covers 24.25–29.5 GHz (millimeter-wave band) antennas. Each group of antennas contains several antennas, and according to the different operating modes, the multiple antennas within the same group can be classified into two categories: a multi-antenna system that operates according to the MIMO mechanism and a multi-antenna system that operates according to the phased array mechanism. Specifically, the first and second groups of antennas work according to the MIMO mechanism, and the third group of antennas work according to the phased array mechanism.

MIMO technology is one of the important technologies in the field of mobile communication in recent decades, and the working principle of the MIMO communication system composed of cell phone multi-antenna and base station multi-antenna is shown in Figure 1.5. Compared with the earlier single-input single-output system, the MIMO

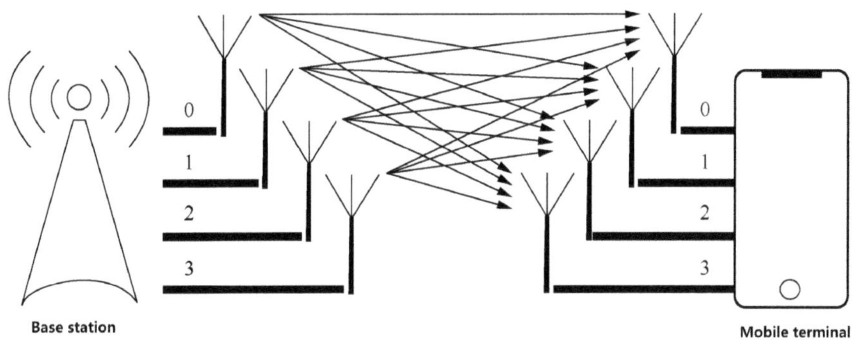

FIGURE 1.5 Working principle of MIMO communication system.

communication system multiplies the channel capacity by arranging multiple antennas at the base station end and the cell phone end, with the advantage that it does not take up more spectrum resources and does not need to increase the antenna transmit power. In addition, compared with 4G, 5G has added some new frequency bands. The operating frequency of these frequency bands is higher, and the corresponding antenna unit is smaller, which is convenient to realize the integrated design of MIMO antenna on a single device.

However, a series of technical difficulties needs to be overcome when applying MIMO technology in cell phone antennas to enhance the transmission rate. For example, due to the limited space inside the cell phone, the increase in the number of antennas working in the same frequency band after the adoption of MIMO technology leads to the close proximity of the antenna units, which usually causes strong coupling, which not only makes it difficult to play the role of increasing the channel capacity, but may even affect the radiation efficiency of the antenna. Thus, decoupling methods are needed to reduce the coupling strength between antenna units. In addition, miniaturization and broadbandization of antennas are always the goals pursued in cell phone design, and the size of the 5G low-band antenna will be much larger than that of the 5G high-band antenna, which is also a difficult problem that needs to be overcome by using the MIMO technology in the low-band.

Considering that the traditional mobile communication bands are concentrated below 6 GHz, which is already very crowded, 5G opens up a new millimeter-wave band on top of the traditional bands to further increase wireless transmission rates. Millimeter wave has a series of unique properties, including the advantages of wide available frequency band, small size of millimeter-wave devices, and good beam directivity, as well as the disadvantages of large path loss and small output power of millimeter-wave devices. How to utilize the advantages of millimeter wave and make up for its shortcomings has stimulated the research interest of many researchers.

Phased array technology is an effective means to overcome the shortcomings of large path loss and has been widely used in radar systems to enhance the antenna gain while realizing the dynamic coverage of large airspace. Phased array technology refers to a number of the same form of antennas in accordance with certain rules uniformly arranged into a linear array or planar array, and each antenna unit is equipped with a phase shifter to change the feed phase of the antenna unit in real time, with different phases of the electromagnetic wave in the space of vector superposition, so that the synthesized signal strength is enhanced or attenuated, thus generating a specific direction of the high-gain beam. Since each antenna unit is switching the beam by electrically controlling the phase, such an antenna is named phased array antenna. Phased array technology has high power consumption, high cost, high complexity, and other problems, and was not applied to the field of mobile communications in the early days. With the increase in the demand for high transmission rates, the decrease in the cost of transceiver components, and the commercialization of millimeter-wave bands, 5G began to introduce phased array technology through Antenna-in-Package (AiP) form to achieve a high degree of integration.

In a typical scenario of 5G, a large-scale antenna is deployed at the base station end, which can generate a corresponding number of beams and beam pointing according to the number and location of users. The mobile terminal antenna has beam scanning capability and can adjust the beam pointing in real time, according to the position and attitude of the terminal, as shown in Figure 1.6. The antenna, as a key component of beam regulation, plays the role of a core device in 5G mobile communication systems. Although the antenna standards for millimeter-wave bands have not been fully finalized, the industry has reached some consensus on 5G millimeter-wave antennas [6,7]. Millimeter-wave antennas for 5G Mobile Terminals need to have three essential characteristics: high gain, wide beam sweep angle, and low cost. The path loss of millimeter waves is high, and to establish a stable communication link under the condition of limited transmit power, millimeter-wave antennas for 5G Mobile Terminals need to have high gain. The increase in gain means that the beam

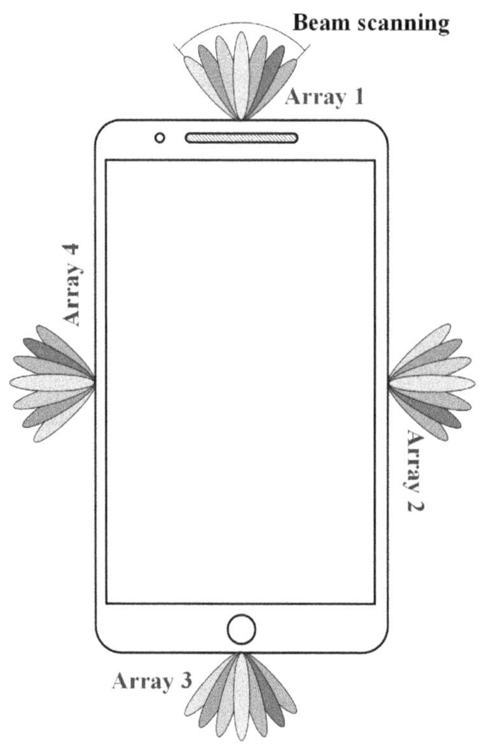

FIGURE 1.6 Cell phone using phased array module for spatial signal coverage.

width becomes narrower and the coverage of the beam becomes smaller. In order to reduce the signal blind spot, the antenna needs to have beam scanning capability to dynamically adjust the beam pointing. In addition, terminal antennas are consumer electronics devices used on a large scale, and any cost reduction will bring significant economic benefits. Therefore, low-cost millimeter-wave antennas with wide-angle scanning are attractive for 5G Mobile Terminals antenna design.

1.5 EXAMPLES OF CELLULAR PHONE PRODUCTS IN INDUSTRY

As a consumer electronic product widely used in people's daily life and production, the cellular phone has a huge market size, which attracts many companies to carry out investment and research and development in this field. Antennas, as one of the most critical components within a cell phone, have received great attention in the R&D process. During the evolution of mobile communication standards, many excellent cell phone companies have emerged around the world, resulting in numerous novel antenna design solutions. According to the different division of labor, the organizations engaged in cell phone antenna R&D in the industrial world are mainly divided into two kinds: one is the whole machine manufacturers, and the other is the professional cell phone antenna manufacturers. The flagships of OEMs often represent the highest level of current cell phone antenna design. Next, we introduce some influential machine manufacturers.

Motorola played a pivotal role in the early stage of mobile communications; it developed a series of high-performance external antennas. However, in the course of the development of mobile communications, Motorola's cellular phone business gradually fell behind in the competition.

Nokia is a benchmark enterprise in the era of feature phones. The company developed a number of best-selling cell phone products in the 2G era. These products are built-in antenna solutions. Sadly, Nokia has gradually declined in the process of shifting from feature phones to smartphones.

Apple is the benchmark enterprise in the smartphone era. Since the launch of the first smartphone in 2007, the company quickly occupied a huge market share and is still the leading cell phone company. Apple has made many innovations in the field of cell phone antenna, and it has taken the lead in launching metal bezel antenna, segmented metal back cover antenna, etc., which has led to the development of the cell phone antenna business.

Starting from the 4G era, domestic cell phone enterprises grew rapidly, seizing the opportunity during the transformation of cell phones from feature phones to smartphones, and started to dominate the cell phone market. Currently, domestic cell phone companies have achieved great success with a global market share of around 50%. Representative cell phone companies are Huawei, Honor, Xiaomi, OPPO, vivo, etc., which have carried out in-depth research in the field of cell phone antennas and have extensive patent layouts.

Figure 1.7 shows the antenna layout of a model of cell phone introduced by Apple. The whole metal bezel is divided into four quadrants; the first quadrant is in the upper right corner, the second quadrant is in the upper left corner, the third quadrant is in the lower left corner, and the fourth quadrant is in the lower right corner. The naming rules for the antenna serial numbers can be based on the quadrant counterclockwise numbering,

Antenna	Location
WiFi_5G Ant1	11#
LMHB+GPS_L1	12#
MHB+N77/N79+WIFI_2.4 G Ant1	21#
GPS_L5	22#
N77/N79	31#
LMHB	32#
MHB+WiFi_2.4G Ant2	41#
N77/N79	42#

FIGURE 1.7 Antenna layout of a model of a cell phone launched by Apple Inc.

e.g., Ant11 indicates the first antenna in the first quadrant. From Figure 1.7, it can be seen that there are six breakpoints in the metal bezel of this model, with mirror symmetry on the left and right sides and rotational symmetry on the top and bottom sides. The Wi-Fi antenna working in the 5 GHz band is the first antenna in the first quadrant, and the second antenna is shared by the low-mid-high band antenna and the L1 band antenna of GPS; the first antenna in the second quadrant is the MHB antenna, the antenna of the N77/N79 frequency bands (the new band of 5G) and the Wi-Fi antenna working in the 2.4 GHz band, and the second antenna is the L5 band antenna for GPS; the first antenna in the third quadrant is the 5G new band antenna and the second antenna is the low MHB antenna; the first antenna in the fourth quadrant is the MHB antenna and the Wi-Fi antenna working in the 2.4 GHz band, and the second antenna is the 5G new band antenna. From this antenna layout, it can be seen that the antenna arrangement is relatively simple, and most of the same frequency band has two antennas for the realization of the MIMO function.

Figure 1.8 shows the antenna layout of a model cell phone introduced by Huawei. From Figure 1.8, it can be seen that there are seven breakpoints in the metal bezel of this model cell phone. The entire metal bezel is similarly divided into four quadrants. The first antenna in the first quadrant covers the MHB and the N78/N41 band, the second antenna covers the low band, the third antenna covers the MHB and the N78/N41 band, and the fourth antenna covers the 5 GHz band of the Wi-Fi antenna; the first antenna in the second quadrant covers the L5 band of the GPS antenna, the second antenna covers the 5 GHz band of the Wi-Fi and the N78 band, the third antenna covers the L1 band of the GPS antenna, the 2.4 GHz band of the Wi-Fi antenna, and the B1/B3/B39 bands, the fourth antenna covers the N78/N41 bands as well as the B38/B40/B41 bands, and the fifth antenna covers the

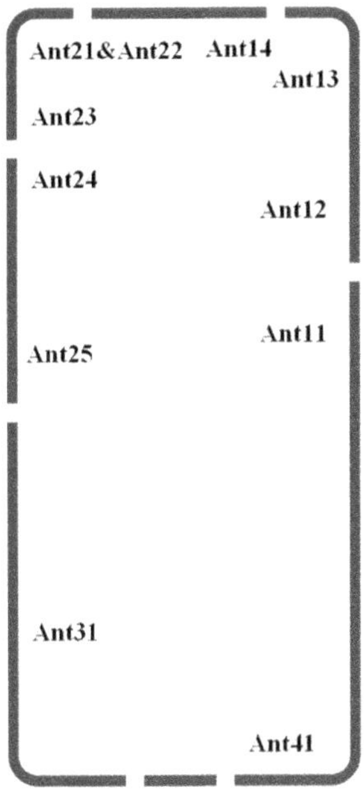

Antenna	Location
MHB+N78_RX+N41_RX	11#
LB	12#
MHB_MIMO+N41_TRX+N78_RX	13#
WiFi_5G	14#
GPS_L5	21#
WiFi_5G+N78_DRX	22#
GPS_L1+WiFi_2.4G+B1/B3/B39	23#
N78_TRX+N41_DRX+B38/B40/B41	24#
WiFi_2.4G+N41_DRX	25#
LB	31#
MHB	41#

FIGURE 1.8 Antenna layout of a model of a cell phone introduced by Huawei.

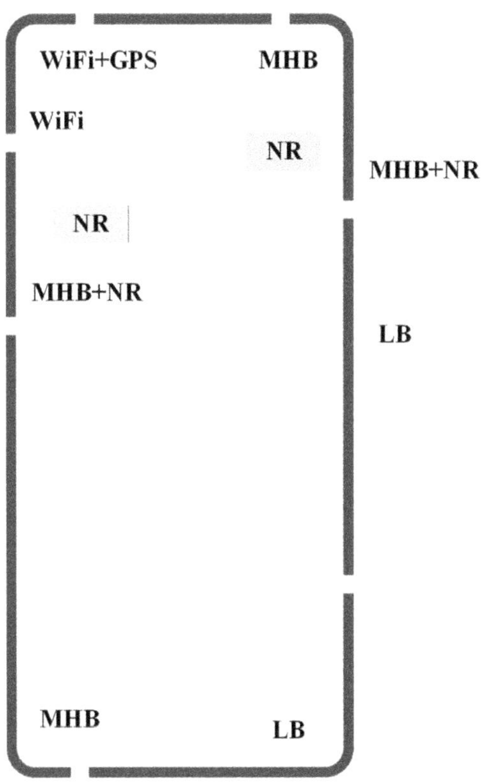

FIGURE 1.9 Antenna layout of a model of a cell phone introduced by vivo company.

2.4 GHz band of the Wi-Fi antenna and the N41 band; there is only one antenna in the third quadrant to cover the low frequency band; the fourth quadrant also has only one antenna for covering the middle and high frequency bands. From this antenna layout, it can be seen that the entire metal bezel includes 11 antennas, with the first and second quadrants being more highly utilized, with four and five antennas laid out, respectively.

Figure 1.9 shows the antenna layout of a cell phone model from vivo. From Figure 1.9, it can be seen that there are seven breakpoints in the metal bezel of this model cell phone. The first quadrant contains three antennas, which are the low-band antenna, the mid-high and 5G new-band antenna, and the MHB antenna; the second quadrant contains three antennas, which are the Wi-Fi antenna and GPS antenna, the Wi-Fi antenna, and the mid-high and 5G new-band antenna; the third quadrant contains one MHB antenna; and the fourth quadrant contains one low-band antenna. There are also two 5G new-band antennas laid out on the inside of the back cover for providing more multi-antenna channels.

REFERENCES

1. Ramsay J. Highlights of antenna history. *IEEE Communications Magazine*, 1981, 19(5): 4–8.
2. Balanis CA. Antenna theory: a review. *Proceedings of the IEEE*, 2002, 80(1): 7–23.
3. Zhang Y, Xing F, Lu C. Development and prospect of wireless communication technology. *Silicon Valley*, 2010, 23: 1.
4. Kraus JD, Marhefka RJ. *Antennas*. 3rd ed. Zhang Wenxun, Translation. Beijing: Electronic Industry Press, 2006.
5. Zhang Z. *Antenna Design for Mobile Devices*. Hoboken: John Wiley & Sons, 2017.
6. Zhang J, Ge X, Li Q, et al. 5G millimeter-wave antenna array: design and challenges. *IEEE Wireless Communications*, 2017, 24(2): 106–112.
7. Hong W, Baek KH, Ko S. Millimeter-wave 5G antennas for smartphones: overview and experimental demonstration. *IEEE Transactions on Antennas and Propagation*, 2017, 65(12): 6250–6261.

CHAPTER 2

Basic Theory of Mobile Terminal Antenna

THIS CHAPTER WILL INTRODUCE the basic theory of mobile terminal antenna, including basic antenna parameters, antenna patterns, typical mobile terminal antenna forms, Multiple-Input Multiple-Output (MIMO) antenna decoupling principle, phased array antenna beam scanning principle, to provide a theoretical basis for mobile terminal antenna design.

2.1 ANTENNA BASIC PARAMETERS

An antenna is a transceiver of electromagnetic waves. The basic parameters of the antenna directly affect the performance of the entire wireless communication system. Figure 2.1 shows a typical wireless communication system Radio Frequency (RF) front-end composition, including antenna, transmission line, duplexer, low noise amplifier (referred to as "low noise amplifier"), power amplifier (referred to as "power amplifier"), and other modules. From the wireless communication link, it can be seen that the antenna is a bridge between free space and circuits [1]. One end of the antenna is connected to the RF microwave circuit through the transmission line, and the processing objects are voltage and current signals; the other end is connected to the free space, and the processing objects are electric field and magnetic field signals. Accordingly, the basic antenna parameters can be divided into two categories: one is the circuit characteristic parameters related to the

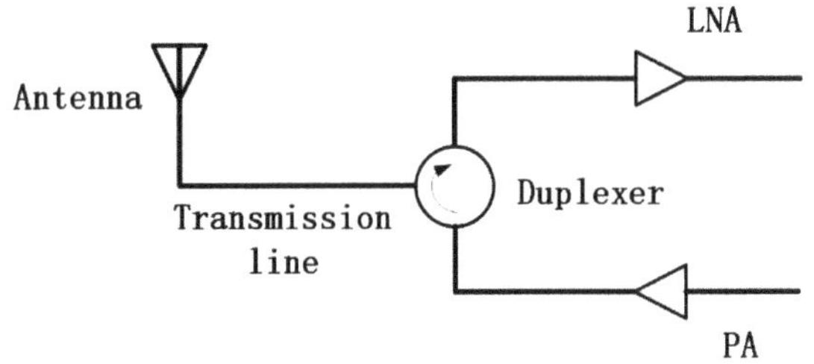

FIGURE 2.1 RF front-end components of a typical wireless communication system.

DOI: 10.1201/9781003770039-2

circuit, and the other is the radiation characteristic parameters related to the free space. In addition, antenna-specific parameters need to be considered in engineering applications.

Circuit characteristic parameters are used to measure how well the transmission line matches the antenna, i.e., how much energy the antenna receives from the transmission line for transmitting or passes to the transmission line for receiving. Specific parameters include reflection coefficient, return loss, Voltage Standing Wave Ratio, frequency band, port coupling coefficient, port isolation, input impedance, characteristic impedance, third-order intermodulation, power capacity, and so on.

The radiation characteristic parameters are used to measure the degree of matching between the antenna and the free space, i.e., the energy exchange relationship between the antenna and the free space; they are also used to measure the beam shape of the antenna in three-dimensional space. The radiation direction diagram is used to measure the beam shape of the antenna in three-dimensional space, and the specific parameters include gain, radiation efficiency, beam pointing, beam width, main lobe, side-lobe, back-lobe, polarization, orifice efficiency, and front-to-back ratio.

2.1.1 Circuit Characteristic Parameters

For mobile terminal antenna, the more commonly used circuit characteristic parameters are reflection coefficient, return loss, voltage standing wave ratio (VSWR), frequency band, port coupling coefficient, and port isolation.

The reflection coefficient is used to measure the ratio of the reflected wave energy to the incident wave energy. As shown in Figure 2.2, when the antenna and the transmission line do not match, i.e., the antenna impedance is not equal to the transmission line characteristic impedance, the antenna cannot be transmitted on the transmission line to absorb all of the high-frequency energy. It can only absorb part of the energy, and a part of the energy of the incident wave is reflected back to the transmission line to form a reflected wave. The ratio of reflected wave voltage $V_{\text{reflected}}$ and incident wave voltage V_{incident} is called the reflection coefficient, which can be expressed by Eq. (2.1).

$$\Gamma = \frac{V_{\text{reflected}}}{V_{\text{incident}}} \tag{2.1}$$

In microwave network theory, the S-parameter is generally used to describe the performance of microwave devices. For a single-port antenna, the reflection coefficient is equivalent to S_{11} (the reflection coefficient of port 1), both of which represent a complex number

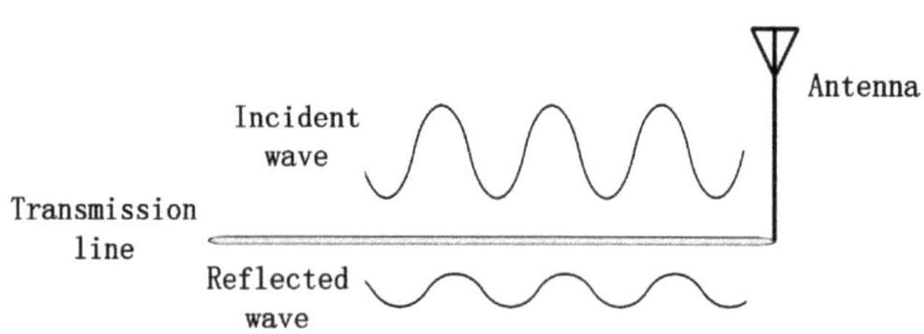

FIGURE 2.2 Definition of reflection coefficient.

containing both amplitude and phase. The amplitude of S_{11} is usually expressed in logarithmic form:

$$S_{11}(\text{dB}) = 20\lg(|S_{11}|) = 20\lg(|\Gamma|) \tag{2.2}$$

Note that 20lg() is taken here instead of 10lg() because $S_{(11)}$ (dB) is used to describe the ratio of the reflected power to the incident power rather than the ratio of the reflected wave voltage to the incident wave voltage, where the power is proportional to the square of the voltage. $S_{(11)}$ (dB) is a real number, not a complex number.

The return loss represents the absolute value of $S_{(11)}$ (dB). Since the reflection coefficient of a passive antenna is always less than 1, $S_{(11)}$ (dB) is always a negative value, and the return loss is always a positive value.

VSWR is used to measure the energy distribution along the transmission path on the transmission line. Typically, there are both incident and reflected waves on the transmission line. The transmission line and the antenna will not match exactly, and the state of the transmission line is not a pure traveling wave state, but a traveling standing wave state that includes both traveling waves and standing waves, or even a pure standing wave state. VSWR represents the ratio of the absolute value of the adjacent wavebelly voltages $V_{\text{wavebelly}}$ to the absolute value of the nodal voltages V_{nodal} on the transmission line, which can quantitatively indicate the different states of the transmission line:

$$\text{VSWR} = \frac{|V_{\max}|}{|V_{\min}|} \tag{2.3}$$

When VSWR = 1, the transmission line is in the traveling wave state; when VSWR is $+\infty$, the transmission line is in the pure standing wave state; and when VSWR is in between, the transmission line is in the traveling standing wave state.

VSWR and reflection coefficient can be converted to each other, and the conversion relationship between the two is:

$$\text{VSWR} = \frac{|V_{\max}|}{|V_{\min}|} = \frac{|V_{\text{incident}}| + |V_{\text{reflected}}|}{|V_{\text{incident}}| - |V_{\text{reflected}}|} = \frac{1+|\Gamma|}{1-|\Gamma|} \tag{2.4}$$

Reflection coefficient, return loss, and VSWR can be used to quantitatively measure the degree of mismatch between the transmission line and the antenna, which can be selected according to individual habits in practical use. The conversion relationship of related parameters is shown in Table 2.1.

The same parameter will have different index regulations in different applications. For example, mobile terminals pursuing antenna miniaturization only require VSWR within 3.0, corresponding to a reflection coefficient of −6.0 dB, which means that 25.0% of the power is reflected back to the transmission line; most of the wireless communication system applications require VSWR within 2.0, corresponding to a reflection coefficient of −9.5 dB, which means that 11.1% of the power is reflected back to the transmission line.

TABLE 2.1 Conversion Relationship of Reflection Coefficient, Return Loss, VSWR, and Other Related Parameters

VSWR	Return Loss (dB)	Reflection Coefficient (dB)	Reflected Power (%)	Reflection Loss (dB)
5.0	3.5	−3.5	44.4	2.55
4.0	4.4	−4.4	36.0	1.94
3.0	6.0	−6.0	25.0	1.25
2.0	9.5	−9.5	11.1	0.51
1.5	14.0	−14.0	4.0	0.18
1.2	20.8	−20.8	0.8	0.04
1.1	26.4	−26.4	0.2	0.01

The base station focuses on antenna performance improvement and requires a VSWR of 1.5 or less, corresponding to a reflection coefficient of −14.0 dB, which means that only 4.0% of the power is reflected back to the transmission line.

Frequency Bandwidth refers to the frequency range in which an antenna's performance parameters meet the specified criteria. The performance of the antenna will generally change with changes in frequency, and engineering needs to define the frequency band of the antenna. In this band, the performance of the antenna should meet the specified standards. Different types of bandwidths can be defined by selecting different parameters. Most applications define the bandwidth in terms of the reflection coefficient or VSWR parameter, i.e., the reflection coefficient or VSWR is less than a specified value in a particular frequency band. For some high gain antennas, such as parabolic antennas, the bandwidth is usually defined in terms of the gain parameter, typically 1 or 3 dB gain bandwidth.

Antenna bandwidth has three commonly used methods of expression, namely absolute bandwidth, relative bandwidth, and ratio bandwidth, which reveal the bandwidth performance from different aspects.

- Absolute Frequency Bandwidth: the actual working frequency range of the antenna, i.e., the difference between the upper and lower limit frequencies.
- Relative Bandwidth: the difference between the upper and lower limit frequencies and the ratio of the center frequency.
- Ratio Bandwidth: the ratio of upper and lower limit frequencies.

The operating frequency of mobile communication is strictly limited to the range, so the definition of absolute bandwidth is generally used. If it is necessary to reveal the radiation capacity of the antenna, the definition of relative bandwidth can be used. For ultra-wideband applications, the definition of ratio bandwidth is more graphic. Figure 2.3 shows the relationship between the reflection coefficient and frequency of a particular antenna. Here the bandwidth of the antenna is defined in terms of −6 dB reflection coefficient or VSWR = 3. As can be seen from Figure 2.3, the absolute bandwidth of this antenna is about 260 MHz (680–940 MHz), the relative bandwidth is about 32.1%, and the ratio bandwidth is about 1.38.

The port coupling coefficient refers to the degree of coupling of energy between multiple feed ports. Multiple antennas have multiple feed ports; the same antenna may also

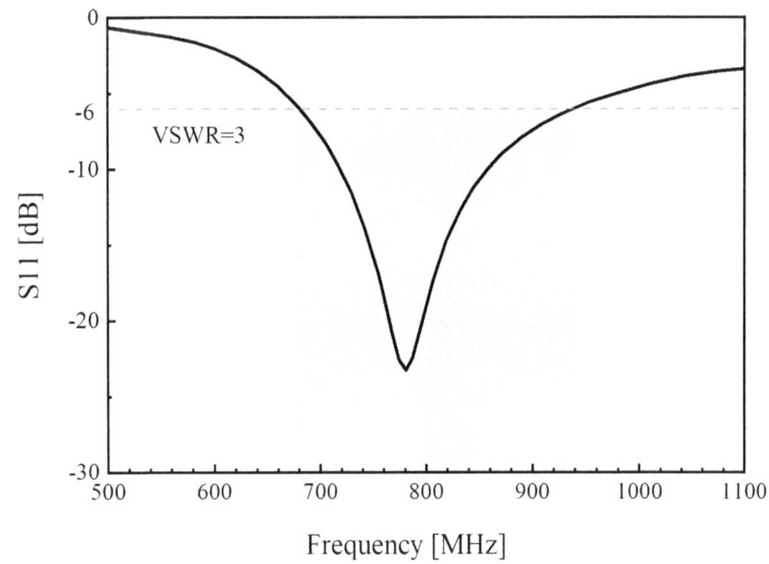

FIGURE 2.3 Reflection coefficient versus frequency for a particular antenna.

TABLE 2.2 Typical Port Coupling Coefficient, Port Isolation, and Coupled Power Conversion Relationships

Port Coupling Factor (dB)	Port Isolation (dB)	Coupling Power (%)
−5	5	31.6
−10	10	10.0
−15	15	3.2
−20	20	1.0
−25	25	0.3

have multiple feed ports, and there is energy coupling between these feed ports. According to the microwave network theory, N feed ports can form an S-parameter matrix of $N \times N$. If $i \neq j$, then $S_{(ij)}$ (dB) denotes the coupling coefficient of the ith port and the jth port, which is generally expressed in logarithmic form.

The port isolation indicates the absolute value of the port coupling coefficient. The port coupling coefficient of the passive antenna is constantly less than 1, the logarithmic form of the port coupling coefficient is negative, and the port isolation degree is constantly positive. The relationship between the port coupling coefficient and the port isolation degree is analogous to the relationship between the reflection coefficient and the return loss.

Generally speaking, we hope that the isolation degree of the two feeder ports tends to infinity, so as to ensure that the two ports work independently. However, in practical engineering, the port isolation degree is constrained by the accuracy of simulation, processing, and testing; there is often an upper limit, and the port isolation degree of mobile terminal antenna is mostly standardized at 15 dB.

For a two-port network, Table 2.2 lists the conversion relationship between typical port coupling coefficient, port isolation degree, and coupling power. If the isolation between two ports is 10 dB, it means that when port 1 is excited, port 2 will receive 10.0% of the energy.

For multi-antenna, both the reflection coefficient and the port isolation need to be taken into account. Just because one antenna has a low reflection coefficient does not mean that all of the energy is being radiated by the antenna; it is possible that the energy is being absorbed by another feed port.

2.1.2 Radiation Characteristic Parameters

The radiation pattern of an antenna is directional, meaning that the field strength radiated varies with the direction in three-dimensional space. The relative relationship between the field strengths radiated in different three-dimensional directions is plotted as a graph, i.e., the radiation direction of the antenna is obtained. Radiation direction diagram is used to characterize the antenna in all directions in space and has the ability to transmit or receive electromagnetic waves. Radiation pattern is a three-dimensional pattern; it can be displayed in different coordinate systems, such as the spherical coordinate system or right-angle coordinate system. The spherical coordinate system facilitates the visualization of beam pointing, while the right-angle coordinate system facilitates the quantitative assessment of metrics. Considering that the drawing of 3D stereograms is more complicated, several typical cut surfaces can be selected in 3D space to form a 2D orientation map. Typical cut surfaces include E-plane and H-plane, which are the cut surfaces composed of electric field and propagation direction, as well as the cut surfaces composed of magnetic field and propagation direction, respectively. Figure 2.4 shows a radiation pattern presented in the right-angle coordinate system, where several parameters can be defined, including the main beam, vice beam (side beam), back beam (rear beam), the width of the main beam, the radiation zero point, and other indicators.

The radiation pattern is generally in the shape of a petal, and these petals are called petals or beams. Among them, the beam containing the maximum radiation direction is called the main beam, the others are called the secondary beams and are divided into the first sidelobe, the second sidelobe, etc., and the beam in the opposite direction to the main beam is called the back lobe. According to the different shapes of the main beam, it can be categorized into broadside radiation pattern, end-fire radiation pattern, and horizontal

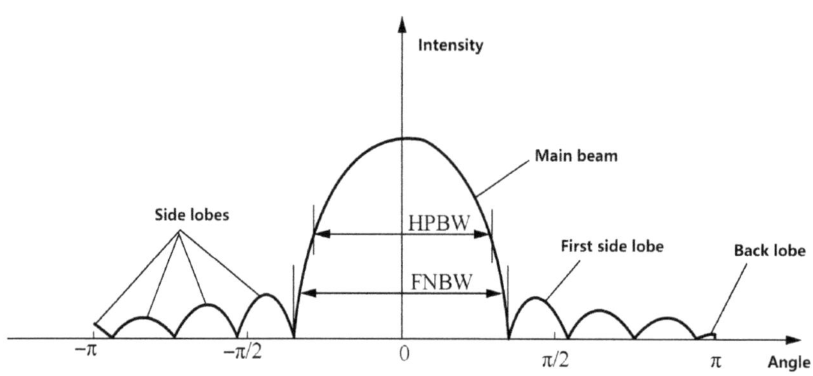

FIGURE 2.4 Radiation pattern in the Cartesian coordinate system. Note: HPBW is Half-Power Beam Width; FNBW is First Null Beam Width.

omnidirectional radiation pattern. In addition to the qualitative description, there are some quantitative descriptions of the radiation pattern parameters.

- Half-Power Beam Width: The horizontal angle corresponding to the time when the radiated power density decreases from the maximum value to half in the plane containing the main lobe.
- First Null Beam Width: The angle between the first zeros on either side of the main lobe in the plane containing the main lobe.
- Sub-Lobe Level: The level of the first sidelobe closest to the main lobe and with the highest level, usually measured in dB.
- Front-to-Back Ratio: The ratio of the level in the direction of maximum radiation (forward) to the level in its opposite direction (backward), usually expressed in dB.
- Directionality Coefficient: The ratio of the power density of the antenna at a point in the far area in the direction of maximum radiation to the power density of an ideal non-directional point source antenna radiating the same power at the same point.

Gain refers to the ratio of the power density of the signal generated by the actual antenna and an ideal non-directional point source antenna at the same point in space. It quantitatively describes the extent to which an antenna concentrates the radiation of the input power. Gain and antenna radiation pattern have a close relationship, in the conditions without considering the loss, the radiation pattern of the narrower the main lobe, the higher the gain. Antenna gain is used to measure the ability of an antenna to send and receive signals in a particular direction. The gain of an antenna under test can be measured by comparing the maximum power density radiated by the antenna under test and a reference antenna of known gain at the same input power.

The antenna gain also describes how many times an actual antenna will amplify the input power in the direction of maximum radiation compared to an ideal non-directional point source antenna. For example, an ideal non-directional point source antenna with an input power of 10 W has the same signal strength in the direction of maximum radiation as an antenna with a gain of 10 dBi and an input power of 1 W. The antenna has a gain of 10 dBi and an input power of 1 W in the direction of maximum radiation. Thus, when the input power is the same, a high-gain antenna can achieve greater radiated power in the maximum radiation direction. However, the antenna is a passive device that cannot generate energy, and the antenna gain only describes the ability to effectively focus the radiation of energy in a particular direction. Increasing the antenna gain increases the point-to-point transmission distance, but at the same timecompresses the beam width, resulting in narrower beam coverage. The selection of antenna gain should be based on the premise that the beam width and the target area are matched, and it is undesirable to compress the beam width excessively in order to improve the gain.

The gain is generally expressed in logarithmic form, and the commonly used symbols are dBi, dBd, and dBiC, in which the reference of dBi is the ideal non-directional point source antenna, the reference of dBd is the half-wave oscillator antenna, and the reference of dBiC is the circularly polarized gain antenna. Since the gain of the half-wave oscillator antenna is 2.15 dBi, 0 dBd=2.15 dBi.

In the actual simulation, it is necessary to pay attention to the difference between different gains in the definition. Taking the full-wave simulation software High Frequency Structure Simulator as an example, the parameters related to gain are the directionality coefficient, gain, and realized gain. The first parameter does not consider antenna loss and matching; the second parameter considers antenna loss but not matching; the third parameter considers both antenna loss and matching. The third parameter should be chosen as the simulation data when comparing with the measured gain.

Radiation efficiency is the ratio of the power radiated out by the antenna (i.e., the power that is effectively converted into the electromagnetic wave portion) and the active power input into the antenna.

The sources of antenna loss are divided into two main categories: one is ohmic loss, and the other is reflection loss. Among them, ohmic loss can be subdivided into conductor loss and dielectric loss. Metal is not a perfect conductor in practice; its conductivity is not infinite when current flows through the metal conductor. The medium used in practice is not an ideal dielectric, and its loss angle tangent value is not 0, such as the FR4 dielectric substrate loss angle tangent value of 0.02; when there is a strong electromagnetic field within the medium, dielectric loss will be produced. Particular attention should be paid to the Dielectric Resonator Antenna "self-resonance" phenomenon, i.e., the antenna in a certain frequency band is well matched, but the radiation efficiency is very low. The reason for this situation is that the dielectric loss in this frequency band rises sharply; the electromagnetic energy is not radiated, but converted into thermal energy loss.

Improvement of antenna radiation efficiency requires improvement in three areas, i.e., selection of metals with high conductivity, dielectric substrates with low loss angle tangent, and improved antenna matching to reduce the reflection coefficient. However, for specific applications, fabrication cost and antenna size constraints also need to be considered.

Radiation efficiency can be expressed in both percentage and logarithmic form. In academia, it is mostly expressed as a percentage, which is more intuitive; in industry, it is mostly expressed in logarithmic form, which makes it easier to evaluate the antenna performance by substituting it into the whole wireless link budget, and the radiation efficiency expressed in logarithmic form can also be called antenna loss. The conversion relationship between the two is shown in Table 2.3.

TABLE 2.3 Conversion Relationship between Radiation Efficiency and Antenna Loss

Radiation Efficiency (%)	Antenna Loss (dB)
79.4	−1
63.1	−2
50.1	−3
31.6	−5
15.8	−8
10	−10

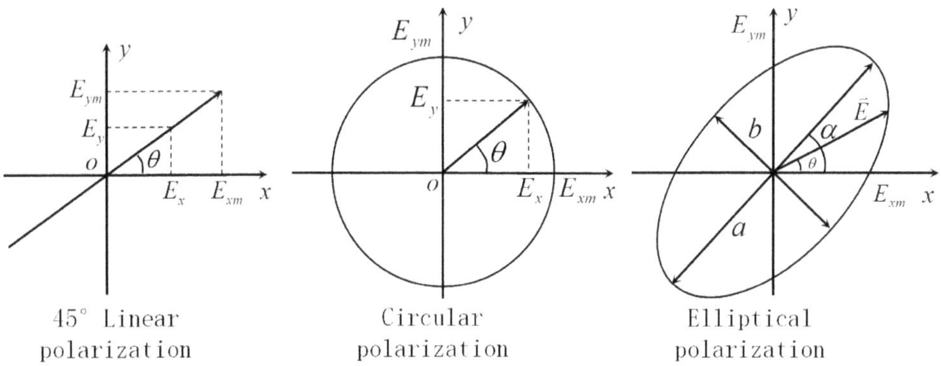

FIGURE 2.5 Typical trajectory curves of electromagnetic wave electric field vector orientation over time (subscript *m* stands for magnitude).

Polarization is the trajectory of the orientation of the electric field vector in space as a function of time at a point in space, viewed along the direction of propagation of an electromagnetic wave. Figure 2.5 shows a typical trajectory curve of the orientation of the electric field vector of an electromagnetic wave as a function of time. A fixed point on the electric field will change with time, if the trajectory of the electric field vector orientation with time is a straight line, it is called line polarization; if the trajectory is a circle, it is called circular polarization; if the trajectory trace is an ellipse, it is called elliptical polarization. (Elliptic) circular polarization can be further subdivided into left-handed (elliptic) circular polarization and right-handed (elliptic) circular polarization, according to the right-handed rule. It should be noted that the polarization is a local feature, which is analogous to the spin angular momentum (i.e. polarization) in optics or the rotation of the Earth. In contrast, vortex electromagnetic waves, which have become more popular in recent years, are macroscopic in nature and can be compared to orbital angular momentum in optics or the Earth's rotation around the Sun.

In a wireless communication system, the system can obtain the maximum transmission power only when the polarization of the receiving and transmitting antennas is matched. Polarization matching of the receiving and transmitting antennas means that the polarization of the receiving and transmitting antennas is consistent when the direction of the maximum electric field is aligned. If the polarization is not consistent and the polarization of the receiving antenna and the polarization of the transmitting antenna cannot be completely matched, then polarization loss will occur. Polarization loss is commonly expressed in logarithmic form, and the polarization loss between different polarization forms is shown in Table 2.4.

For cell phone antennas, polarization is not a key parameter, which is mainly due to two reasons: on the one hand, the size of the cell phone antenna is strictly limited, and the antenna mostly adopts the bent structure, which makes it difficult to ensure a purer polarization component; on the other hand, the base station antenna usually has ±45° dual-polarized transmit/receive capability, and it can make up for the loss of polarization through the reception of the dual-polarized electromagnetic wave. Even if the polarization

TABLE 2.4 Polarization Loss between Different Polarization Forms

Transmitting Antenna Polarization	Receiving Antenna Polarization	Polarization Loss (dB)
Vertical/horizontal	Vertical/horizontal	0
Vertical/horizontal	Vertical/horizontal	∞
Vertical/horizontal	Circular polarization	3
Left/right rotation	Left/right	0
Left/right	Right rotation/left rotation	∞

of the mobile phone antenna changes arbitrarily, a stable communication link can be established.

2.1.3 Mobile Terminal Antenna Specific Parameters

In the application of mobile terminal antenna, in addition to the conventional circuit and radiation characteristic parameters, there are also some specific parameters defined, which are convenient to visually measure the performance of the antenna in the carrier of a cell phone.

Total Radiated Power (TRP) is calculated as the integral of the power radiated in all directions over the spherical surface, representing the total power averaged across the sphere. It reflects the transmit power of the whole cell phone and is related to the transmit power and antenna radiation performance of the cell phone in the conduction situation. The definition of TRP is given in the standard of the Wireless Communications and Internet Association, which requires the measurement of the omnidirectional radiated power in spherical coordinates:

$$\mathrm{TRP}=\frac{1}{4\pi}\int_{\theta=0}^{\pi}\int_{\varphi=0}^{2\pi}\left[\mathrm{EIRP}_{\theta}(\theta,\varphi)+\mathrm{EIRP}_{\varphi}(\theta,\varphi)\right]\sin\theta d\theta d\varphi \tag{2.5}$$

where effective isotropic radiated power (EIRP) denotes the effective omnidirectional radiated power in θ*-component* or φ*-component.*

EIRP can be directly measured by the TRP measurement platform shown in Figure 2.6, and the TRP of the antenna can be obtained by data processing based on Eq. (2.5). Usually, the peak EIRP cannot be a direct measure of the good or bad performance of the cell phone antenna. For example, if the radiation pattern of the cell phone antenna system is highly directional, the peak EIRP is very large, and the EIRP of other directions is smaller instead, and the signal coverage of the whole space is poor; the closer the EIRP of the antenna in all directions is, the more uniform the signal coverage is, the bigger the calculated TRP is, and the better the performance of the transmitter is.

Total Isotropic Sensitivity (TIS) is calculated as the integral of the receiver sensitivity over the spherical surface, and is used to evaluate the performance of the receiver. A typical TIS measurement platform is shown in Figure 2.7. A higher measured TIS indicates better receiver performance. A low TIS can degrade the quality of the signal received by the user and even cause the user to lose base station information, resulting in call termination.

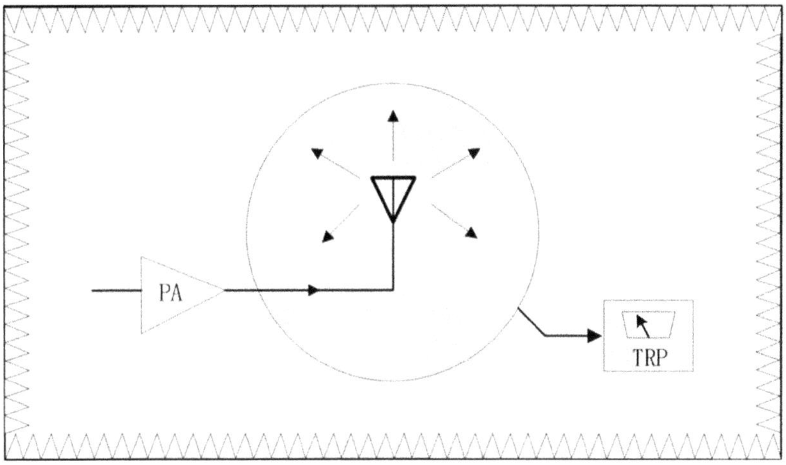

FIGURE 2.6 TRP measurement platform.

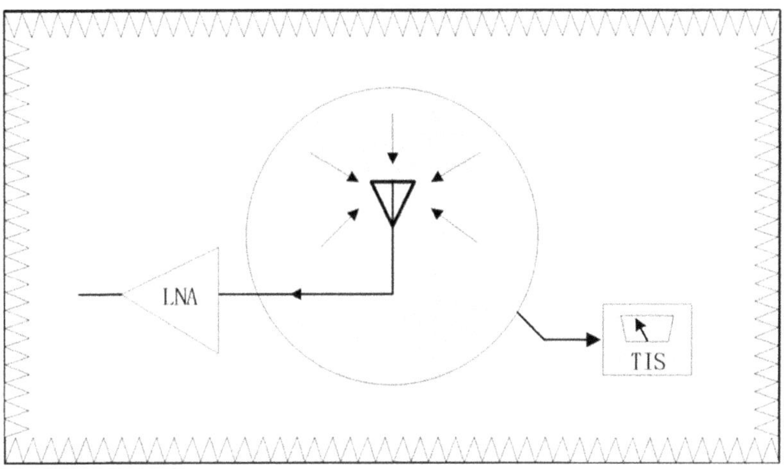

FIGURE 2.7 TIS measurement platform.

The Wireless Communications and Internet Association of America standard requires that the receiver's TIS be measured when the transmitter has maximum transmit power and that the TIS be measured in spherical coordinates:

$$\mathrm{TIS} = \frac{4\pi}{\int_{\theta=0}^{\pi}\int_{\varphi=0}^{2\pi}\left[\frac{1}{\mathrm{EIS}_{\theta}(\theta,\varphi)}+\frac{1}{\mathrm{EIS}_{\varphi}(\theta,\varphi)}\right]\sin\theta d\theta d\varphi} \tag{2.6}$$

where EIS denotes the effective isotropic sensitivity of *the θ-component* or *the φ-component*, which is used to measure the reception capability at a point, corresponding to the EIRP, which is used to measure the transmission capability.

TRP characterizes the transmitting performance of a cell phone, while TIS characterizes the receiving performance of a cell phone, both of which can comprehensively measure the transmitting and receiving capabilities of a cell phone.

Specific absorption rate (SAR) is used to assess the effects of cell phone radiation on the user. The known effects of electromagnetic radiation exposure at the operating frequencies of mobile devices such as cell phones are based on the heating of human tissue as a starting point. In the presence of an external electromagnetic field, the human body will generate an induced electromagnetic field. Since all organs of the human body are consumptive media, the induced EMF in the human body will generate an electric current, which in turn absorbs and dissipates the electromagnetic wave, which is the phenomenon of heating of human tissues by EMF. SAR is commonly used in biological dosimetry to characterize this physical heating process, and SAR is defined as the electromagnetic power absorbed or dissipated per unit mass of human tissue, with the unit of W/kg, and the calculation formula is shown in Eq. (2.7):

$$\mathrm{SAR} = \frac{\sigma E^2}{\rho} \tag{2.7}$$

where ρ represents density, σ represents conductivity, and E represents electric field strength. The international standard for SAR compliance: timed at 6 min, the electromagnetic power absorbed by each kilogram of brain tissue should not exceed 2 W, i.e., 2 W/kg (any 10 g of tissue fluid). In addition, the U.S. SAR compliance standard is 1.6 W/kg (any 1 g of tissue fluid); relatively, the U.S. SAR standard is more stringent.

SAR is related to the conductivity of the body medium. The smaller the conductivity, the smaller the induced high-frequency current, the less electromagnetic waves absorbed, and the smaller the SAR. The conductivity of the body medium is related to the frequency of the electromagnetic wave; in the range of 900–2400 MHz, the conductivity increases with increasing frequency.

For the TRP metric of over the air testing, it is generally desirable to have a larger TRP so that the power entering the antenna from the amplifier can be effectively radiated, and the better the performance of the transmitter. In the SAR test, it is hoped that the TRP is smaller, so that the electromagnetic power absorbed or dissipated by the human body tissues is smaller, to ensure that it can pass the SAR standard. Therefore, TRP and SAR are a pair of contradictory indicators, and how to ensure that both indicators in the antenna design meet the relevant standards needs to be considered comprehensively at the beginning of the antenna design.

2.2 ANTENNA MODES

Modes are the basic basis for analyzing the working principle of an antenna. Essentially, the radiation field of any antenna can be regarded as the electric field solution or magnetic field solution of Maxwell's equation system under specific boundary and excitation. Different solutions correspond to different current or electric field distributions in the antenna structure. The current distribution or electric field distribution in the antenna structure is called a mode.

2.2.1 Theoretical Basis of Modes

Starting from Maxwell's system of equations, the fluctuation equations, also known as Helmholtz equations, can be derived in the region of a passive homogeneous medium containing only electric or magnetic fields. The time-varying electric field satisfies Eq. (2.8):

$$\nabla^2 \boldsymbol{E} = \mu\sigma \frac{\partial \boldsymbol{E}}{\partial t} + \mu\sigma \frac{\partial^2 \boldsymbol{E}}{\partial t^2} \tag{2.8}$$

where $\boldsymbol{E}$ represents the electric field, and σ represents the electric conductivity. For a sinusoidal electromagnetic field, the complex vectors of the electric and magnetic fields under passive conditions satisfy the vector chi-square Helmholtz equation, and the expression can be simplified as:

$$\nabla^2 \boldsymbol{E} + k^2 \boldsymbol{E} = 0 \tag{2.9}$$

where k represents the wave number. In the right-angle coordinate system, the components of the complex vector of the electromagnetic field satisfy the scalar chi-square Helmholtz equation, taking *the x* component as an example:

$$E_x = f(x)g(y)h(z) \tag{2.10}$$

Solve using the method of separated variables:

$$\begin{cases} \dfrac{d^2 f(x)}{dx^2} + k_x^2 f(x) = 0 \\ \dfrac{d^2 g(y)}{dy^2} + k_y^2 g(y) = 0 \\ \dfrac{d^2 h(z)}{dz^2} + k_z^2 h(z) = 0 \end{cases} \tag{2.11}$$

The generalized solutions of each of the above three equations are:

$$\begin{cases} f(x) = A_1 \cos(k_x x) + A_2 \sin(k_x x) \\ g(y) = B_1 \cos(k_y y) + B_2 \sin(k_y y)\,. \\ h(z) = C_1 \cos(k_z z) + C_2 \sin(k_z z) \end{cases} \tag{2.12}$$

Substituting the boundary conditions into the expressions, the corresponding solutions can be found theoretically. However, in practical engineering, the boundary conditions of

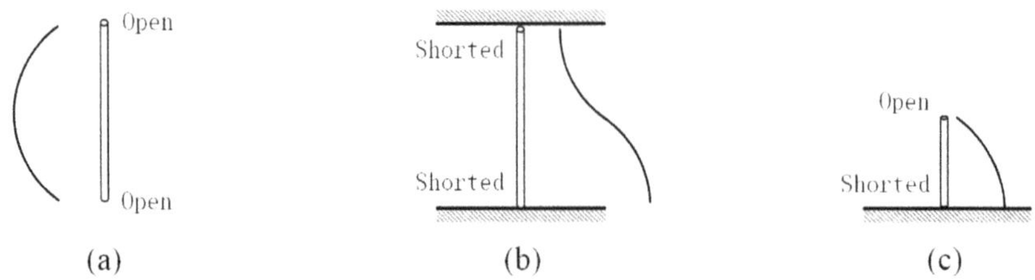

FIGURE 2.8 Three cases of wire antenna forms: (a) both ends open at the same time, (b) both ends shorted at the same time, and (c) one end open and one end shorted.

the sky line are often more complicated and difficult to solve by analytical methods, which are more commonly solved with the help of numerical analysis algorithms in full-wave simulation software. However, field analysis based on the symmetry of the structure helps to understand the working principle of the antenna.

Antennas can be classified into linear antennas, planar antennas, and volumetric antennas based on whether their structural distribution is primarily one-dimensional, two-dimensional, or three-dimensional, respectively. Each type supports distinct operational modes and radiation characteristics.

2.2.2 Wire Antenna Modes

A wire antenna only has an electric field or a current change in one dimension; this kind of antenna has small wind resistance and is lightweight because of its one-dimensional distribution, especially suitable for low frequency bands such as meter waves. Wire antennas are simple in structure and can be reduced in size by bending.

As shown in Figure 2.8, according to the one-dimensional boundary conditions, there are the following three cases of the wire antenna form, which are open at both ends at the same time, shorted at both ends at the same time, open at one end and shorted at one end, and the corresponding modes are the one-half-wavelength mode, one-half-wavelength mode, and one-quarter-wavelength mode, respectively.

2.2.3 Surface Antenna Modes

In a surface antenna, the electric field or current changes only in two dimensions. The antenna is a flat or curved surface distribution and can be used in the planar circuit board process for batch production. This antenna is the mainstream form of realization of the microwave band.

As shown in Figure 2.9, according to the two-dimensional boundary conditions, the surface antenna form has the following three cases: the boundary conditions in each dimension are similar to those of the line antenna, and the corresponding modes are also divided into one-half wavelength mode and one-quarter wavelength mode. Due to the mirroring principle, the third case requires the presence of a floor, and the antenna can be short-circuited to the floor via a metal surface or a metal post. For miniaturized designs, the quarter-wavelength mode can be used in both dimensions simultaneously. Since surface antennas are typically printed on dielectric substrates, the physical size of the antenna

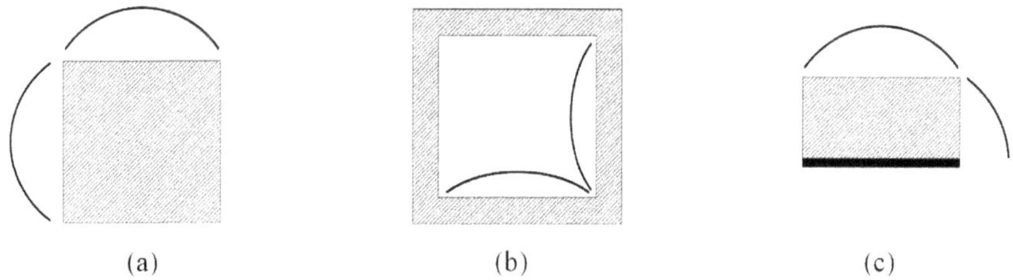

FIGURE 2.9 Three cases of faceted antenna forms: (a) both dimensions satisfy the open circuit boundary condition, (b) both dimensions satisfy the short circuit boundary condition, and (c) one dimension satisfies the open-circuit boundary condition and the other dimension satisfies the open-short-circuit boundary condition.

FIGURE 2.10 Two cases of the body antenna forms: (a) all faces satisfy the open circuit boundary condition and (b) five faces satisfy the open circuit boundary condition.

can be reduced by loading it with a high dielectric constant while keeping the electrical dimensions constant.

2.2.4 Body Antenna Modes

The body antenna has electric field or current changes in all three dimensions, because the antenna can be adjusted in all three dimensions. Therefore, the degree of freedom of the design is larger, and it is commonly used in Dielectric Resonator Antenna, but the antenna is more complicated to fabricate in a three-dimensional structure.

As shown in Figure 2.10, according to the three-dimensional boundary conditions, the body antenna form has the following two cases: one is that all the faces satisfy the open-circuit boundary conditions and work in one-half wavelength mode in all three dimensions; the other is that one face is placed on the infinite metal floor, and the remaining five faces satisfy the open-circuit boundary conditions. Due to the mirroring effect of the floor, the antenna operates in the quarter-wavelength mode in the dimension perpendicular to the floor, and the size can be reduced by half.

2.2.5 Multi-Mode Synergy

After analyzing the inherent properties of the antenna using various mode analysis methods, we have a qualitative understanding of the full range of modes that the antenna may support, i.e., we understand the passivity of the antenna. Based on this, we can further analyze the performance of the antenna under active excitation to produce a specific radiation pattern.

The type and location of the feed source are determined by the field distribution of the mode to be excited. Generally, placing a capacitive feeder at the voltage maximum or an inductive feeder at the current maximum maximizes the coupling of the mode to the feeder, thus effectively exciting a specific mode.

Currently, on the one hand, antennas are developing in the direction of miniaturization and multifunctionality, and it is difficult for a single mode to satisfy the multifunctional requirements of antennas; the performance of antennas will be further deteriorated when the antenna size is reduced. On the other hand, the antenna itself supports multiple modes, and the performance of the antenna can be further improved if multiple modes are simultaneously excited by a suitable feed. Therefore, multi-antenna cooperative work provides a new idea for the design of functional small antennas.

For the modes in antenna design, the synergy is mainly reflected in utilizing the commonality or difference between multiple modes to find ways to improve the antenna performance. After summarization, the enhancement effect of multi-antenna synergy is mainly reflected in three aspects [2]: expanding the operating bandwidth, increasing the orthogonal patterns, and improving the characteristics of the radiation pattern.

The difficulty of multi-mode synergy lies in finding or constructing multiple modes with commonality or difference. For example, in order to extend the impedance bandwidth of the antenna, it is necessary to find multiple modes with similar current distributions, and through certain means to make these modes coincide with the location of the current maximum or voltage maximum, and then place the feed source at this coincident point to simultaneously stimulate these modes, and the combination of multiple modes can effectively extend the impedance bandwidth of the antenna. For example, if the antenna has multiple resonant modes at the same frequency point, and the current directions of these modes are orthogonal to each other, these orthogonal modes can be utilized to increase the number of channels and realize a multi-antenna design. In addition, the current distribution of intrinsic modes determines their radiation pattern characteristics. The radiation patterns of different modes vary greatly, with some radiating in the side-shooting direction, some in the horizontal direction, some with high gain, and some with wide beam. Therefore, a single mode cannot achieve both high gain and wide beam at the same time. Multi-mode synergy, on the other hand, provides a new option to improve the characteristics of the radiation pattern and to realize dynamic wide coverage in the airspace through reconfiguration.

2.3 TYPICAL MOBILE TERMINAL ANTENNA FORMS

According to the different working principles of the pattern, the typical mobile terminal antenna can be divided into several typical forms, including monopole antenna, Planar Inverted-F Antenna (PIFA), slot antenna, loop antenna, and so on.

2.3.1 Monopole Antenna

Monopole antenna is one of the most widely used mobile terminal antenna forms, and the earliest commercial cellular phones use monopole antenna as the radiating antenna. The monopole antenna adopts a wire antenna structure, with one end open and the other end

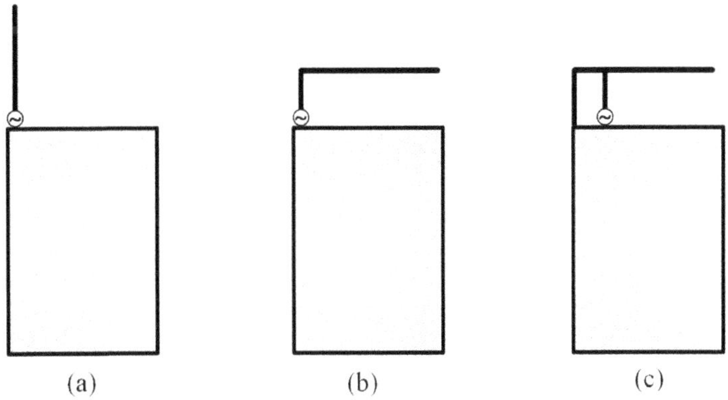

FIGURE 2.11 Basic structure of monopole antenna: (a) straight monopole, (b) inverted L-shaped monopole, and (c) inverted F-shaped IFA structure.

connected to the cell phone floor through a feed port. Due to the mirroring effect of the floor, the monopole antenna conforms to the boundary condition of an open circuit at one end and a short circuit at the other and thus operates in the quarter-wavelength ($\lambda/4$) mode at its lowest resonant frequency. Other resonant modes include higher sub-modes such as three-quarter wavelength, five-quarter wavelength, and seven-quarter wavelength.

The basic structure of a monopole antenna is shown in Figure 2.11. Early external antennas with low antenna size requirements could be in the form of a straight monopole. Since some of the mirror current on the floor also participates in the radiation, the floor size needs to be optimized during design. As the antenna evolved from an external to an internal form, the straight monopole structure could be bent into an inverted L-shape in order to meet the size requirements, and the antenna still operated in the quarter-wavelength mode, but the price of miniaturization was a degradation of antenna performance. A simple explanation is that the horizontal arm of the inverted L-shaped monopole, being bent close to the floor, induces a current of opposite phase, and the radiated fields generated by the two parts of the current cancel each other out in space, with the effective radiated length being only the vertical arm portion. In order to improve the impedance matching of the feed port, the inverted F-shaped Inverted-F Antenna (IFA) structure can be used, i.e., a short-circuit branch is connected in parallel near the feed point of the inverted L-shaped monopole, which is used to act as a shunt inductor, and good impedance matching can be achieved in a single frequency band by adjusting the length, width, and position of the short-circuit branch. In the actual project, inductors or capacitors can also be added to build a matching network to extend the impedance bandwidth of the monopole antenna.

2.3.2 PIFA

The three-dimensional structure of the PIFA is shown in Figure 2.12, which contains a piece of cell phone floor, a suspended metal sheet, a short-circuiting metal strip, and a feeder wire. There is a certain distance between the metal sheet and the cell phone floor to prevent the mirrored current on the floor from completely canceling the current on the metal sheet. The PIFA can be transformed from the IFA structure by simply rotating the

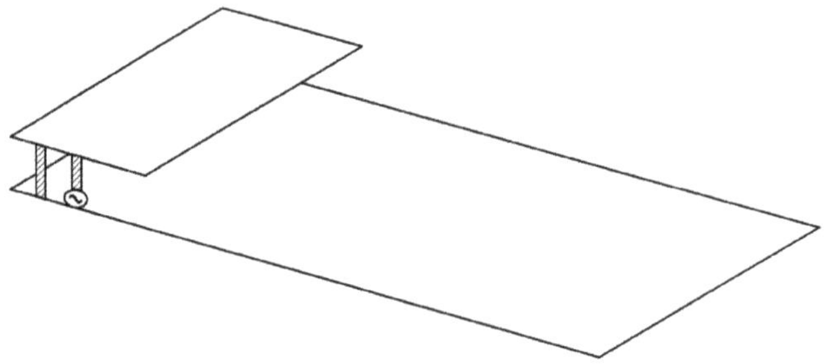

FIGURE 2.12 3D structure of PIFA.

IFA from the end-fire direction of the cell phone floor to the broadside direction and then stretching the wire antenna into a planar antenna. Built-in PIFAs were widely used in cellular phones in the 1990s, such as Nokia's 2100 model. However, with the development of cellular phones toward the trend of flattening, the disadvantage of the high profile of PIFA has limited its application range.

The metal sheet of the PIFA is open-circuited on one side and short-circuited to the ground plane via a vertical metal strip on the other side, satisfying the boundary conditions for the quarter-wavelength mode. The shorter side of the metal sheet is relatively short, with a constant field distribution along it, and the electric field is oriented perpendicular to both the ground plane and the metal sheet. Therefore, the fundamental operating mode of the PIFA is the $TM_{\frac{1}{2}0}$ mode. The bandwidth of the PIFA is directly related to the distance between the metal sheet and the floor, and the larger the distance, the larger the bandwidth. In addition, the width of the metal strip also has a certain effect on the bandwidth. In practical applications, the size of PIFA is often reduced by bending the metal strip, and the number of resonant branches is increased, so as to realize the miniaturization and multi-antenna design of cell phone antenna.

2.3.3 Slot Antenna

Slot antenna or slit antenna is a commonly used form of cell phone antenna, especially in the era when metal bezels and metal back covers are more popular. The use of a metal bezel or a metal back cover can enhance the texture of the cell phone and also increase the mechanical strength of the cell phone. However, RF signals cannot penetrate metal. In order to reduce this radiation obstruction, slot antenna technology is gradually applied to the antenna design of cell phones with metal bezels or metal back covers. According to the Babine principle, the gap on the infinite metal plate can be equated to a magnetic dipole in free space, so the typical gap length of slot antenna is one-half of the working wavelength, and the slot antenna is structurally complementary to the half-wave electric dipole. As shown in Figure 2.13, slot antennas commonly used in cell phone antenna design are divided into closed slot antennas and open slot antennas, in which the open slot antenna is complementary to the monopole antenna with a quarter-wavelength mode of operation, which is more widely used.

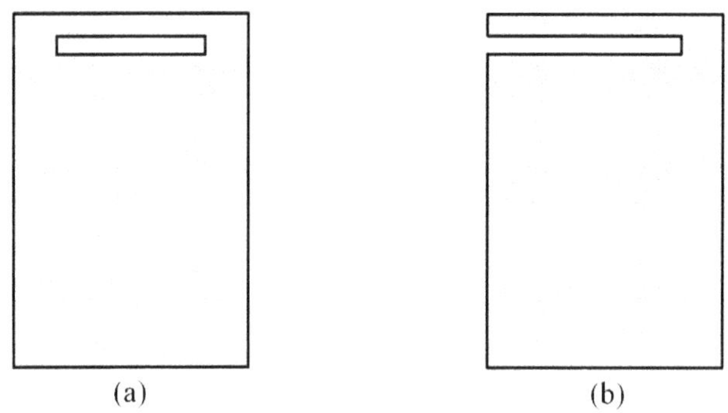

FIGURE 2.13 Slot antenna structure: (a) closed slot antenna and (b) open slot antenna.

FIGURE 2.14 loop antenna structure.

2.3.4 Loop Antenna

A loop antenna is a type of wire antenna. When the physical size of the ring is much smaller than the resonance wavelength, this antenna is called an electric small ring antenna, and its radiation ability is weak; when the physical size of the ring is close to or larger than the resonance wavelength, this antenna is called an electric large ring antenna, and its radiation ability is strong. The electric large ring antenna is used in cell phone antenna design; the antenna circumference is about a working wavelength.

A conventional loop antenna consists of a circular metal ring. In the process of cell phone antenna design, in order to reduce the area of the ring antenna, the ring antenna is usually bent into the structure shown in Figure 2.14, and the ring antenna has two connection points with the floor, which can be used to realize the ring antenna's differential-mode and common-mode excitation. The differential mode signal makes the antenna work in one wavelength mode, and the common mode signal makes the antenna work in two parallel one-half wavelength modes. In the low-frequency band, the cell phone floor has less effect on the differential mode performance of the antenna, and the ring antenna can achieve a large operating bandwidth, but the physical size of the ring antenna is larger than that of a monopole antenna.

2.4 MULTIPLE-INPUT MULTIPLE-OUTPUT ANTENNA DECOUPLING PRINCIPLE

In order to meet the demand for high wireless transmission rates, Multiple-Input Multiple-Output (MIMO) technology can be used to increase the transmission rate exponentially in the case of limited bandwidth resources, and MIMO technology is also one of the important technologies in the field of mobile communication in recent decades. Compared with the early single-input single-output system, the MIMO system arranges multiple antennas at the transceiver end at the same time, which can greatly improve the spectrum utilization, but the antennas must be independent and uncorrelated with each other.

In carriers such as cell phones, where the installation space of antennas is strictly limited, integrating multiple antennas faces great challenges: on the one hand, the development trend of cell phones, such as high screen-to-body ratio, use of metal backshells, and thinness, makes the ground clearance of antennas smaller and smaller, and the space of individual antennas is severely compressed; on the other hand, there is strong coupling between multiple antennas in close proximity, and the signals received or transmitted by antennas affect each other, while the MIMO technology theoretically requires that there is no correlation between antennas, and the stronger the correlation, the weaker the ability of MIMO technology to enhance channel capacity. How to reduce the coupling strength between multiple antennas under the limit of small ground clearance and small cell spacing is a topic worthy of in-depth study.

2.4.1 Antenna Coupling Sources

There are various coupling relationships between antennas in a mobile terminal. Taking the MIMO system containing two antennas as an example, the typical coupling can be categorized into four types, including surface wave coupling between antenna units, current coupling, spatial propagation coupling, spatial scattering coupling, etc., as shown in Figure 2.15.

Surface wave coupling refers to the case where two antennas are deployed on the same dielectric substrate, where surface wave propagation is supported by the dielectric substrate, resulting in electromagnetic wave transmission through the dielectric substrate, forming a surface wave coupling path. Generally, the thicker the dielectric substrate, the larger the relative dielectric constant, the easier it is to form surface waves.

Current coupling refers to the case where two antennas share the same metal floor, and current is transmitted through the floor from one antenna port to the other antenna port, forming a current coupling path.

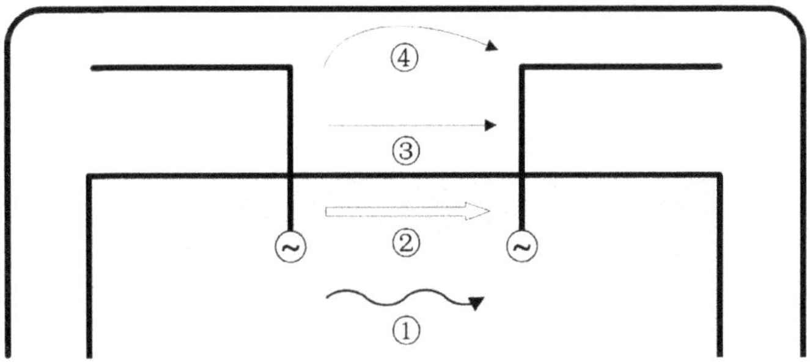

FIGURE 2.15 Source of coupling between two antennas of a mobile terminal.

Spatial propagation coupling refers to the existence of a direct propagation path between two antennas, where electromagnetic waves are transmitted directly through three-dimensional free space, forming a spatial propagation coupling path. In general, the closer the two antennas are, the greater the strength of the spatial propagation coupling.

Spatial scattering coupling refers to the existence of additional propagation paths between two antennas besides the direct path, such as those caused by reflection or scattering from metal bezels, cameras, or other components, which form spatial scattering coupling paths.

The above coupling has different coupling mechanisms, coupling paths, and coupling strengths, which are superimposed on each other, constituting the coupling of a multi-antenna system, which is macroscopically manifested as the transmission coefficient in the scattering matrix.

In order to reduce the coupling strength of a multi-antenna system, multiple decoupling methods can be used to remove the correlation between antennas. According to the decoupling principle, the relevant decoupling methods can be classified into three categories, which are adding resonant structures to form a stop band, introducing controlled coupling to realize energy pair cancellation, and decoupling through mode quadrature. Taking the dual antenna of a mobile terminal with a metal bezel as an example, the realization principles of various decoupling methods are introduced next.

2.4.2 Adding Resonant Structure to Form a Stop Band

The basic idea of the decoupling method of adding resonant structures to form a blocking band is to block the coupling path between the two antennas. Typical resonant structures include defected ground structure (DGS), parasitic resonant branch structure, and artificial electromagnetic materials [such as electromagnetic bandgap and artificial magnetic conductor (AMC)], and the working principle of forming a blocking band with these structures will be introduced next.

A dual antenna model based on defect ground decoupling is shown in Figure 2.16. Two monopole antennas are placed back-to-back, both spaced a certain distance apart, and the feed ports are located on the floor of the cell phone, generally connecting the antennas to the feed ports via microstrip or coaxial lines. An inverted T-slot is carved on the top of the floor, which is generally symmetric about the central axis of the two antennas, and the

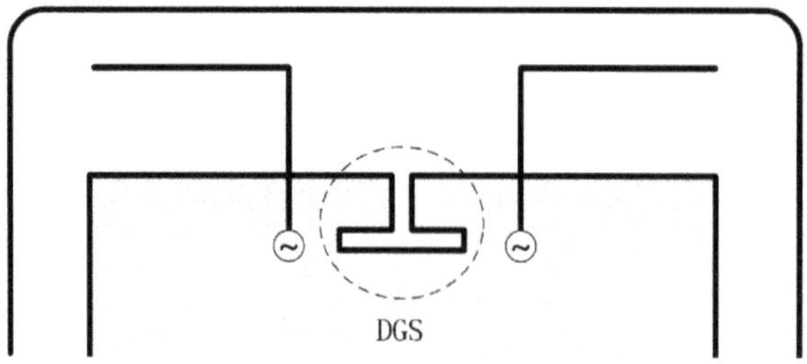

FIGURE 2.16 Dual antenna model based on defect ground structure decoupling.

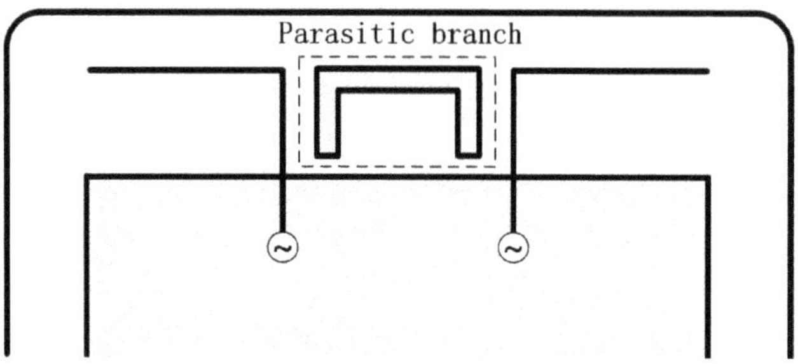

FIGURE 2.17 Dual antenna model based on parasitic resonant branch decoupling.

total length of the T-slot is about one quarter of the operating wavelength, which meets the boundary conditions of open circuit at one end and short circuit at the other end, and the decoupled structure is in a resonant state.

The basic principle of defected ground decoupling: Defected ground technology introduces a new resonant structure on the floor, adding a controlled resonator between two antennas. By adjusting the size of the DGS, the coupling coefficients between the three resonators can be changed, making the coupling coefficients between the two antennas as small as possible, thus reducing the amount of energy that the two antennas are coupled through the floor path. In addition, a similar effect can be achieved by mirroring the DGS above the floor to form a T-shaped raised structure according to the principle of complementarity.

A dual antenna model based on parasitic resonant branch decoupling is shown in Figure 2.17, which adds a suspended metal strip between two back-to-back monopole antennas to form a parasitic resonant branch. In order to adjust the length of the parasitic resonant branch more flexibly, the metal strip is bent into an inverted U shape. The total length of the inverted U-shaped parasitic resonant branch is about one-half of the operating wavelength, which meets the boundary condition of an open circuit at both ends, and this decoupled structure is also in a resonant state.

The basic principle of parasitic resonant branch decoupling: Introduce a new resonant structure between two antennas; by adjusting the length of the parasitic resonant branch and the spacing between the parasitic resonant branch and the two antennas, the coupling coefficients between the three resonators can be changed, making the coupling coefficients between the two antennas as small as possible, so as to reduce the energy of the two antennas through the spatial path of coupling.

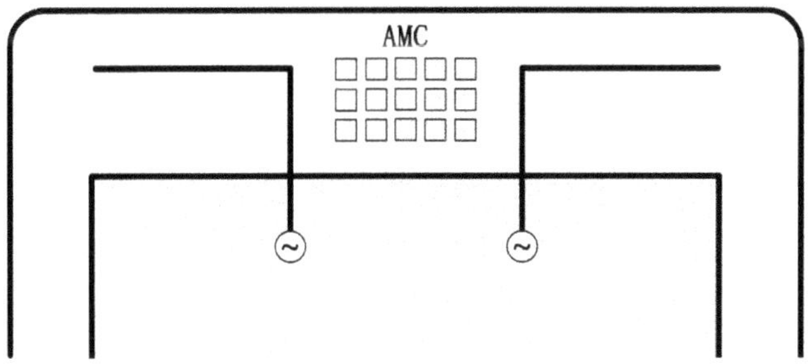

FIGURE 2.18 Dual antenna model based on AMC decoupling.

Figure 2.18 shows a dual-antenna model based on AMC decoupling, with periodic rows of AMC added between the two back-to-back monopole antennas. AMC cell size belongs to the sub-wavelength range, i.e., the cell size is much smaller than the wavelength of the medium. A large number of sub-wavelength cells can effectively regulate the direction of the electromagnetic wave.

The basic principle of AMC decoupling: The sub-wavelength units arranged periodically between two antennas can generate multiple space harmonics, and by adjusting the unit size, unit spacing, and the number of units of AMC, the spatial coupling coefficient between space harmonics and the two antennas can be controlled, so as to reduce the energy of the two antennas coupled through the spatial path.

2.4.3 Introducing Controlled Coupling to Realize Energy Cancellation

Different from the working principle of adding a resonant structure to form a stop band to block the coupling path between two antennas, the decoupling method of introducing controlled coupling to realize energy pair cancellation introduces controlled coupling paths, so that the signals on the new coupling paths are equal in amplitude to the signals on the existing coupling paths with a phase difference of 180°, so as to realize the pair cancellation of signals on the two coupling paths, and the specific techniques include the introduction of neutralization lines and the introduction of decoupling circuits. The decoupling method based on introducing controlled coupling differs from the resonant stop-band approach. Instead of blocking the original coupling path, it creates a controlled coupling path. By making the signal on this new path equal in amplitude but opposite in phase (180° out of phase) to the signal on the existing path, the two signals cancel each other out. The specific techniques include the introduction of neutralization lines and the introduction of decoupling circuits.

Figure 2.19 shows a dual-antenna model based on neutralization line decoupling, which adds a bent connecting line between two back-to-back monopole antennas, and the connecting line is generally made of a metal strip. The length and width of the neutralization line and its connection position to the two antennas can be flexibly adjusted.

The basic principle of neutralization line decoupling: The neutralization line introduces a controlled coupling path between two antennas. By adjusting the line width of the neutralization line, the amplitude of the signal on the coupling path can be adjusted, and by adjusting the length of the neutralization line, the delay and phase of the signal on the

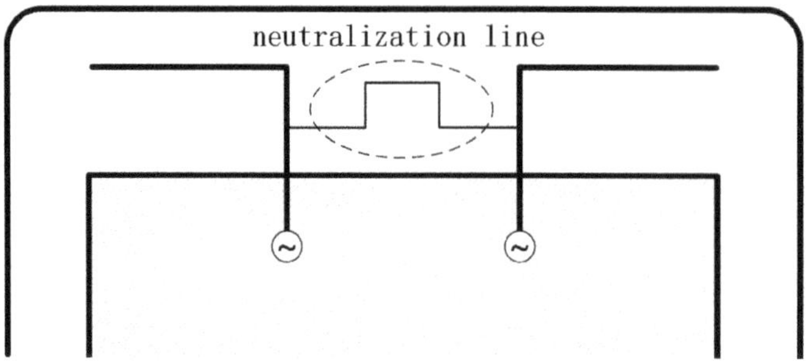

FIGURE 2.19 Dual antenna model based on neutralization line decoupling.

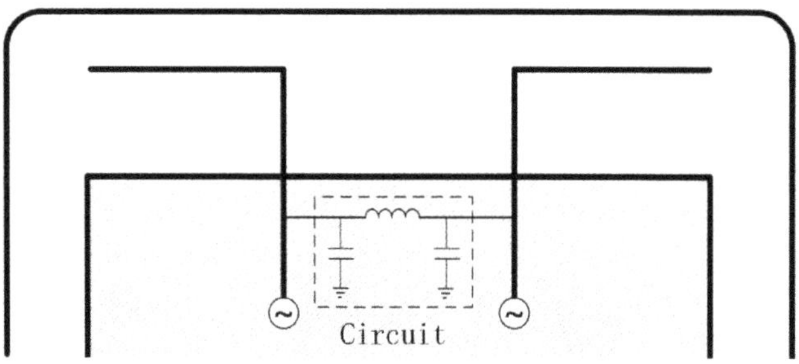

FIGURE 2.20 Dual antenna model based on decoupling circuit.

coupling path can be adjusted, so that the signals on the controlled coupling path can realize the current cancellation with the coupled signals between the two antennas. In addition, in order to realize wide-band decoupling, multiple neutralization lines can be introduced to construct a cascaded decoupling network to realize current cancellation in a wide frequency band.

Figure 2.20 shows a dual antenna model based on decoupling-circuit decoupling, which adds a decoupling circuit constructed by the lumped components between the feed ports of two back-to-back monopole antennas. The realization form of the circuit, the number of lumped components, and the connection position between the circuit and the feed line can be flexibly adjusted.

The basic principle of circuit decoupling: Decoupling circuit introduces a controlled coupling path between two antennas; by adjusting the parameters of the inductance, capacitance, and other devices in the decoupling circuit, it can change the strength and phase of the current on the controlled coupling path, so that the signals on the controlled coupling path and the coupled signals between the two antennas can realize the current pair cancellation. Compared with neutralization line decoupling, decoupling-circuit decoupling has a greater degree of design freedom and greater versatility, and is especially suitable for decoupling problems under strong mutual coupling conditions.

2.4.4 Decoupling by Mode Orthogonality

According to the eigenmode theory, the mobile terminal platform itself supports multiple modes, which may have similar or complementary field distributions, and the orthogonality

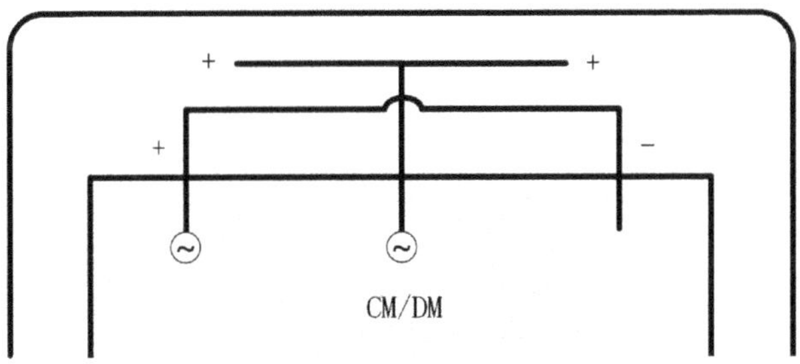

FIGURE 2.21 Dual antenna model based on common mode/differential mode theoretical decoupling.

between the modes can be realized by selecting a suitable external feeder to stimulate a specific mode. The advantage of this decoupling method is that no additional decoupling structure is required and port isolation is high. Two typical decoupling methods are common mode/differential mode theory and polarization quadrature.

Figure 2.21 shows a dual-antenna model based on common mode/differential mode theory decoupling, which has two antennas realized in different forms, a T-shaped monopole structure and a U-shaped dipole structure, with the feed ports of the monopole antenna on the centerline and the dipole antenna at the corners of the floor. The left and right ends of the monopole antenna are open-circuited and center-fed, and the length from the center to the left or right end is approximately one-quarter of the operating wavelength, which meets the boundary conditions for a quarter-wavelength resonator. The dipole antenna is open at the right end and has a length of approximately one-half the working wavelength, meeting the boundary conditions for a one-half wavelength resonator.

The basic principle of common-mode/differential-mode theory decoupling: The monopole antenna works in the common-mode state due to its symmetrical structure and center-fed; the dipole antenna is fed from one side, and the one-half-wavelength resonator can be excited to work in the differential-mode state. The currents on the two arms in the common-mode and differential-mode states behave differently (i.e., the currents on the two arms in the common-mode state are in opposite phases, and the currents on the two arms in the differential-mode state are in the same phase), and the radiated electric fields of the two modes are orthogonal to each other, so as to achieve high isolation of the two antennas. In addition, to improve the symmetry of the differential mode antenna, two feed ports can be used to form a differential signal.

Figure 2.22 shows a dual antenna model based on polarization orthogonality decoupling, which has two monopole antennas with the same implementation form, which are placed at the long and short edges of the phone floor. The relative placement of the two antennas has various forms, including simultaneous clockwise placement, simultaneous counterclockwise placement, one clockwise placement and one counterclockwise placement.

The basic principle of polarization orthogonality decoupling: Two monopole antennas excite the currents on the long side and short side of the cell phone floor, respectively, and since the two sides are orthogonal, the corresponding antenna polarization is also orthogonal, and there exists a natural isolation between the orthogonal polarizations, which ensures a high isolation of the two antennas.

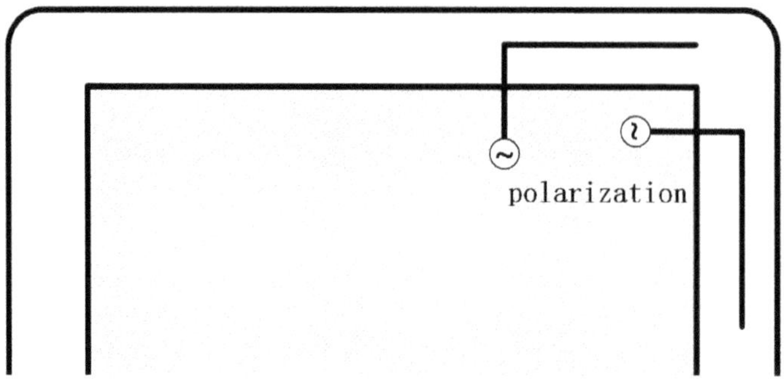

FIGURE 2.22 Dual antenna model based on polarization orthogonality decoupling.

2.5 PHASED ARRAY ANTENNA BEAM SCANNING PRINCIPLE

Phased array antenna is the current mainstream form of beam modulation realization, which has a wide range of applications in the field of 5G mobile communication beam fouling, new system radar target detection, and so on. By forming an array of units of the same form, and then applying a specific feed phase to each unit to form the desired phase gradient distribution, a specific pointing beam can be generated. One of the most basic and typical types of phased array architecture is represented by a uniformly excited one-dimensional equally spaced linear array.

2.5.1 One-Dimensional Equally Spaced Linear Array Radiation Pattern

According to the directional map product theorem, it is known that the directional map of an antenna array composed of similar cells is determined by the product of the cell directional map and the array factor. The array structure of a one-dimensional equally spaced linear array is shown in Figure 2.23, with N isotropic point sources instead of the actual cells and lined up along the z-axis co-linearly, and the spacing of the neighboring cells is d. If there is a linear gradient phase of excitation current between the cells, the excitation current I_n can be expressed as:

$$I_n = A_n e^{jn\alpha}, \quad n = 0, \cdots, N-1 \tag{2.13}$$

where A_n denotes the amplitude of the excitation current of the nth cell and α denotes the phase difference of the excitation current between neighboring cells. Considering the case

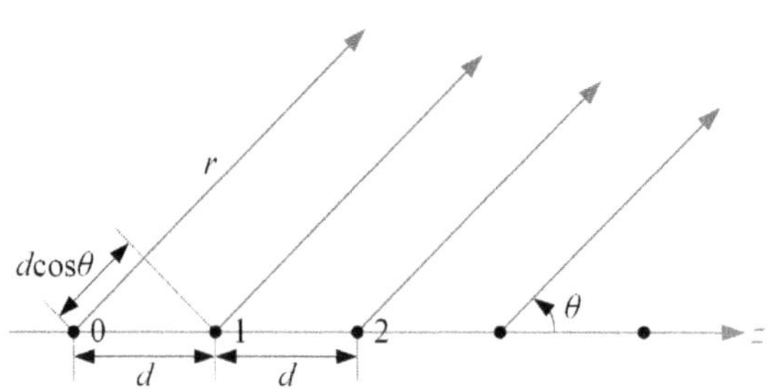

FIGURE 2.23 Structure of a one-dimensional equally spaced linear array.

of uniform excitation of the array, i.e., the amplitude of the excitation currents of all cells is equal to A_0, the array factor AF is:

$$\mathrm{AF} = A_0 + A_0 e^{j\psi} + \cdots + A_0 e^{j(N-1)\psi} = A_0 \sum_{n=0}^{N-1} e^{jn\psi} \tag{2.14}$$

$$\psi = \beta d \cos\theta + \alpha \tag{2.15}$$

where β is the phase constant in free space and ψ is the far-field phase difference generated by any neighboring cell, including the wave range difference introduced by cell spacing and α.

Summing Eq. (2.14) in an isoparametric series and omitting the common phase factor terms present gives the normalized array factor:

$$f(\psi) = \frac{\sin(N\psi/2)}{N\sin(\psi/2)} \tag{2.16}$$

In order to obtain the maximum value of the normalized array factor, it is necessary to satisfy:

$$\theta_m = \arccos\left[\frac{1}{\beta d}(-\alpha \pm 2m\pi)\right],\ m = 0,1,2,\cdots \tag{2.17}$$

When $m = 0$, θ_0 corresponds to the maximum value of the main beam of the array factor, and the phase difference of the excitation current between neighboring cells should be satisfied:

$$\alpha = -\beta d \cos\theta_0 \tag{2.18}$$

Accordingly, controlling the step phase of the excitation current can make the maximum radiation direction of the array factor point to a specific angle and realize the electronically controlled scanning of the beam. In particular, when θ_0=90°, the broadside radiation pattern is formed; when θ_0=0° or 180°, the end-fire radiation pattern is formed.

When m takes a non-zero value, the array factor produces additional grating lobes. In order to avoid the grating lobes falling into the visible region, when the maximum scanning angle is θ_m, the cell spacing d needs to satisfy the suppression condition:

$$d < \frac{\lambda}{1+|\cos\theta_m|} \tag{2.19}$$

2.5.2 Scattering Characteristics of Phased Array Antenna Parameters

The phased array system requires several excitation sources of the antenna to work simultaneously, and the mutual coupling between the units will inevitably affect the input impedance of each unit. Conventional passive S-parameters only consider the scattering characteristics of a feed port when it works independently, but they cannot be used to evaluate the design performance of the phased array antenna under different feed excitations. Especially when the amplitude and phase of the excitation currents of each unit of the antenna are significantly different, such as in the wide-angle beam scanning scenario, the severe mutual coupling will significantly deteriorate the impedance matching performance of the units.

The active S-parameter reflects the scattering characteristics of the port when other ports are excited simultaneously, contains the coupling effect of the port with other ports or units, and is an effective description of the actual working state of the active phased array antenna. Usually, the active S-parameters can be expressed as a linear combination of the passive S-parameters [3]. When the excitation a_n of the nth cell is not zero, the following relation exists:

$$\text{active } S_{mn} = \sum_{i=1}^{N} S_{mi}\frac{a_i}{a_n} = \sum_{i=1}^{N} S_{mi}\frac{A_i e^{j\varphi_i}}{A_n e^{j\varphi_n}} \quad m,n=1,\cdots,N \tag{2.20}$$

where S_{mi} is the passive coupling coefficient of the ith and mth cell, A_n and φ_n are the excitation amplitude and phase of the nth cell, and N is the number of cells or feed ports.

In particular, the active S-parameters are obtained when $m = n$:

$$\text{active } S_{mm} = \sum_{i=1}^{N} S_{mi}\frac{a_i}{a_m} = \sum_{i=1}^{N} S_{mi}\frac{A_i e^{j\varphi_i}}{A_m e^{j\varphi_m}} \quad m=1,\cdots,N \tag{2.21}$$

From this, the reciprocal relationship can be established:

$$\text{active } S_{mm} = \frac{a_n}{a_m}\text{active } S_{mn} \quad m,n=1,\cdots,N \tag{2.22}$$

Other relevant active parameters can be obtained based on the active S-parameter transformation. For example, for the active VSWR, the following definition is given:

$$\text{active VSWR}_{mm} = \frac{1+|\text{active } S_{mm}|}{1-|\text{active } S_{mm}|} \quad m=1,\cdots,N \tag{2.23}$$

Based on the above definition, for N-element phased arrays, the extraction of the complete active S-parameters requires the completion of $N(N-1)/2$ vector network analyzer-based passive tests at both ports, and the test complexity will rise significantly with the increase of array size. In this regard, the testing process can be simplified within a reasonable error range by using the center cell [4] or by means of directional coupler [5] in practical engineering.

2.5.3 Typical Cell Forms of Phased Array Antennas

2.5.3.1 Patch antenna

As one of the simplest realization forms of planar printed antenna, the patch antenna always plays an important role in the field of wireless communication by virtue of the advantages of small mass, low profile, and easy commonality. Figure 2.24 shows the geometric

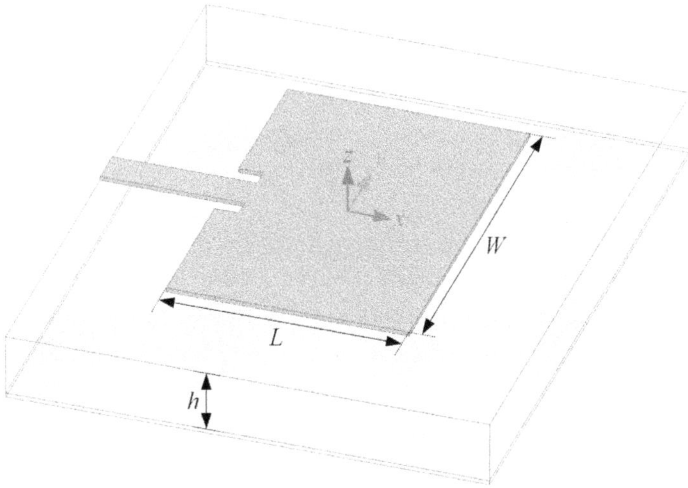

FIGURE 2.24 Schematic geometry of side-fed microstrip patch antenna.

structure of a side-fed microstrip patch antenna, in which the dielectric substrate has two conducting planes on its upper and lower surfaces, serving as the radiating patch (referred to as "patch") and the ground plane, respectively. The thickness of the dielectric substrate is usually much smaller than the wavelength. According to the analytical method of transmission line model, the geometric structure can be regarded as a low impedance transmission line with length L and two parallel narrow slots radiating into the half-space. Based on the cavity model, the conductor-covered dielectric substrate region can in turn be viewed as a one-half wavelength loaded resonant cavity with open terminals, and the two radiating slits form an equal-amplitude, inverted-phase, binary array with a spacing *of* L, generating directional radiation directed into the upper half of space.

Due to the presence of the patch edge field, the electromagnetic wave exists simultaneously in a layered medium consisting of a dielectric substrate and air, so the effective dielectric constant ε_{eff} can be introduced to describe the propagation of the electromagnetic wave, and the initial static value of ε_{eff} at low frequencies is given by Eq. (2.24) [6]:

$$\varepsilon_{\text{eff}} = \frac{\varepsilon_r + 1}{2} + \frac{\varepsilon_r - 1}{2}\left(1 + 12\frac{h}{W}\right)^{-1/2}, \quad W/h > 1 \tag{2.24}$$

where ε_r is the relative dielectric constant, W is the width of the patch, and h is the thickness of the substrate. Meanwhile, the edge effect makes the equivalent electrical length of the patch larger than the actual physical length, and the empirical estimation of the extended length ΔL at the ends of the patch is [7]:

$$\Delta L = 0.412h\frac{(\varepsilon_{\text{eff}} + 0.3)(W/h + 0.264)}{(\varepsilon_{\text{eff}} - 0.258)(W/h + 0.8)} \tag{2.25}$$

Accordingly, the length and width of the patch with operating mode TM_{01} and resonant frequency f_r can be calculated based on the following equation:

$$L = L_{\text{eff}} - 2\Delta L = \frac{1}{2f_r\sqrt{\varepsilon_{\text{eff}}}\sqrt{\mu_0\varepsilon_0}} - 2\Delta L \tag{2.26}$$

$$W = \frac{1}{2f_r\sqrt{\mu_0\varepsilon_0}}\sqrt{\frac{2}{\varepsilon_r + 1}} \tag{2.27}$$

where L_{eff} is the effective length, μ_0 is the vacuum permeability, and ε_0 is the vacuum dielectric constant. In this resonant mode, the radiated field of the patch antenna can be calculated from the equivalent surface magnetic flow [8]. Assuming that the fringe electric field is $\boldsymbol{E}_a$, the surface magnetic flow $\boldsymbol{M}_S$ can be obtained according to the mirror theory:

$$\boldsymbol{M}_S = 2\boldsymbol{E}_a \times \boldsymbol{n} \tag{2.28}$$

where $\boldsymbol{n}$ represents the normal unit vector. The far-field component is obtained from the magnetocurrent radiation:

$$E_\theta = E_0 \cos\varphi f(\theta,\varphi) \tag{2.29}$$

$$E_\varphi = -E_0 \cos\theta \sin\varphi f(\theta,\varphi) \tag{2.30}$$

where E_0 is the electric field amplitude, θ is the pitch plane angle, and φ is the azimuth plane angle.

$$f(\theta,\varphi) = \frac{\sin\left(\frac{\beta W}{2}\sin\theta\sin\varphi\right)}{\frac{\beta W}{2}\sin\theta\sin\varphi}\cos\left(\frac{\beta L}{2}\sin\theta\cos\varphi\right) \tag{2.31}$$

where β is the phase constant. When φ takes the values of 0° and 90°, the radiation pattern of the E-plane and H-plane of the antenna can be obtained, respectively.

In engineering, the feeding techniques often used for patch antennas include microstrip line side-feeding, coaxial probe bottom-feeding, aperture coupling, and proximity coupling. By selecting specific patch shapes or operating modes and adding adaptive element loads, patch antennas can be flexibly designed for optimal tuning and reconfigurable design of frequency, polarization, and radiation pattern [9].

2.5.3.2 Dipole Antennas

In the evolution of antenna technology for more than a hundred years, the wire antenna represented by dipole antenna is the oldest and most widely used type of antenna, which is often used as the basis for the design of various new types of antennas [10]. The basic structure of a dipole antenna and the current distribution of a half-wave dipole antenna are shown in Figure 2.25. The dipole antenna uses a set of balanced two-wire transmission lines to feed the excitation from the center, and it is usually approximated that the current

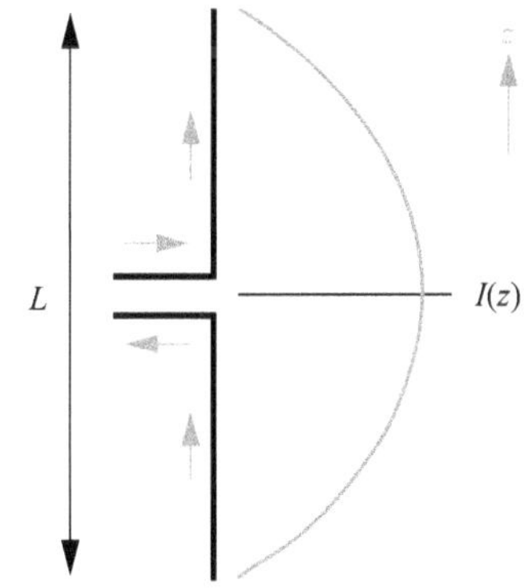

FIGURE 2.25 Basic structure of a dipole antenna and current distribution of a half-wave dipole antenna.

distribution on the symmetric two arms is sinusoidally distributed, and the currents at the two ends are 0, which can be written as:

$$I(z) = I_m \sin\left[\beta\left(\frac{L}{2} - |z|\right)\right],\ |z| < \frac{L}{2} \tag{2.32}$$

where β is the phase constant, L is the total length of the antenna, I_m is the maximum current on the antenna, and z is the coordinate of a point on the z-axis. The electrical length of a dipole antenna directly determines its current distribution and hence the shape of the far-field radiation pattern. In particular, when $L = \lambda/2$, the currents on the two arms of the dipole antenna always remain in phase, producing the strongest radiation in the direction perpendicular to the dipole antenna, called the half-wave dipole antenna. When $L > \lambda$, anti-phase currents will occur every one-half wavelength. Considering the dipole antenna as a continuous line source distributed along the z-direction, the radiated electric field can be obtained by integrating the current distribution.

As the length of the dipole antenna increases, the beam width gradually becomes narrower and the directionality increases. When the electric length is larger than one wavelength, the presence of inverse phase currents will cause a radiation canceling effect, resulting in multi-petal radiation. In contrast, the half-wave dipole antenna is more commonly used, and its radiation resistance is about 73 Ω, which is easy to achieve good impedance matching with the feed line.

REFERENCES

1. Kraus JD, Marhefka RJ. *Antennas*. 3rd ed. Zhang Wenxun, Translation. Beijing: Electronic Industry Press, 2006.
2. Deng CJ. *Application of Multi-Antenna Collaborative Analysis Method in Small Antenna Design*. Beijing: Tsinghua University Press, 2019.
3. Kahn W. Active reflection coefficient and element efficiency in arbitrary antenna arrays. *IEEE Transactions on Antennas and Propagation*, 1969, 17(5): 653–654.
4. Kalfa M, Halavut E. A fast method for obtaining active S-parameters in large uniform phased array antennas. *IEEE International Symposium on Phased Array Systems and Technology*. Piscataway, USA: IEEE, 2013: 684–688.

5. Zhang C, Lai Q, Gao C. Measurement of active S-parameters on array antenna using directional couplers. *IEEE Asia Pacific Microwave Conference (APMC)*. Piscataway, USA: IEEE, 2017: 1167–1170.
6. Balanis CA. *Advanced Engineering Electromagnetics*. Hoboken: John Wiley & Sons, 1989.
7. Hammerstad EO. Equations for microstrip circuit design. *5th European Microwave Conference*. Piscataway, USA: IEEE, 1975: 268–272.
8. Stutzman WL, Thiele GA. *Antenna Theory and Design*. 3rd ed. Hoboken: John Wiley & Sons, 2012.
9. Balanis CA. *Antenna Theory: Analysis and Design*. 3rd ed. Hoboken: John Wiley & Sons, 2005.
10. King RWP. The linear antenna—Eighty years of progress. *Proceedings of the IEEE*, 1967, 55(1): 2–16.

CHAPTER 3

Single Antenna Design for 5G Mobile Terminals

THIS CHAPTER INTRODUCES THE impedance matching of a single antenna for 5G Mobile Terminals, which lays the foundation for multi-antenna design. Impedance matching is an important index for single antenna design in 5G Mobile Terminals. There are various methods to achieve antenna impedance matching, which are mainly divided into two categories: optimizing the antenna structure and adding matching circuits. Among them, matching circuits are widely used in engineering practice, and there are already more mature theories and devices, while the optimization of antenna structure is related to ground clearance and specific antenna forms, which is a research hotspot in the academic community. In the process of impedance matching, observing the impedance matching law only from the reflection coefficient amplitude curve in the Cartesian coordinate system often does not get enough information, and it is easy to misjudge the number and location of resonance points. The Smith Chart diagram is an intuitive tool for observing impedance matching, which can reflect the effect of impedance matching in a visual way. Therefore, this chapter will first introduce the Smith Chart diagram, then introduce the role of aggregate elements in impedance matching based on the Smith Chart diagram, and finally introduce several typical single antenna design examples for 5G Mobile Terminals.

3.1 SMITH CHART DIAGRAM

In microwave engineering, the most basic operation is the conversion between reflection coefficient and input impedance, which is performed on the basis of known characteristic parameters (load impedance, characteristic impedance, propagation constant, and length).

The Smith Chart diagram is a special tool that combines the characteristic and operating parameters and solves them by graphing. Still widely used today with the popularity of computer software, the Smith Chart diagram provides an easy way to visualize transmission lines.

DOI: 10.1201/9781003770039-3

The Smith Chart diagram is applicable to any transmission line type as well as any characteristic impedance, and for this reason, the idea of normalization can be used to normalize the transmission line length (l) as well as the impedance, removing the effect of the specific operating wavelength and characteristic impedance.

- Define the electrical length (θ):

$$\theta = \beta l = 2\pi \frac{l}{\lambda} \Leftrightarrow 360° \frac{l}{\lambda} \tag{3.1}$$

where β is the phase constant and λ is the operating wavelength.

- Define the normalized impedance (Z):

$$Z = \frac{Z_{\text{in}}(Z)}{Z_c} = \frac{1+\Gamma(Z)}{1-\Gamma(Z)} \tag{3.2}$$

where Γ is the reflection coefficient, Z_{in} is the input impedance, and Z_c is the characteristic impedance.

With pre-normalized transmission line lengths and impedances, the Smith Chart diagram is universal, i.e., it is applicable to all operating wavelengths and all characteristic impedances.

3.1.1 Components of the Smith Chart Diagram

The basic idea of the Smith Chart diagram is to nest the resistance circle and reactance circle inside the reflection coefficient circle, so that one circle diagram reflects the parameter relationship corresponding to all operating wavelengths of the transmission line.

According to the transmission line principle, the following relationship is satisfied between Γ and Z:

$$Z = \frac{1+\Gamma(Z)}{1-\Gamma(Z)} = r + ix \tag{3.3}$$

Γ is generally a complex number with real and imaginary parts:

$$\Gamma = \Gamma_r + i\Gamma_i \tag{3.4}$$

Then, the real and imaginary parts of Z can be expressed by the real and imaginary parts of Γ to obtain the resistance plot function r and the reactance plot function x, respectively:

$$r = F_1(\Gamma_r, \Gamma_i) \tag{3.5}$$

$$x = F_2(\Gamma_r, \Gamma_i) \tag{3.6}$$

can also be written as:

$$Z=\frac{1+\Gamma}{1-\Gamma}=\frac{1+\Gamma_r+i\Gamma_i}{1-(\Gamma_r+i\Gamma_i)}=r+ix \tag{3.7}$$

For the real part (resistance):

$$r=\frac{1-\left(\Gamma_r^{\,2}+\Gamma_i^{\,2}\right)}{(1-\Gamma_r)^2+\Gamma_i^{\,2}} \tag{3.8}$$

can be obtained after transformation:

$$\left(\Gamma_r-\frac{r}{r+1}\right)^2+\Gamma_i^{\,2}=\left(\frac{1}{r+1}\right)^2 \tag{3.9}$$

This equation is the equation of a circle with center coordinates ($\frac{r}{r+1}$, 0) and radius 1/(r + 1).

For the imaginary part (reactance):

$$x=\frac{2\Gamma_i}{(1-\Gamma_r)^2+\Gamma_i^{\,2}} \tag{3.10}$$

This is obtained after a transformation:

$$(\Gamma_r-1)^2+\left(\Gamma_i-\frac{1}{x}\right)^2=\left(\frac{1}{x}\right)^2 \tag{3.11}$$

This equation is also an equation for a circle with center coordinates (1, 1/x) and radius $|1/x|$.

According to Eqs. (3.9) and (3.11), using the tracing method, a resistance circle diagram and a reactance circle diagram can be drawn, and the combination of the two is the Smith Chart diagram.

Using the equal resistance circle as an example, Table 3.1 lists the normalized resistance values versus radius in the coordinates of the center of several typical circles. Depicting these typical values on the reflection coefficient circle results in a resistance circle diagram, as shown in Figure 3.1.

Each curve in Figure 3.1 corresponds to an equal resistance circle. The physical significance of the equal resistance circle is as follows:

- The resistive portion of the normalized impedance is a fixed value.
- The reactance portion can be an arbitrary value.

TABLE 3.1 Normalized Resistance Values and Radii at Several Typical Circle-Centered Coordinates

	Circle Center Coordinates		
Normalized Resistance Value	**Real Part**	**Imaginary Part**	**Radius**
$r=0$	0	0	1
$r=1$	1/2	0	1/2
$r=2$	2/3	0	1/3
$r=\infty$	1	1/3	0
$r=1/2$	1/3	0	2/3
$r=1/4$	1/5	0	4/5

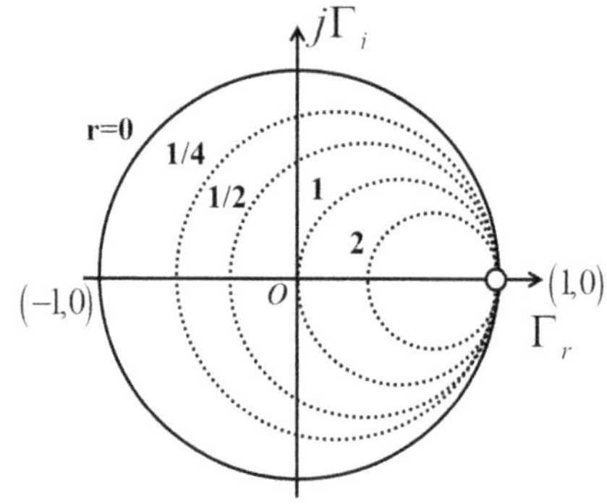

FIGURE 3.1 Resistance circle diagram.

- Each point corresponds to a different value of reactance and a different value of the corresponding reflection coefficient.

The resistance circles are characterized as follows:

- All equal resistance circles are tangent to the (1, 0) open-circuit point.
- The centers of the circles are all on the real axis.
- The scale corresponding to r on the real axis is not uniform, with the left half of the real axis $r<1$ (small resistors), the right half of the real axis $r>1$ (large resistors), and the center of the circle $r=1$.
- The larger (smaller) the radius, the smaller (larger) the corresponding normalized resistance value.
- The normalized resistance value at infinity corresponds to the point (1, 0) (open circuit point), and the normalized resistance value equal to 0 corresponds to the outermost unit circle (pure reactance point).

Similarly, the equal reactance circle can be represented by tracing the points. The reactance circle is shown in Figure 3.2.

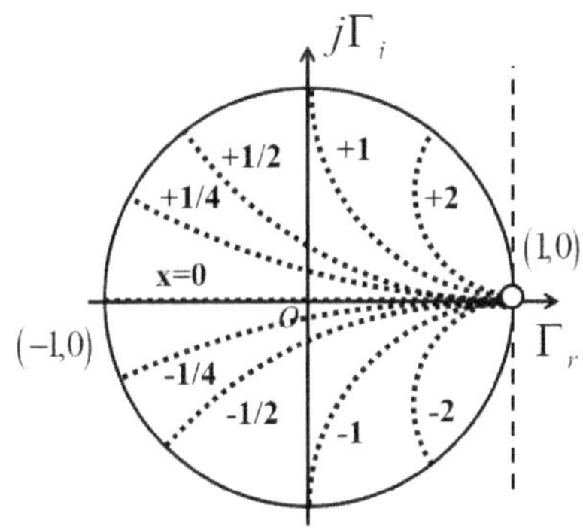

FIGURE 3.2 Reactance circle diagram.

The physical significance of the equal reactance circle is as follows:

- The reactance portion of the normalized impedance is a reflection coefficient trace of a certain value.
- The resistance portion can be any value.
- Each point corresponds to a different value of reactance and a different value of reflection coefficient.

The reactance circles are characterized as follows:

- All equal reactance circles are tangent to the (1, 0) open circuit point.
- The centers of the circles are all on the line with $\Gamma_r = 1$.
- The scale corresponding to x on the unit circle is not uniform.
- The larger (smaller) the radius, the smaller (larger) the corresponding normalized reactance value.
- The value of normalized reactance at infinity corresponds to the point (1, 0), and the value of normalized reactance equal to 0 corresponds to the point on the real axis that is purely resistive.
- The upper half plane is the inductive reactance region, and the lower half plane is the capacitive reactance region.

The two circle diagrams can be combined to form the Smith Chart diagram, as shown in Figure 3.3. The equal resistance circle intersects the equal reactance circle, and the only thing determined is a normalized impedance value $r + ix$ and a corresponding reflection coefficient value. Any point on the circle diagram corresponds to four parameters, r, x, $|\Gamma|$, and φ (phase), and either the first two or the last two parameters are known to determine the position of the point on the circle diagram.

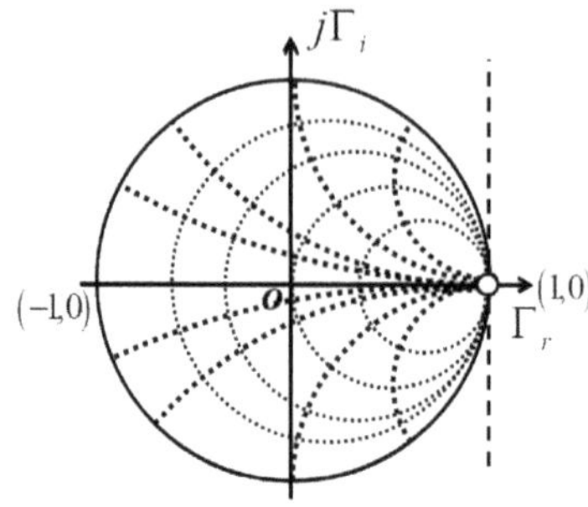

FIGURE 3.3 Smith Chart diagram.

3.1.2 Properties of the Smith Chart Diagram

The properties of the Smith Chart diagram can be summarized in three levels: "Point-Line-Surface".

First, there are three special points on the Smith Chart diagram, namely the rightmost open point, the leftmost short point, and the matching point in the center.

Second, there exist two special lines on the Smith Chart diagram, the pure resistance line on the real axis and the pure reactance line on the outermost unit circle, and the points on the lines have only real and imaginary parts, respectively. The pure resistance line on the negative real axis is the voltage wave node, and the pure resistance line on the positive real axis is the voltage wave belly point.

Finally, there are two special surfaces on the Smith Chart diagram: the upper half plane is the inductive load plane, and the lower half plane is the capacitive load plane.

3.1.3 Why the Smith Chart Plot Provides More Information than Return Loss

Junior antennae are often accustomed to observing return loss, or S_{11}, in logarithmic form, because rectangular plots in a Cartesian coordinate system can be used to visualize resonance points and bandwidths. However, there are often limitations and misconceptions in this observation. For example, judging the number of modes based on the number of resonance points and the bandwidth based on the resonance depth can be erroneous. This is illustrated by an example below.

The reflection coefficient (here S_{11} is equivalent to the reflection coefficient) curves for two antennas are shown in Figure 3.4. The impedance matching of both antenna 1 and antenna 2 is poor; antenna 1 has only one resonance point with deeper resonance depth, and antenna 2 has two resonance points with shallower resonance depth. Judging only from the reflection coefficient curves, the performance of the two antennas is close to each other, and neither of them has good impedance matching. However, if the Smith Chart plot is analyzed, a significant difference in the bandwidth potential of the two antennas is found.

The actual structures of the two antennas are shown in Figure 3.5 with Smith Chart diagrams. It can be seen that antenna 1 is an L-shaped monopole and antenna 2 is a straight monopole. Analyzing from the mode perspective, both antennas work in

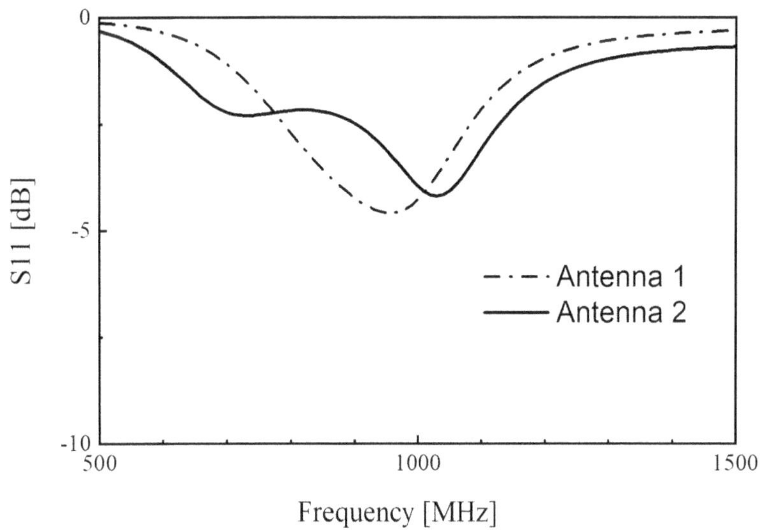

FIGURE 3.4 Reflection coefficient curves of the two antennas.

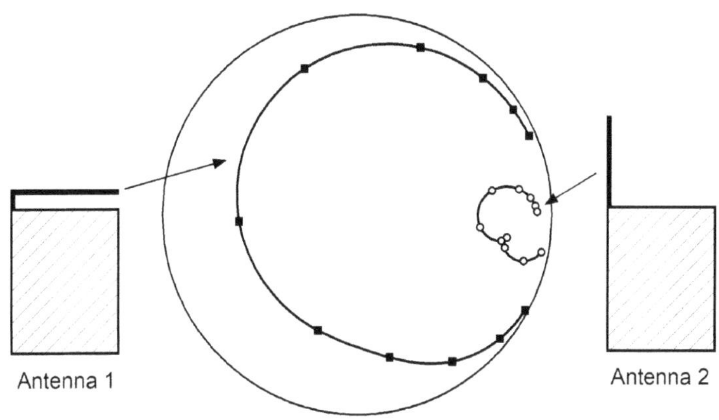

FIGURE 3.5 Actual structure of the two antennas vs. Smith Chart plot.

quarter-wavelength monopole mode, so the two resonance points of antenna 2 actually correspond to the same mode. Judging from the size occupied by the antenna, antenna 2 should have a larger bandwidth. Here, for illustration, the feed port impedance of antenna 2 is deliberately changed to 20 Ω, which makes it severely mismatched and presents the effect shown in Figure 3.4 on the reflection coefficient curve. However, if analyzed based on the Smith Chart diagram, it can be seen that the curve of antenna 1 spans a very large arc distribution within the Smith Chart diagram, which indicates that broadband matching is very difficult, while the curve of antenna 2 has a more centralized distribution within the Smith Chart diagram, with only the curve shifted to the right side, and impedance matching can be easily achieved by appropriate rotation. This information cannot be read from the reflection coefficient curve.

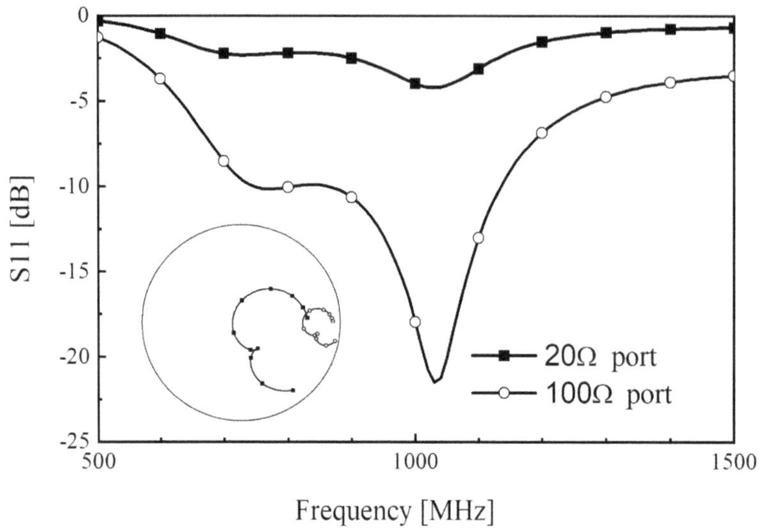

FIGURE 3.6 Changing the feed port impedance of antenna 2.

Figure 3.6 compares the reflection coefficients of antenna 2 with feed port impedances of 20 Ω and 100 Ω, respectively. When the feed port impedance is increased from 20 to 100 Ω, the curve shifts from the right side of the Smith Chart diagram to the center region, the reflection coefficient is significantly reduced, and the antenna bandwidth is greatly improved. Thus, the Smith Chart plot can visualize the shift in the impedance curve and bandwidth potential and can provide more information than the reflection coefficient or Voltage Standing Wave Ratio curves.

3.2 IMPEDANCE MATCHING

Impedance matching is one of the most important factors to consider for cell phone antennas. A well impedance-matched cell phone antenna can effectively radiate the fed electromagnetic signal into free space. There are various techniques to achieve impedance matching, which can be roughly divided into two categories: one is to adjust the antenna impedance by optimizing the antenna structure, and the other is to add a lumped component at the feed port to form a matching circuit. Although the realization of the two types of methods takes different forms, the basic idea is the same, i.e., let the input impedance be as close as possible to the center of the Smith Chart diagram. Since the workings of the lumped element are more intuitively presented with the help of the Smith Chart diagram, this section will explain impedance matching with the help of the lumped element.

3.2.1 The Role of Lumped Components

Inductors, capacitors, and resistors are commonly used lumped components. For impedance matching of antennas, resistive elements are generally not used because resistance introduces losses, making the antenna radiation efficiency significantly lower. Inductors and capacitors are commonly used inductive and capacitive components to adjust the imaginary reactance of the input impedance, and the amount of adjustment is determined

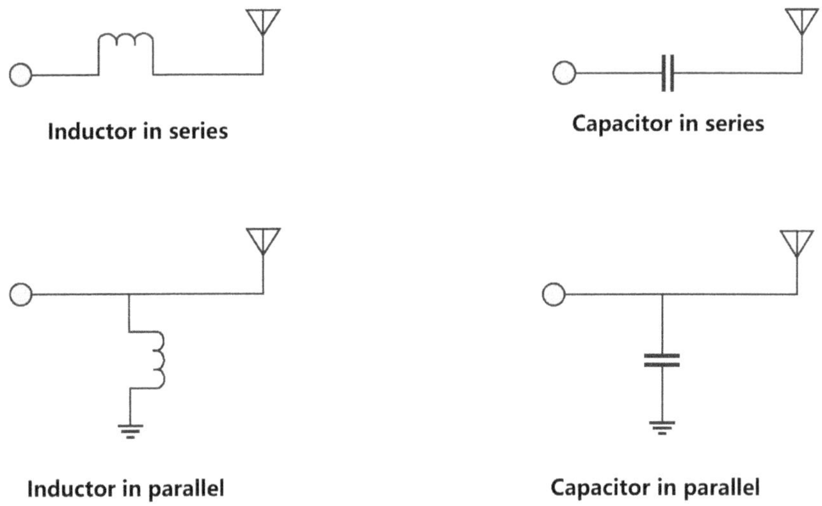

FIGURE 3.7 Four ways of accessing a single lumped component.

by the size of the inductor and capacitor. The relationship between input impedance and inductance L and capacitance C is shown below:

$$Z_L = i\omega L \tag{3.12}$$

$$Z_C = \frac{1}{i\omega C} \tag{3.13}$$

where ω is the frequency. Considering that the lumped components are connected in both parallel and series, there are four possible ways of accessing a single collector element, which are series inductance, parallel inductance, series capacitance, and parallel capacitance, as shown in Figure 3.7. The matching effect of these four ways can be visualized on the Smith Chart diagram.

Figure 3.8 shows the performance of a single lumped component on the Smith Chart diagram. When an inductive element is added, the input impedance of the antenna moves toward the upper inductive region, where the series form moves the input impedance clockwise up along the iso-resistive circle, while the parallel form moves the input impedance counterclockwise up along the iso-conductive circle. When the capacitive element is added, the input impedance of the antenna moves toward the lower capacitive impedance region, in which the series form makes the input impedance move counterclockwise downward along the equal resistive circle, while the parallel form makes the input impedance move clockwise downward along the equal admittance circle. Four types of matching circuits provide four tuning paths, allowing the input impedance to be transformed within a certain range. However, the degrees of freedom offered by a single component are too limited, resulting in blind spots in impedance transformation. This often makes it difficult to move the antenna impedance to the central region. There are blind zones, and it is often difficult to move the impedance of the antenna to the center region.

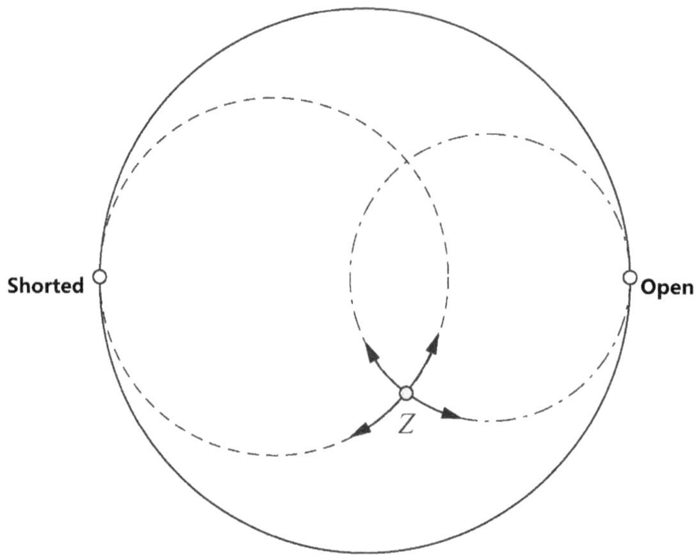

FIGURE 3.8 Performance of a single lumped component in the Smith Chart diagram.

3.2.2 Single Band Impedance Matching

A single lumped component can only rotate along the equal resistance circle or the equal conductance circle, and the range of impedance transformation is limited. In order to realize arbitrary impedance matching, except for extreme cases such as short circuit and open circuit, two lumped components can generally be used. Considering the two ways of series and parallel connection, and the two elements of inductance and capacitance, 16 kinds of matching circuits can be formed theoretically, and 10 kinds of matching circuits are left after eliminating the duplicated topological networks. Of these, four types of matching circuits are more widely used, as shown in Figure 3.9.

These matching circuits provide flexible and versatile options for impedance matching of antennas. There are multiple matching forms for the same antenna impedance, and even for the same matching form, there may be multiple choices of different lumped component values. Taking Figure 3.10 as an example, assuming that the input impedance of the antenna is in the lower right corner of the Smith Chart diagram, there is the option of either connecting an inductor in parallel first or connecting an inductor in series first. Moreover, after an inductor is connected in parallel, either a series capacitor or a series inductor can be selected. If an inductor is connected in series first, you can also choose to connect a capacitor in parallel or an inductor in parallel. All four matching circuits can move the antenna impedance to the center area of the Smith Chart diagram to achieve perfect matching. In practical engineering, it is generally necessary to determine the appropriate matching circuit based on the commercially available components.

3.2.3 Dual Band Impedance Matching

Dual-band impedance matching is a little more complicated than single-band impedance matching, and it has a certain relationship with the frequency interval between

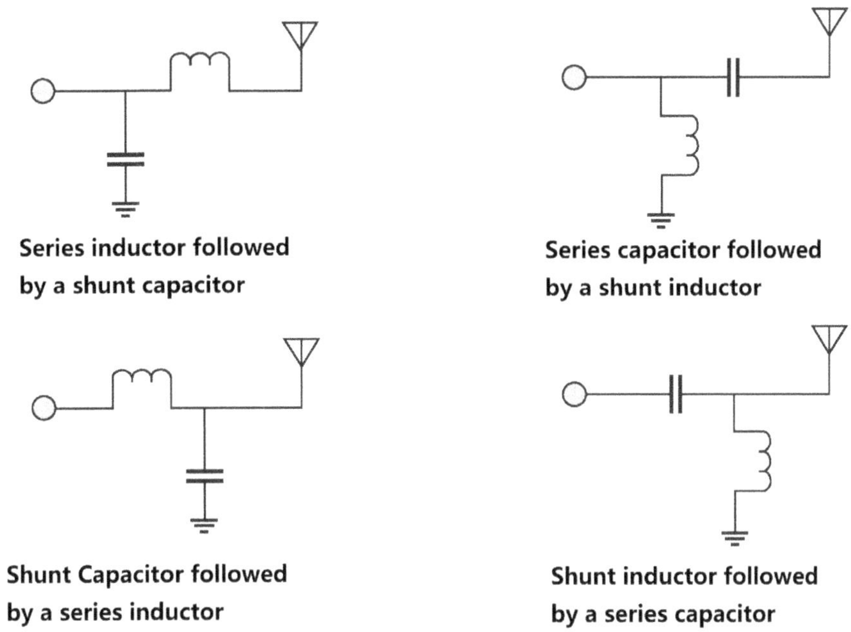

FIGURE 3.9 Four kinds of matching circuits for two lumped components.

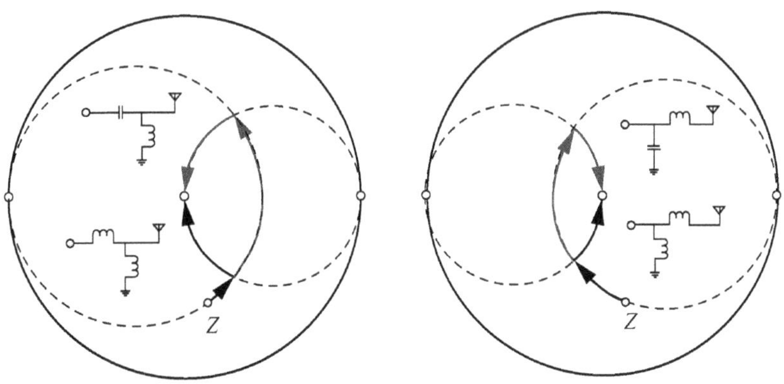

FIGURE 3.10 Two matching methods (left is parallel inductance first and right is series inductance first).

the two bands. The basic principle of dual-band impedance matching is: first match the impedance of a frequency band, and then select the aggregate components that do not affect the band to match the impedance of another frequency band. This method can be regarded as a cascade of two single-band impedance matching. For example, if you choose to match the impedance of the low band before matching the impedance of the high band, you need to select the high-pass element to match the impedance of the low band first and then cascade the low-pass element to match the impedance of the high band. Since the high-pass element is approximately straight through to the low-frequency band and the low-pass element is approximately straight through to the high-frequency band, the two frequency bands can be adjusted relatively independently.

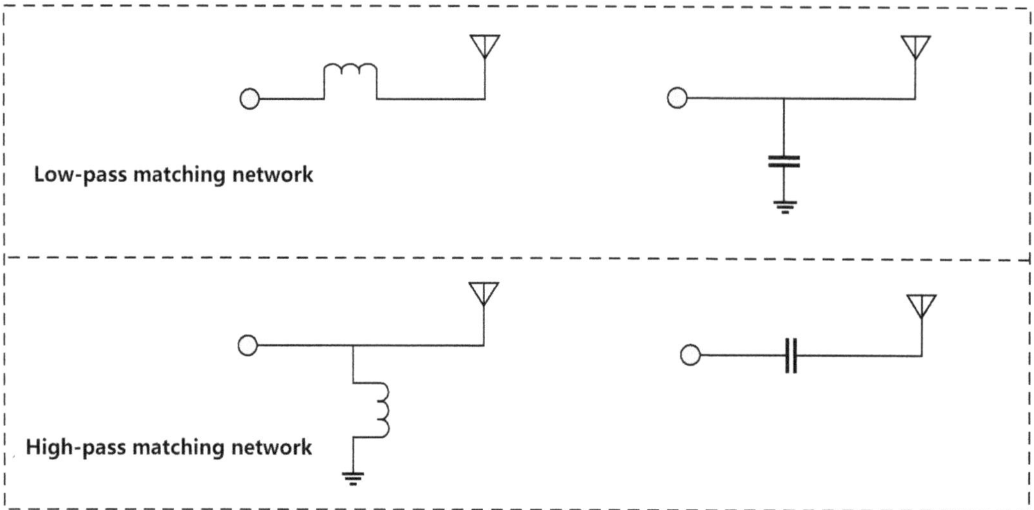

FIGURE 3.11 Commonly used low-pass element and high-pass element.

Commonly used low-pass and high-pass elements are shown in Figure 3.11. Series capacitors and parallel inductors can be regarded as high-pass components for matching the impedance of the low-frequency band; series inductors and parallel capacitors can be regarded as low-pass components for matching the impedance of the high-frequency band. If you choose to match the impedance of the low-frequency band before matching the impedance of the high-frequency band. Then you need to select the shunt inductor or series capacitor to match the impedance of the low-frequency band before selecting the series inductor or shunt capacitor to match the impedance of the high-frequency band.

3.3 SINGLE ANTENNA RESEARCH STATUS

Mobile terminals in the frequency bands below 3 GHz usually include a primary antenna and an auxiliary antenna. Among them, the primary antenna covers a wide frequency band, which can generally cover 698–960 MHz (low band) and 1710–2690 MHz (medium-high band), while the auxiliary antenna covers only a part of the band to improve transmission rate and enhance robustness. In addition to the demand for wide-band coverage, antenna miniaturization is also an important research direction. Essentially, wide bandwidth and miniaturization are two different expressions of the same goal, with the former focusing on bandwidth expansion under fixed size and the latter on size reduction under fixed frequency. In this section, we will investigate the broadband and miniaturization techniques in single antenna design, including element loading, multi-antenna, capacitively coupled feed, frequency reconfigurable, and other techniques.

3.3.1 Element Loading Technology

Element loading technology refers to the antenna radiation structure or feed circuit loading components to enhance the antenna performance or reduce the size of the antenna technology. The specific forms of realization are divided into two kinds: lumped component loading and distributed component loading. The lumped element includes inductance and capacitance, and the resistance is not used for antenna matching due to the loss; the distributed element refers to the equivalent inductance or capacitance constructed by a microstrip line. Element loading provides additional degrees of freedom for antenna matching, especially inductors and capacitors, which are more flexible in dual-band impedance matching by virtue of frequency responses such as high-pass or low-pass.

Lumped elements are used in a variety of ways, either loaded in the antenna radiating structure to change the path and length of the radiating current, or loaded in the feed circuit to improve impedance matching. Some typical element-loaded terminal antenna designs are given in the reference, including the use of lumped element loading [1–3] and distributed element loading [4–6]. For example, a 27 nH lumped inductor [1] loaded on an L-shaped monopole pulls the resonant frequency of the monopole's third mode from triple to doublet, and the length of the fundamental mode is reduced from 0.25 to 0.17 wavelengths, creating two broad bands at 900 and 1900 MHz; a stereo-bent monopole loaded with a 1.8 pF lumped capacitor [2] on a sterically bent monopole to reduce the size and lower the resonance frequency, and the capacitor loading is used to generate a double resonance in the band of interest, and a single dipole can cover the bands at 690–1070 MHz and 1710–2750 MHz; a lumped element [3] is loaded on a matching circuit at the feed port for the construction of the matching network, and the radiators are of a collapsed-ring structure, with two capacitors and an inductor forming a Π-shaped high-pass circuit, which can generate an additional resonance point in the low-frequency band without affecting the high-frequency band, and can cover the 740–960 MHz and 1600–2860 MHz bands; distributed C-shaped coupler lines are constructed between the two arms of the folded monopoles to introduce distributed capacitance [4], and the capacitance can be varied by controlling the spacing of the cross-toe shaped structure, which can cover the 733–1150 MHz and 1705–2680 MHz bands; a distributed inductor [5] is introduced between the parasitic monopole and the floor, and the size of the inductor can be varied by controlling the length of the microstrip line, and a single monopole can cover the 690–1180 MHz and 1670–2890 MHz bands; a distributed inductor [6] is given to the monopole in parallel near the feed port. By controlling the number of coils to change the inductor size, the same wide band coverage is realized.

3.3.2 Multi-Resonator Technique

Multi-resonator technique refers to the technique of constructing multiple branches or adding multiple resonant structures in one antenna structure to extend the bandwidth by combining multiple resonant modes. Generally, two resonant modes can be combined to form a double resonance in the low frequency band, and the higher modes of the resonant structure of the low frequency band and the fundamental modes of other resonators can be used to form a multi-resonance in the high frequency band, so as to extend the bandwidth through the multi-mode synergy of the two frequency bands.

Mobile terminal antennas have various forms of resonators, which are combined for a more flexible design. References [7–10] gives common forms of combining multiple resonators. For example, using two quarter-wavelength open slots [7], the quarter-wavelength mode of the long slot operates in the 900 MHz band, the three-quarter-wavelength mode in the 1800 MHz band, and the quarter-wavelength mode of the short slot in the 1800 MHz band, and the combination of three modes extends the bandwidth; combining a quarter-wavelength monopole with a one-half-wavelength ring [8], the combination of two resonant modes extends the bandwidth; the use of a multibranch drive monopole and a parasitic branch [9] generates two and four resonances in the low and high frequency bands, respectively, and the combination of these modes extends the bandwidth to cover the 698–960 MHz and 1710–2690 MHz bands; by combining a drive branch with a ground branch [10], the fundamental modes of the two branches form dual resonances in the lower frequency band, while the third-order modes form dual resonances in the higher frequency band.

3.3.3 Capacitively Coupled Feed Technology

Capacitive coupling feed technology means that the feed line is not directly connected to the main radiator, but is fed through capacitive coupling. This feeding method introduces distributed capacitance, and the capacitance value can be adjusted by the coupling area and spacing between the feedline and the main radiator, which additionally increases the optimization degree of freedom.

References [11–14] gives specific designs for capacitively coupled feeds, where the feed line is in the form of a monopole, and the main radiator structure includes forms such as Planar Inverted-F Antenna (PIFA) and monopole. For example, the PIFA structure is used for folded cell phone antenna design. There is an L-shaped monopole below the planar patch of the PIFA, which plays a role of capacitive coupling feed in the low frequency band, and can be used as a radiating structure to extend the bandwidth in the high frequency band, which can achieve good impedance matching both in the folded and open state; two open slots are fed through an L-shaped monopole, and the slot antenna is bent to form a three-dimensional structure. The two slot antennas generate two resonances in the low and high frequency bands, and the monopole generates another resonance in the high frequency band, and the bandwidth after the combination of the three resonances can cover the frequency bands of 890–1000 MHz and 1710–2993 MHz; the driving antenna is a short monopole, and the parasitic antenna is a two-branched monopole, and the parasitic branch is tucked in a three-dimensional bending way into a mobile terminal, and the structure is fed through the capacitive coupling to cover the frequency bands of 698–960 MHz and 1710–2690 MHz bands; T-shaped driving monopole and two long parasitic monopoles form a capacitively coupled feed, and the parasitic monopole uses distributed inductance to increase the degree of freedom of regulation, and the fundamental and cubic modes of the two parasitic monopoles can be resonated in the low-frequency band and high-frequency band, respectively.

3.3.4 Frequency Reconfigurable Techniques

Reconfigurable technology refers to the technology to improve the electrical performance of an antenna through reconfigurable devices such as switches, including frequency reconfigurable, polarization reconfigurable, radiation pattern reconfigurable, and other technologies. Since mobile terminals are mainly concerned with frequency band coverage, this section focuses on the current research status of frequency reconfigurable mobile terminal antennas.

Frequency reconfigurable technology dynamically adjusts the operating frequency of the antenna by introducing state-adjustable devices such as varactor diodes or Radio Frequency switches to realize the time-sharing coverage of ultra-wide band. According to the position of adjustable devices, reconfigurable techniques include antenna reconfigurable techniques and circuit reconfigurable techniques. The basic idea of the two reconfigurable techniques is similar, i.e., to dynamically change the impedance of the radiating structure or matching circuit, so that the input port realizes impedance matching at the designed frequency. The frequency reconfigurable technique is more maturely developed and is now widely used in mobile terminals such as cell phones. Several typical antenna designs for frequency reconfigurable mobile terminals are given in the references [15–18]. For example, a P-type - Intrinsic - N-type (PIN) diode is loaded on a folded ring structure, and the antenna operates in the ring mode when the PIN diode is on, and in two quarter-wavelength monopole modes when the PIN diode is off, which have different resonant frequencies and can be used to cover different frequency bands [15]. If two PIN diodes are loaded on a horizontal metal strip, the antenna operates in ring mode when both diodes are switched on. When the left diode is activated while the right diode remains off, the antenna functions in PIFA mode, and the two modes and their higher sub-modes can dynamically cover the 880–960 MHz and 1710–2170 MHz bands [16]. There is a gap with a width of 0.5 mm between the metal bezel and the floor, and there is a switching circuit loaded on the gap, which can be switched between two paths in order to select different matching elements; by switching to different states, the structure can dynamically cover the 695–979 MHz band and the 1710–2780 MHz band [17]. The driving short branch and the parasitic long branch are fed by capacitive coupling. The varactor diode is loaded on the driving short branch, and the parasitic long branch can be selected to be shorted or not by switching state. Because the capacitance value of the varactor diode can be continuously varied, the resonant frequency can be continuously adjusted between 700–960 MHz [18]. Frequency reconfigurable technique exponentially extends the antenna bandwidth without increasing the antenna size, and it is suitable for non-resonant electrically small antennas with a distinct size advantage.

3.4 EIGHT-BAND PLANAR MOBILE TERMINAL ANTENNA

In this section, we design an eight-band planar mobile terminals antenna for 5G Mobile Terminals with an antenna size of 72 mm×8 mm, which mainly consists of a driving band (including three branches) and a parasitic band (including two branches). By appropriately combining the quarter and three-quarter wavelength patterns of these branches, the antenna can cover the 5G low band (698–960 MHz) and the mid-high band (1710–2690 MHz).

3.4.1 Antenna Geometry

The eight-band planar type mobile terminal antenna is in the form of a planar monopole, and its geometric structure is shown in Figure 3.12. The antenna is a double-surface plate structure; the upper layer is a radiating metal strip, the lower layer is the floor, and the upper and lower metal structures are printed on the FR4 dielectric substrate. The relative

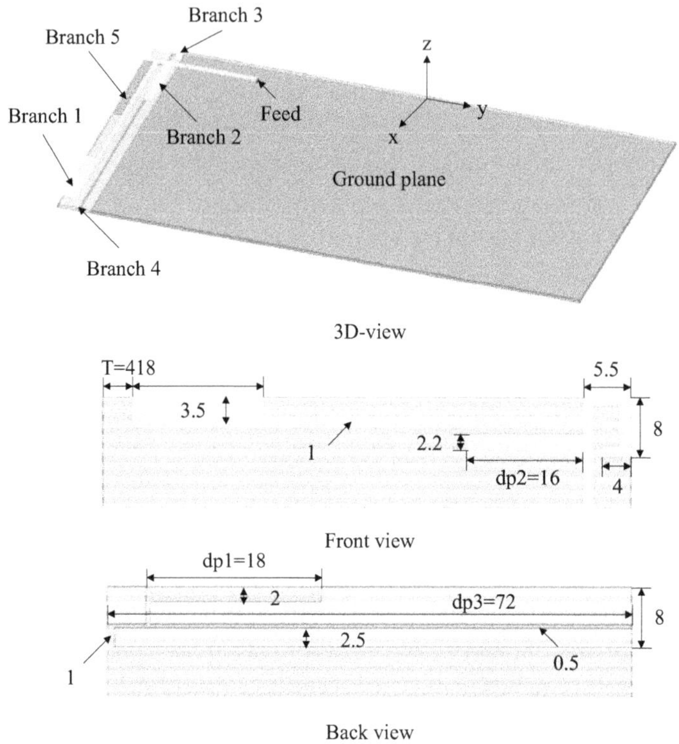

FIGURE 3.12 Antenna geometry.

permittivity of the dielectric substrate is 4.4, the tangent of loss angle is 0.02, the planar size is 72 mm × 140 mm, and the thickness is 0.8 mm. The antenna is printed on the edge of the floor, and this layout can effectively improve the radiation efficiency of the antenna as well as extend the operating bandwidth.

The antenna consists of two parts: one part is a multi-antenna directly connected to the feed port. The direct-feed part has three monopole branches, namely, branch 1, branch 2, and branch 3. The other part is a short path connected to the floor. This short path part has two branches, namely, branch 4 and branch 5, which are connected to the floor to form an open-slot structure. The upper feed line is excited by a quarter-wavelength mode around 730 MHz to form a resonance point, and the quarter-wavelength monopole mode (900 MHz) is excited by branch 1. The quarter-wavelength mode, which forms a resonance point around 730 MHz, and the quarter-wavelength monopole mode (900 MHz) stimulated by branch 1 are combined to realize the coverage of the low-frequency band of 698–960 MHz. In the middle and high frequency bands, branch 5 works in the quarter-wavelength monopole mode, forming a resonance point around 1840 MHz; branch 3 and branch 2 work in the quarter-wavelength monopole mode, forming a resonance point around 2200 and 2420 MHz, respectively; and the cubic mode of branch 1 forms a resonance point around 2690 MHz. The above modes jointly complete the coverage of the low and middle-high frequency bands. These modes can be defined by observing the current distribution of each branch at each resonance point.

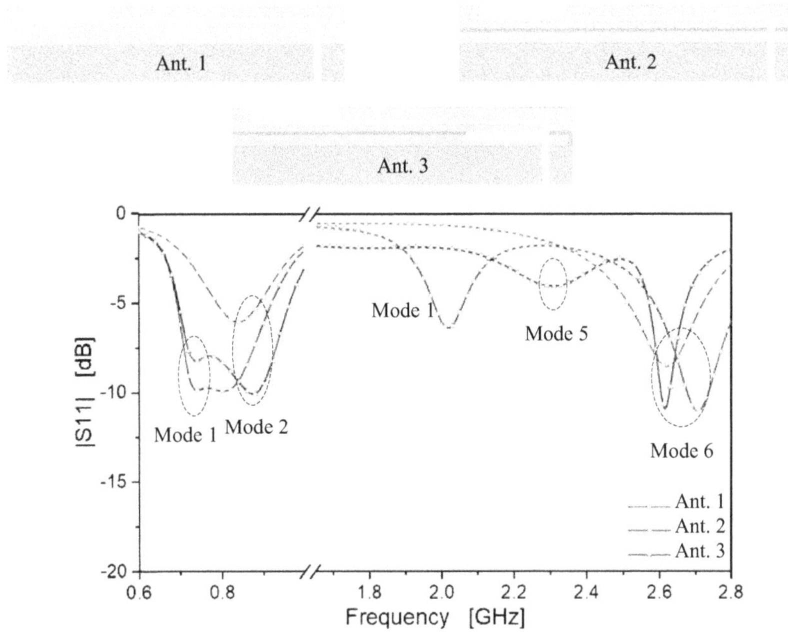

FIGURE 3.13 Structure and simulation results of antenna 1, antenna 2, and antenna 3.

3.4.2 Design Process Analysis

The structure of the designed antenna and the simulation results are shown in Figures 3.13 and 3.14. In order to reduce the size of the cell phone antenna, the design of a single branch is carried out first, and the structure is shown as antenna 1 in Figure 3.13. The simulation of antenna 1 was carried out using HFSS software. From the simulation results, it was observed that antenna 1 has two resonance points near 850 MHz (mode 2) and 2600 MHz (mode 6), which corresponds to the quarter-wavelength mode and three-quarter-wavelength mode of the monopole structure, respectively.

Based on the principle of covering the low-frequency band first, in the design of antenna 2, the grounded branch 4 is added, and this grounded short branch forms an open slot with the floor. From the simulation results, the quarter-wavelength mode of the open slot operates around 730 MHz (mode 1) and generates a high sub-mode. At this time, there are two resonance points in the low-frequency band, and it is expected to realize the coverage of the 698–960 MHz band by adjustment, but the middle and high frequency bands are poorly matched, and the resonance points in the middle and high frequency bands need to be increased subsequently by the method of adding branches to the monopole antenna. In the design of antenna 3, branch 2 is added next to branch 1, and at this time, observing the simulation results, it can be seen that there are still two resonance points in the low-frequency band, which indicates that branch 2 does not affect the operating mode of the open slot in the low-frequency band and generates mode 5 in the middle and high-frequency bands, which originates from the quarter-wavelength mode of branch 2. A bent branch 3 is then added on top of the direct-feed branch to add more resonance

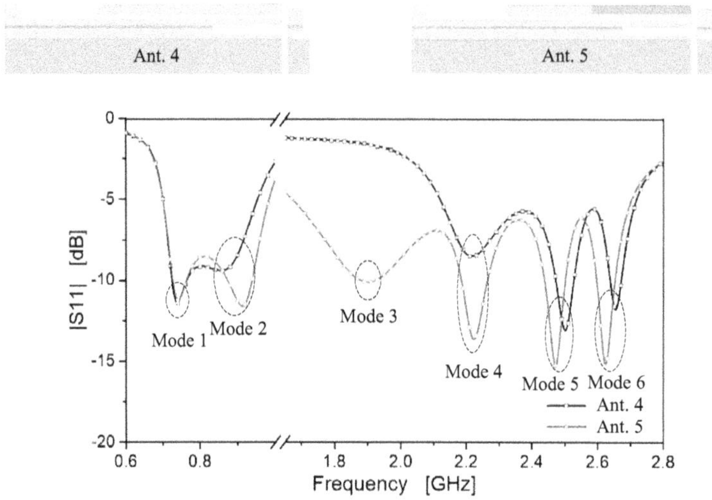

FIGURE 3.14 Structure and simulation results of antenna 4 and antenna 5.

points in the mid- and high-frequency bands, and the structure is shown as multi-antenna 4 in Figure 3.14. It can be observed that branch 3 does not affect the resonance points in the low band, but generates modes in the mid and high bands 4. Branch 5 is added on top of the grounded branch 4 to form the final antenna model, i.e., antenna 5. Comparing the simulation results of antenna 4 and antenna 5, it can be seen that the branch 5 adds one resonance point, i.e., mode 3, which suggests that the branch 5 can generate resonance points near 1840 MHz. At this point, the antenna realizes the coverage of the 5G low band (698–960 MHz) as well as the middle and high band (1710–2690 MHz).

3.4.3 Analysis of Key Parameters

Antenna performance is closely related to the physical size of the antenna. For the multi-antenna size parameters in this design, only two representative physical parameters are scanned and analyzed in this section, and in the process of parameter analysis, the corresponding parameters are analyzed using the control variable method.

First, analyze the parameter T. The parameter T is the distance from branch 1 to the edge of the long side of the phone and determines the length of branch 1. The length of branch 1 directly affects its quarter-wavelength monopole and three-quarter-wavelength monopole modes, i.e., the resonance of the antenna at mode 2 and mode 6. Theoretically, the larger the value of T, the smaller the length of branch 1, and the larger the resonant frequency corresponding to each mode of the monopole. By changing only the size of parameter T (i.e., changing the length of branch 1) while keeping other original dimensions unchanged, it can be observed from the reflection coefficient curves shown in Figure 3.15 that the resonance frequency of the quarter-wavelength monopole mode (mode 2) increases as parameter T increases and that the resonance frequency of the three-quarter-wavelength monopole mode (mode 6) also increases correspondingly. Branch 1 is parallel and close to

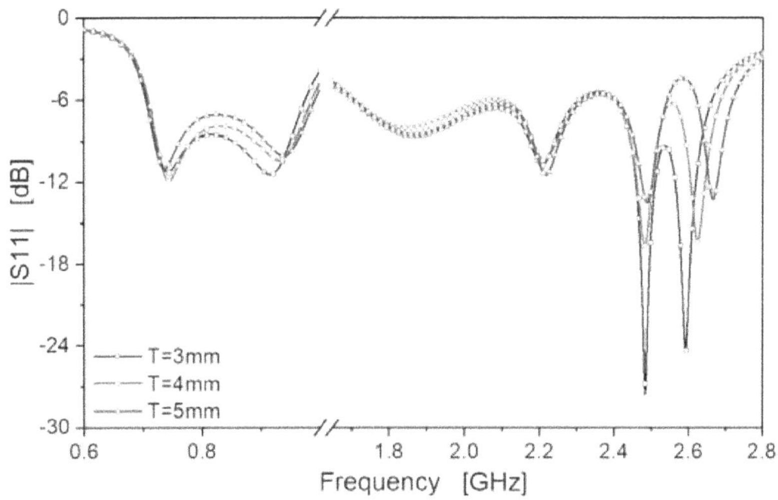

FIGURE 3.15 Parameter T scanning results.

the floor opening slot, so changing the length of branch 1 also has an effect on the resonant frequency of mode 1 generated by the opening slot. The frequencies corresponding to the other modes are not affected by the parameter *T*. The results of the parameter *T* scan show that the quarter-wavelength monopole mode and the three-quarter-wavelength monopole mode generated by branch 1 correspond to mode 2 and mode 6 of the antenna, respectively.

Next, the parameter dpl is analyzed. Parameter dpl is the length of branch 5. From the previous antenna design process, it can be seen that branch 5 will generate a quarter-wavelength monopole mode with a center frequency around 1840 MHz, i.e., mode 3 of the antenna. Theoretically, dpl becomes larger to increase the length of the branch 5, which prolongs the current distribution path of the mode, and so its corresponding resonance frequency decreases, i.e., the center frequency of mode 3 decreases when dpl becomes larger. Keeping the other dimensions of the antenna unchanged, only the size of the parameter dpl (i.e., the length of branch 5) is varied. As it increases from 22.5 to 24.5 mm, the reflection coefficient curves in Figure 3.16 show that the resonant center frequency corresponding to mode 3 decreases.

Due to the proximity of the branches, there is a certain coupling effect, so the two resonant modes of the low frequency will be affected to a certain extent, but this effect can be counteracted by adjusting other parameters. The center frequencies of the modes in the middle and high frequency bands are unchanged except for mode 3, indicating that changing the length of branch 5 has less effect on the other resonant modes, and the resonant frequency of mode 3 can be adjusted individually by changing the parameter dpl in the actual adjustment process.

3.4.4 Experimental Results

In this section, the designed antenna is processed and tested, and the assembled antenna model is shown in Figure 3.17. Figure 3.18 shows the comparison between the measurement results and the simulation results, and the data of the two are more consistent. In the

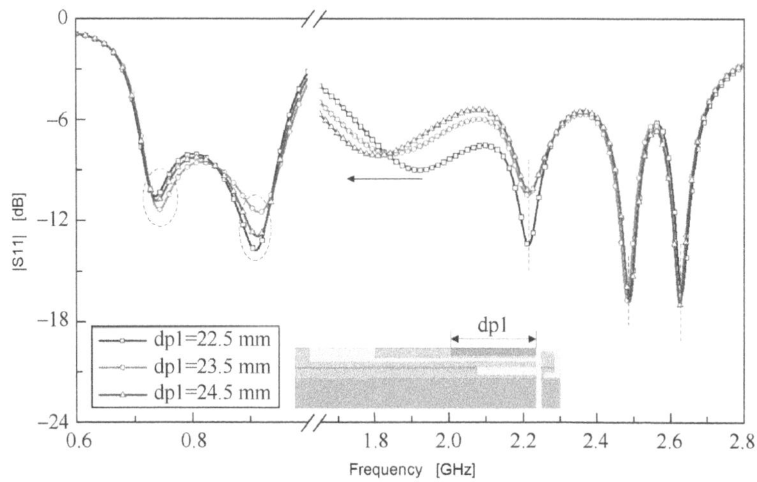

FIGURE 3.16 Parameter dpl scanning results.

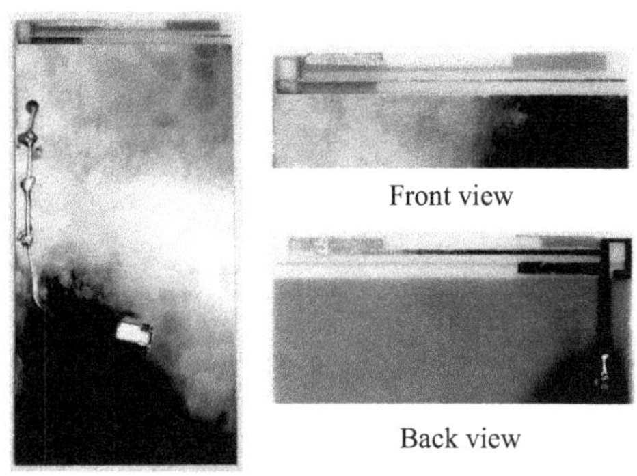

FIGURE 3.17 Assembled antenna model.

actual measurement, the antenna is able to realize the LTE700 (704–787 MHz), GSM850 (824–894 MHz), GSM900 (880–960 MHz), GSM1800 (1710–1880 MHz), GSM1900 (1850–1990 MHz), UMTS (1920–2170 MHz), LTE2300 (2300–2400 MHz), and LTE2500 (2500–2690 MHz), and other 8 bands are covered. However, some resonant frequencies in the middle and high frequency bands are shifted, which may be due to two reasons: First, the processing precision is not enough, for example, there are processing errors in the width and length of the metal strips, and the permittivity of the FR4 dielectric substrate is unstable; second, the gap of the FR4 dielectric substrate cannot be filled in the coaxial line welding process, and there may be errors at the feed power.

The radiation efficiency curve of the antenna is shown in Figure 3.19, from which it can be observed that the radiation efficiency of the antenna in the 698–960 MHz band is generally higher than 60%. In the 1710–2000 MHz band, the radiation efficiency of the antenna is higher than 70%. However, in the 2000–2690 MHz band, the radiation efficiency of the antenna fluctuates around 45%, and the lower radiation efficiency of the antenna in this band should be the result of the interaction between modes 4 and 5, and the stronger

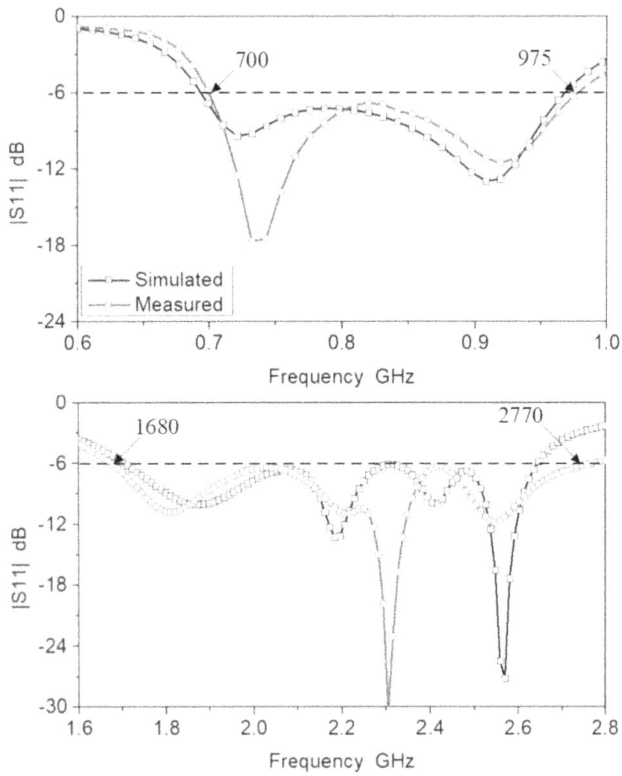

FIGURE 3.18 Comparison between measured and simulated results.

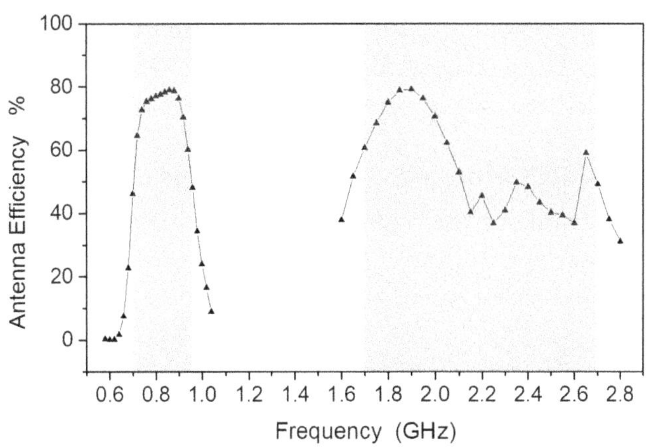

FIGURE 3.19 Radiation efficiency curve of the antenna.

coupling between the corresponding branches leads to the electromagnetic wave not being able to be effectively radiated.

Table 3.2 shows the performance comparison between this design and the same type of cell phone antenna. The antenna designed in the references [9,10,19] occupies a floor width of more than 10 mm, which is not in line with the trend of miniaturization of cell phone antennas.

TABLE 3.2 Performance Comparison between this Design and the Same Type of Cell Phone Antenna

Design	Antenna Size (mm×mm)	Frequency Band Coverage	Whether Loaded with lumped Elements or not
Reference [19]	13.5×35.5	Seven-band	Yes
Reference [20]	10×60	Eight-band	Yes
Reference [9]	15×60	Eight-band	No
Reference [21]	8×60	Seven-band	No
Reference [10]	15×64	Eight-band	No
This design	8×72	Eight frequency band	No

Note: Seven frequency bands represent GSM850, GSM900, GSM1800, GSM1900, UMTS, LTE2300, and LTE2500; Eight frequency bands represent LTE700, GSM850, GSM900, GSM1800, GSM1900, UMTS, LTE2300, and LTE2500.

The antenna in reference [20] is relatively small in size and can also cover eight frequency bands of 5G mobile communication, but the antenna is loaded with lumped components, which increases the loss and cost of the antenna.

3.5 SEVEN-BAND ANTENNA BASED ON UNBROKEN METAL BEZEL

The metal bezel of a cell phone is widely used for antenna design because the size of the bezel is comparable to the wavelengths in the mobile communication bands that can generate resonance points. The combination of a metal bezel and the floor of a cell phone comes in a variety of configurations and designs. The bezel can either be suspended on the outer edge of the floor or attached to the floor. Of these, the latter design is more attractive and avoids static electricity problems. In addition, an unbroken metal bezel without breaks provides better mechanical strength and also suppresses the effect of user behavior on the cell phone signal. This section introduces a cell phone antenna design based on a complete metal bezel, which uses the metal bezel to act as a radiator, removing the traditional radiating structure and greatly reducing the ground clearance occupied by the antenna.

3.5.1 Eigenmode of the Metal Frame

Based on the eigenmode theory, the resonance characteristics of the metal bezel itself can be analyzed without interference from external excitation sources. In this theory, λ_n and J_n denote the eigenvalue and eigencurrent of the nth eigenmode, respectively. If λ_n is equal to 0, then the eigenmode n is resonant. The eigenmode n can be identified by observing the eigencurrent at the resonant frequency.

Figure 3.20 shows the metal bezel model. The model uses an unbroken, uncut metal bezel, and there is a gap of a certain width between the cell phone floor and the metal bezel, with a gap width of 2 mm along both sides of the long side of the floor, and 2 and 6 mm along both sides of the short side of the floor, respectively. The two metal connecting strips connecting the metal bezel and the cell phone floor are located near the midpoint of the long side of the floor. The cell phone floor is flush with the bottom of the metal frame. The dimensions of the floor are 120 mm×65 mm (ignoring the thickness of the floor), and the dimensions of the metal frame are 128 mm×69 mm×6 mm.

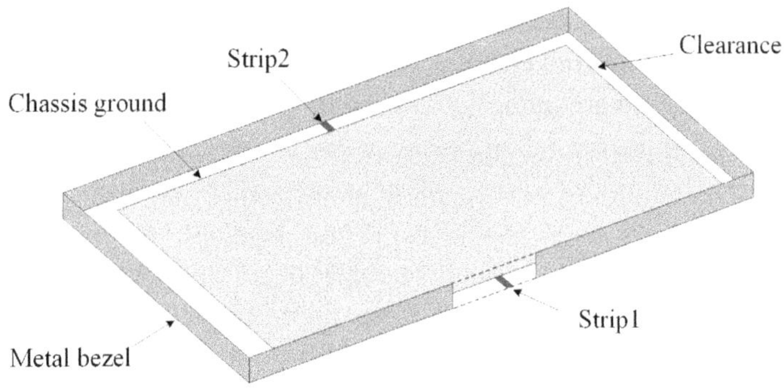

FIGURE 3.20 Metal bezel model.

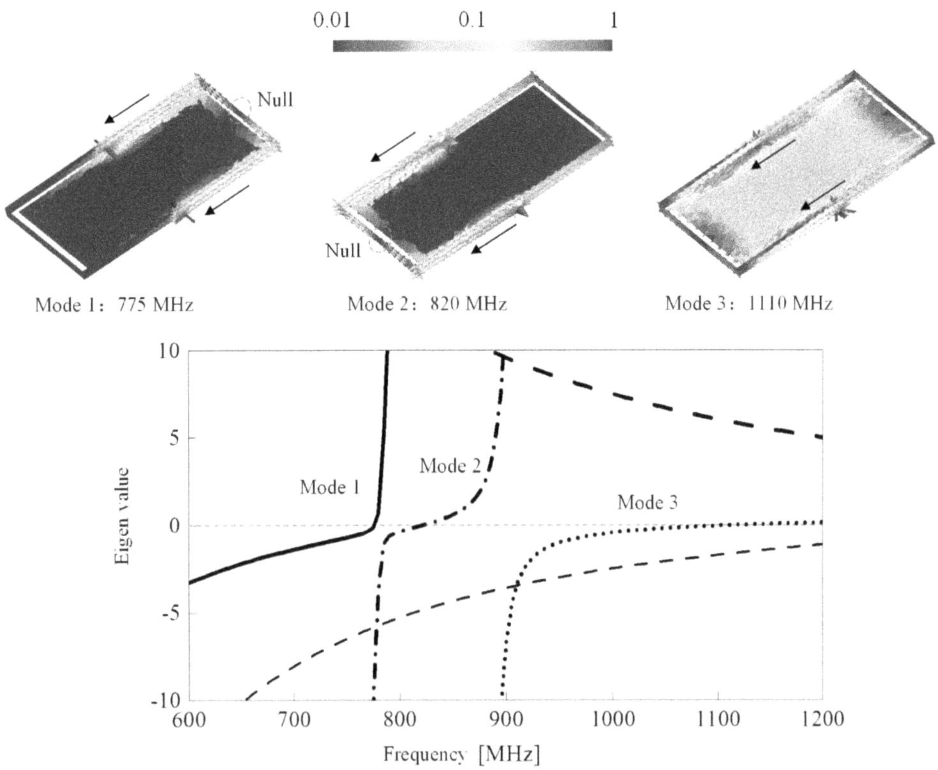

FIGURE 3.21 Normalized eigencurrent distribution of the metal-bezel model for the eigenmode distribution at 600–1200 MHz.

Considering that the 600–3000 MHz band of interest covers a very wide frequency range, it is divided into two bands, 600–1200 MHz and 1200–3000 MHz, to analyze the intrinsic mode of the metal bezel. The distribution of the intrinsic modes of the metal bezel model at 600–1200 MHz is shown in Figure 3.21. There are three resonant modes in the 600–1200 MHz band (labeled as mode 1-3), and the resonance frequencies of the three modes are approximately 775, 820, and 1110 MHz, respectively.

Next, the eigencurrents are analyzed to identify the corresponding radiation modes. Figure 3.21 shows the normalized eigencurrent distributions for the three resonant modes at their respective resonant frequencies. The metal bezel has strong currents at the first two resonance points, while the cell phone floor has strong currents at the third resonance point. This indicates that modes 1 and 2 are generated by the metal bezel, while mode 3 is generated by the cell phone floor. From the current distribution, both mode 1 and mode 2 are one-half wavelength loop modes of the metal bezel, and mode 3 is a one-half wavelength mode along the long side of the cell phone floor. If the current magnitude is ignored and only the current direction is considered, modes 1 and 2 can also be regarded as even and odd full-wavelength modes of the metal bezel because the current direction on both sides of the connecting strip is opposite in mode 1 and the same in mode 2.

The distribution of intrinsic modes of the metal-bezel model at 1200–3000 MHz is shown in Figure 3.22. Considering the large number of eigenmodes in the 1200–3000 MHz band, only the resonant modes are plotted in Figure 3.22, with a total of six resonant modes (labeled as mode 4–9). The resonant frequencies of these modes are approximately 1375, 1550, 2070, 2360, 2630, and 2900 MHz, respectively. The normalized eigencurrent distributions of these modes at their respective resonant frequencies are also shown in Figure 3.22. Based on the relative strengths of the currents on the cell phone floor and the metal bezel, mode 4–8 is generated by the metal bezel, and mode 9 is generated by the cell phone floor. Observing the current distributions, modes 4 and 5 are twofold wavelength loop modes of the metal bezel, modes 6 and 7 are threefold wavelength loop modes of the metal bezel, and mode 8 is a fourfold wavelength loop mode of the metal bezel. These modes are the high sub-modes of the metal bezel. Similar to the fundamental mode, the higher secondary modes have even and odd states. Comparing the current directions on both sides of the connecting strip, it can be concluded that modes 4, 6, and 8 are even states and modes 5 and 7 are odd states. In addition, the current distribution indicates that mode 9 is a full-wavelength mode along the long side of the cell phone floor.

From the eigenmode analysis, it is clear that there are multiple resonant modes in the 600–3000 MHz band, and both the metal bezel and the cell phone floor have the ability to radiate. The classification of the eigenmodes is shown in Figure 3.23, where seven and two resonant modes are available for the metal bezel and cell phone floor, respectively, in the 600–3000 MHz band. Due to the introduction of connecting strips, each metal bezel mode has even and odd states, thus increasing the number of available modes.

3.5.2 Excitation of Metal Bezel Models

Inductive feed and capacitive feed are two typical ways to excite the eigenmodes. Since the inductive feed occupies a small antenna size, this design uses this method to excite the eigenmode. The inductive excitation source should be placed at the point where the current appears to be extremely large in order to obtain the maximum coupling to the corresponding mode. According to the intrinsic current distribution, in the metal-bezel mode, the

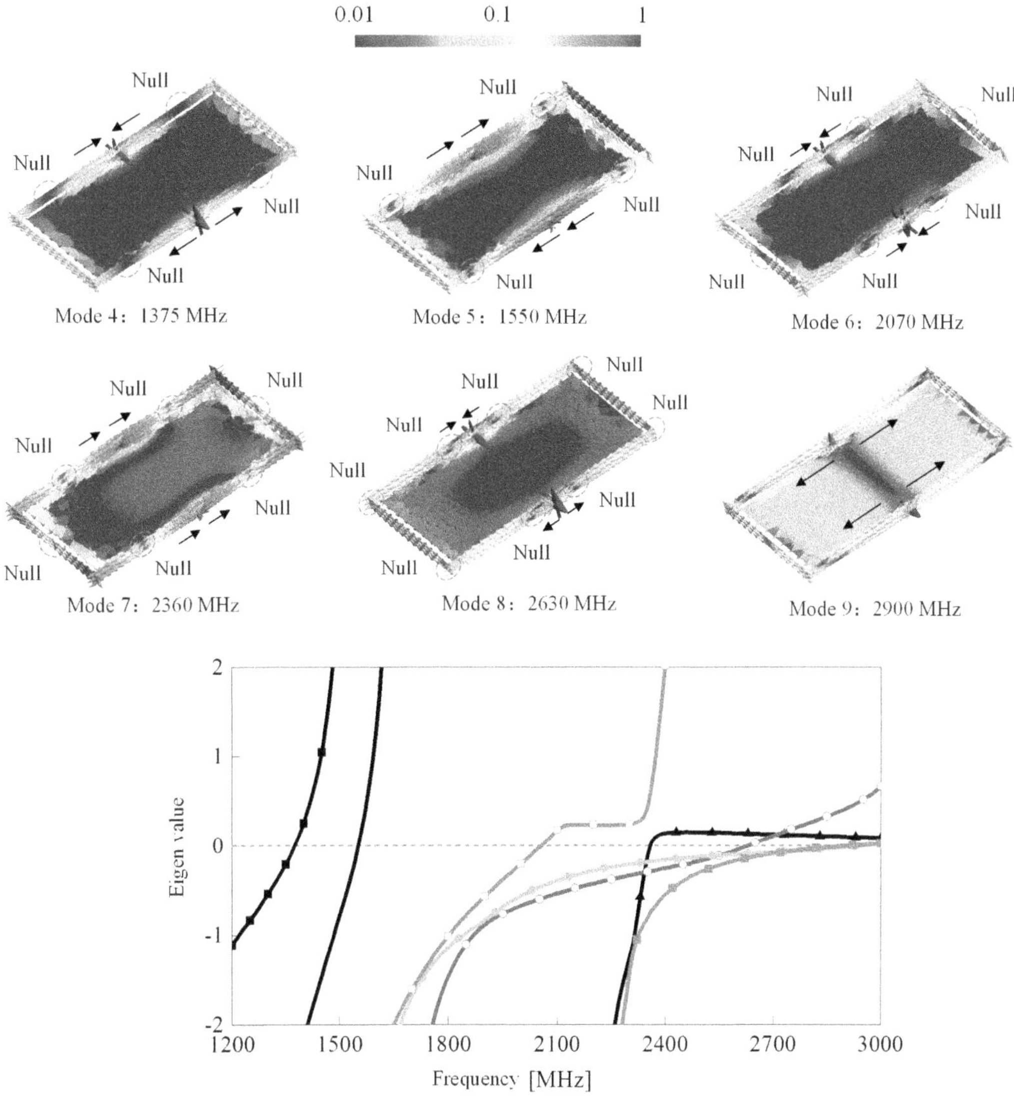

FIGURE 3.22 Normalized eigencurrent distribution of the metal bezel model at 1200–3000 MHz eigenmode distributions.

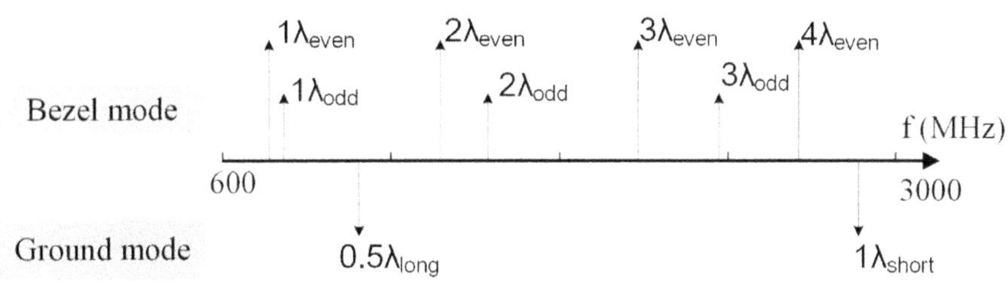

FIGURE 3.23 Eigenmode classification.

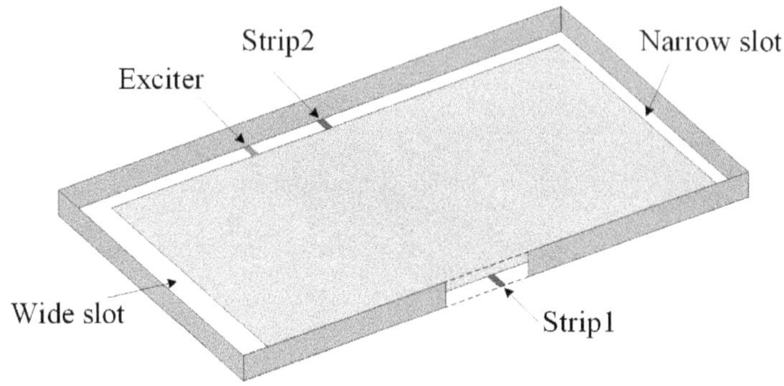

FIGURE 3.24 Metal bezel model with external excitation source added.

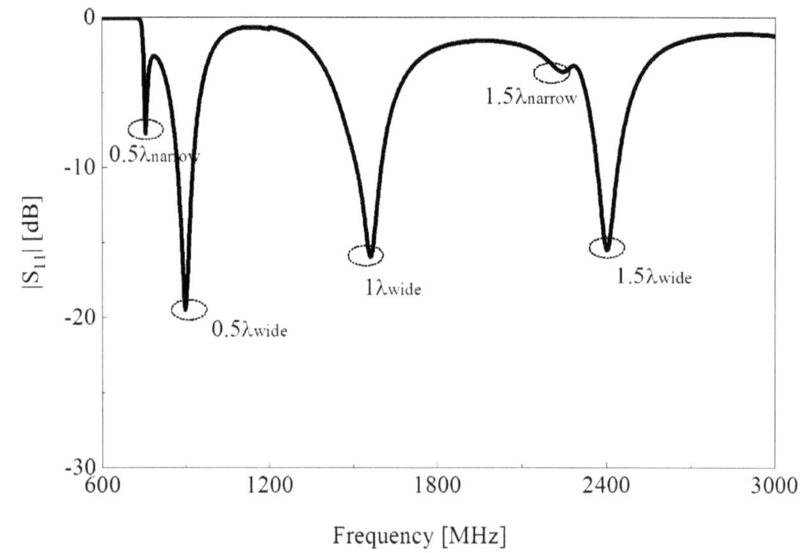

FIGURE 3.25 Reflection coefficient curve.

current is concentrated on the two connecting strips. Therefore, the inductive excitation source should be placed around the connection bars. Figure 3.24 shows the metal bezel model with the addition of an external excitation source. The two connection strips divide the gap between the metal bezel and the floor into two parts: a wide slot and a narrow slot. The inductive excitation source is placed in the wide slot to directly excite the modes in the wide slot, while the modes in the narrow slot are excited by coupling.

Figure 3.25 shows the reflection coefficient curves when the inductive excitation source is placed in the wide slot and the distance from the connecting strip is 5 mm. Two modes in the low frequency band and three modes in the high frequency band can be observed. In the low-frequency band, the distribution of modes at 746 and 868 MHz is similar to the distribution of even and odd full-wavelength modes in the intrinsic current distribution of the metal border, which can also be considered as one-half wavelength modes in the narrow and wide slots. In the high frequency band, the currents at 1470 and 2330 MHz are concentrated on the metal bezel of the wide slot and are generated by the full-wavelength and three-half-wavelength modes of the wide slot. At 2200 MHz, the currents on both sides of the connecting strip are in opposite directions, and observation of the current zeros reveals that this mode corresponds to the three-half wavelength mode of the narrow slot.

The impedance matching of the metal-bezel model remains poor even when multiple modes are excited. To further expand the bandwidth, the parameters of the metal bezel model were investigated. Considering that the dimensions of the metal bezel and the phone floor are difficult to change, the inductive excitation source and the positions of the two connecting strips are the three key parameters to optimize the bandwidth. By scanning these three parameters, the one-half wavelength narrow slot mode and the one-half wavelength wide slot mode were effectively fused, which provided a large bandwidth for the low frequency band. However, the 1710–2690 MHz band is not fully covered, and further bandwidth expansion is required.

Due to the limited number of available parameters, additional degrees of freedom are needed to cover the middle and high frequency bands. The improved metal bezel model is shown in Figure 3.26. Connection strip 2 is replaced by a 7 nH collector inductance. This adjustment mainly affects the model with narrow slots and has little effect on the model with wide slots. In addition, the inductor acts as a low-pass filter, which mainly affects the higher-order modes in the high-frequency band. With this adjustment, the low-frequency band is hardly affected, while the performance of the high-frequency band can be improved.

The reflection coefficient curves with and without inductance are shown in Figure 3.27. The introduction of inductance increases the two resonance points at 2310 and 2960 MHz. The resonant frequencies produced by the one-half wavelength, full wavelength, and three-half wavelength modes of the wide slot are virtually unaffected. By combining the full-wavelength and three-half-wavelength modes of the wide slot, as well as the two modes introduced by the inductor, the bandwidth of the high-frequency band is significantly broadened. In order to identify the characteristics of the 2 newly introduced operating modes, the current distributions at these two resonance frequencies are observed, and 3 and 4 current zeros are found along the narrow slot, respectively, which suggests that the two resonances are generated by the three-fourths wavelength and twice wavelength

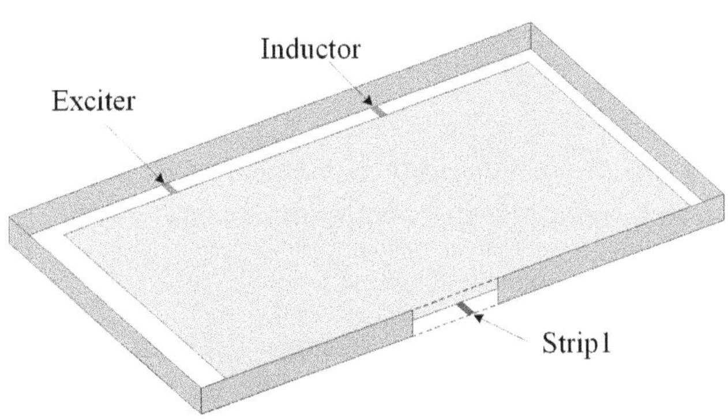

FIGURE 3.26 Improved metal bezel modeling.

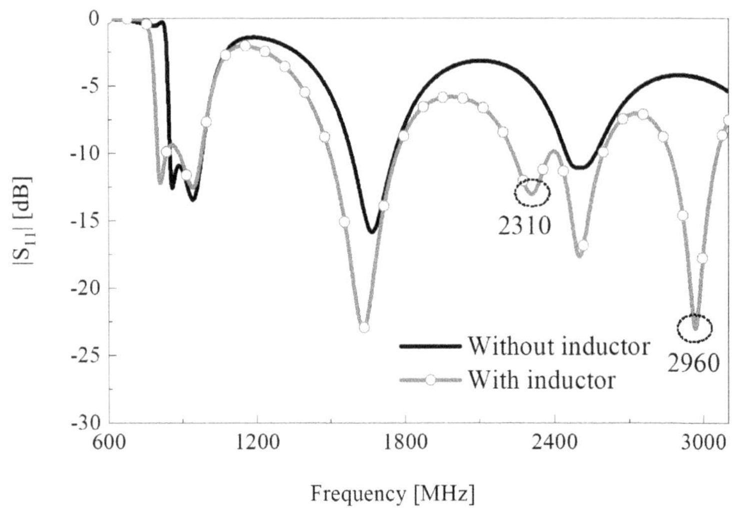

FIGURE 3.27 Reflection coefficient curves with and without inductance.

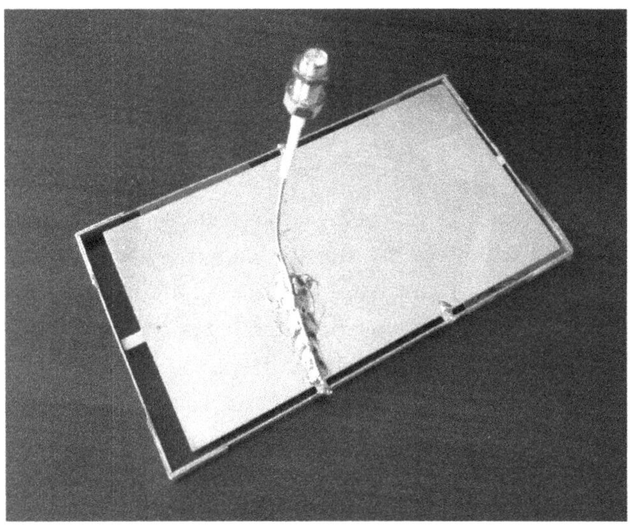

FIGURE 3.28 Physical photo of the final model.

modes of the narrow slot. This implies that the number of resonance points in the higher frequency band can be increased by introducing an inductor.

3.5.3 Physical Model Validation

The physical photograph of the final model is shown in Figure 3.28. The cell phone floor is printed on a 1 mm thick FR4 dielectric substrate (ε_r=4.4, tanδ=0.02) with a floor size of 120 mm×65 mm. The metal bezel is made of 0.6 mm thick copper strips with a metal bezel size of 128 mm×69 mm×6 mm. A 50 Ω coaxial cable is used to feed the antenna. All parameters in the final model are re-optimized to take into account the dielectric loading effect of

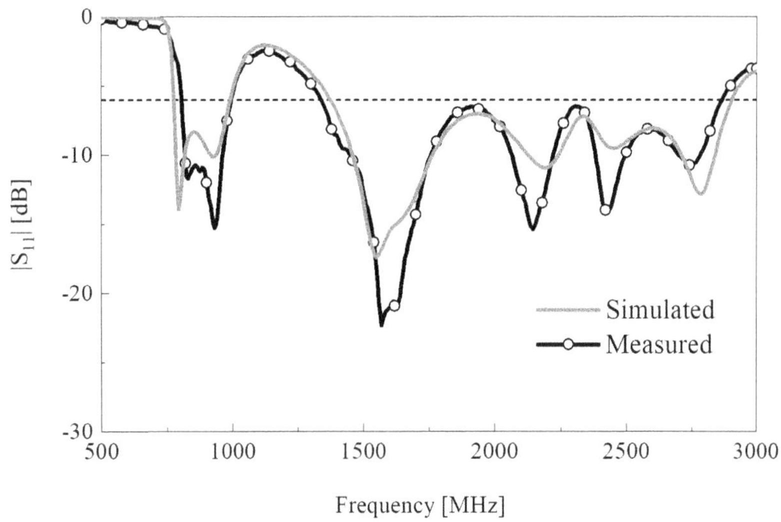

FIGURE 3.29 Simulated and measured reflection coefficient curves for the model.

the FR4 dielectric substrate. Four copper bars are soldered together to form an unbroken metal bezel. The lumped inductor is soldered where the connecting strip 2 is located. The gap between the cell phone floor and the metal bezel is precisely controlled by five short branches extending from the FR4 dielectric substrate. The inner and outer conductors of the coaxial cable are soldered to the metal bezel and the cell phone floor, respectively.

The simulated and measured reflection coefficient curves for the designed model are shown in Figure 3.29. The measured results are in good agreement with the simulation results. Two resonance points and four resonance points are observed in the low and high frequency bands, respectively. The measured reflection coefficient at −6 dB in the low band has a bandwidth of 190 MHz, which covers the GSM850 and GSM900 bands. The bandwidth measured in the high band with a reflection coefficient of −6 dB is 1530 MHz, covering the GSM1800, GSM1900, UMTS, LTE2300, and LTE2500 bands.

The radiation performance was measured in a microwave chamber. The simulated and measured radiation efficiency of the designed model is shown in Figure 3.30. The measured data is slightly lower than the simulated data due to processing errors and cable losses. The measured radiation efficiency is above 80% in the low frequency band (left gray area) and at 65%–85% in the high frequency band (right gray area). The design has high radiation efficiency because there is no strong electric field inside the dielectric substrate. The radiation efficiency in the high-frequency band can be further improved by reducing the losses caused by impedance matching.

A performance comparison of this design with other full metal-bezel designs is shown in Table 3.3 [22–25]. All designs have similar bandwidths covering the 824–960 MHz and 1710–2690 MHz bands. Compared to other designs, the present design has significant advantages in terms of reduced top and bottom ground clearance. For example, compared with the reference [24], the top clearance is reduced from 7 to 2 mm, and the bottom clearance is reduced from 7 to 6 mm. This means that the present design has a smaller ground clearance and is more practical in mobile terminals. In addition, the radiation efficiency of this design is better than that of other designs in the low-frequency band.

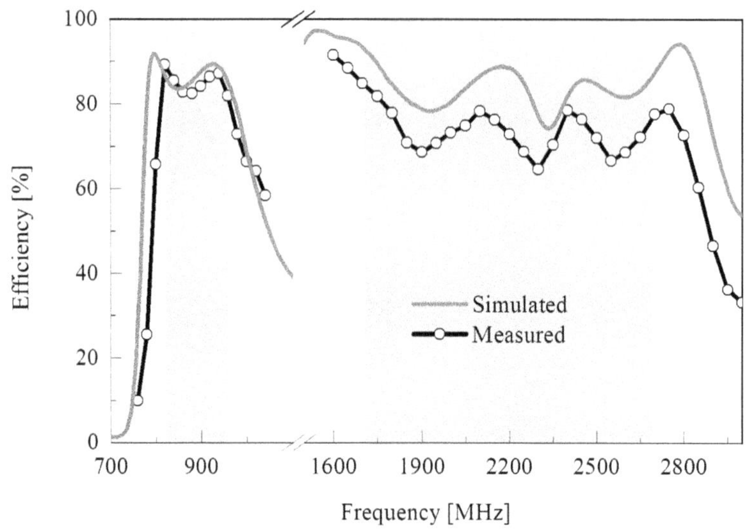

FIGURE 3.30 Simulated and measured radiation efficiency of the model.

TABLE 3.3 Comparison of the Performance of this Design with other Unbroken Metal Bezel Designs

Design	Top Ground Clearance (mm×mm)	Bottom Ground Clearance (mm×mm)	Left and Right Clearance (mm)	−6 dB Bandwidth (Frequency Range, MHz)	Radiation Efficiency (%)
Reference [22][a]	3×60	8×60	3	760–960, 1510–2720	54–86, 50–74
Reference [23]	7×70	12×70	2	770–1130, 1612–3000	62–79, 60–82
Reference [24]	7×60	7×60	2	798–968, 1440–2950	50–65, 60–78
Reference [25]	7×70	22×70	2	800–1150, 1700–3000	65–90
This design	2×65	6×65	2	805–995, 1340–2870	80–90, 65–85

[a] Is a conventional monopole antenna.

3.5.4 Influence of Head and Hand Modeling

This section investigates the effect of the head-hand model on the performance of the metal bezel antenna. The Specific Absorption Rate (SAR) evaluation model (head model) is shown in Figure 3.31, which is provided by the CST software version 2016. The metal bezel antenna is located close to the ear (at a distance of 1 mm) and tilted 60° to the vertical line. For the GSM850 and GSM900 bands, the input power for the SAR test is 24 dBm; for the other bands, the input power for the SAR test is 21 dBm. The simulated SAR for the 1 g head tissue is shown in Table 3.4. For the 1 g head tissue, the SAR at all frequencies is below the SAR threshold of 1.6 W/kg, which is in compliance.

Figure 3.32 shows the human hand model, where the human hand directly holds the metal bezel at different locations. Since a non-adjustable generic model was chosen for the human hand model, the part of the model that overlaps with the antenna was scooped out. Three positions, top, middle, and bottom, of the metal bezel are used to represent the

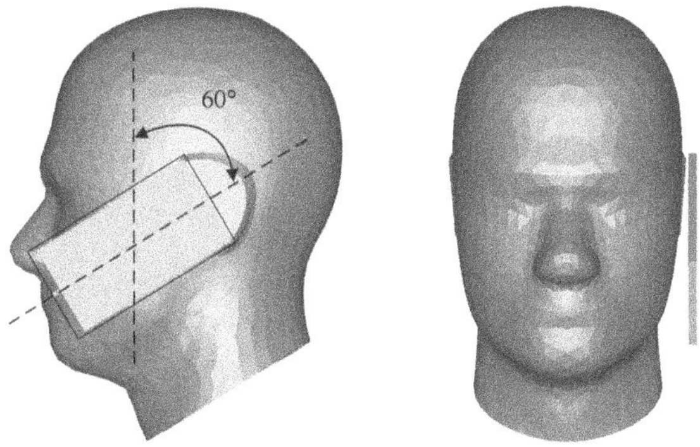

FIGURE 3.31 SAR evaluation models.

TABLE 3.4 Simulated SAR for 1 g Head Tissue

Frequency (MHz)	SAR (W/kg)	S_{11} (dB)	Radiation Efficiency (%)
859	1.19	−5.9	27.6
925	1.59	−7.9	33
1795	0.37	−9.3	57
1920	0.58	−8.6	49
2045	0.88	−12.1	45
2350	1.23	−12.4	44
2595	0.68	−6.9	49

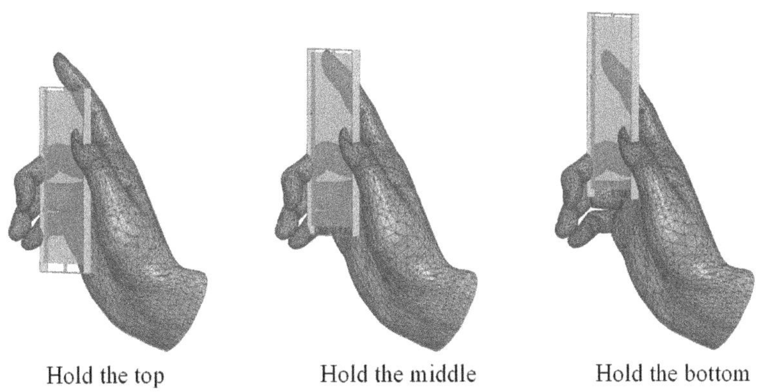

FIGURE 3.32 User hand model.

typical hand behavior of the user. The relative permittivity and conductivity of the human hand model vary with frequency.

The radiation efficiency after adding the human hand model is shown in Figure 3.33. At all three gripping positions, the radiation efficiency appears to be reduced to a great extent. The main reason for this situation is that the human hand is a lossy medium that absorbs a significant portion of the power, resulting in a lower quality factor. The average radiation efficiency of these three models of the human hand is about 30%. About 50% of the power is absorbed by the human hand compared to in free space.

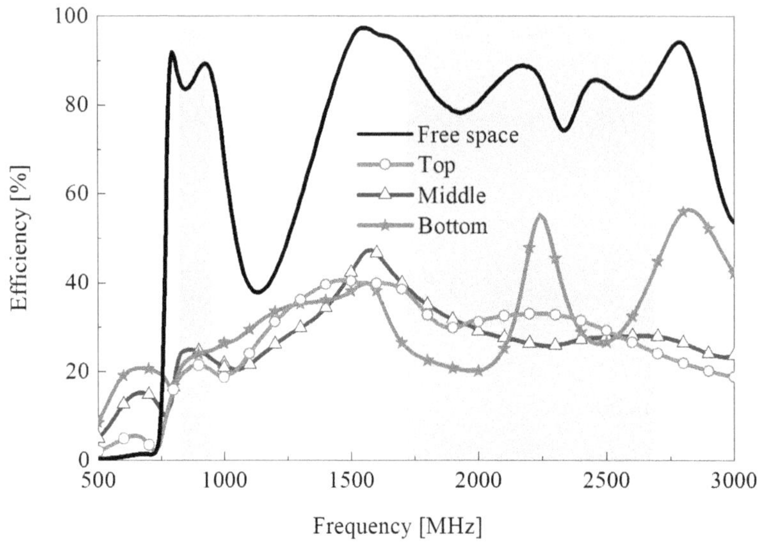

FIGURE 3.33 Radiation efficiency with the addition of the manual hand model.

3.6 FREQUENCY RECONFIGURABLE MOBILE TERMINAL ANTENNA

This section briefly introduces a frequency reconfigurable mobile terminal antenna design [26], which dynamically adjusts the resonance frequency of the monopole antenna through a varactor diode, so that the fundamental and high secondary modes can cover the low frequency band and the medium-high frequency band, respectively. It should be noted that in practice, the varactor diode can be replaced by a set of parallel capacitors with different capacitances and a single-pole, multi-throw switch, thus reducing the complexity of the control voltage.

3.6.1 Reconfigurable Antennas Based on Varactor Diodes

A simple monopole antenna ref1 is first analyzed, the structure of which is shown in Figure 3.34a. The antenna is fed by connecting a 50 Ω microstrip line directly to the metal bezel, and the fed monopole (open slot) antenna is shorted to the floor near the end of the antenna, and a varactor diode is connected in series with the shorted line [its equivalent circuit model is shown in Figure 3.34b], and the ground clearance between the metal bezel and the floor has a clearance width of 1.5 mm. After analyzing the antenna's characteristics as well as several simulation experiments, the final varactor diode model chosen is SMV2019-079LF, and the capacitance value will change from 2.1 to 0.3 pF when the DC voltage loaded on both ends of the varactor diode changes from 0 to 20 V. According to the datasheet, the varactor diode of this model has an equivalent resistance of R_v=4.5 Ω and an equivalent inductance *of* L_v=0.7 nH.

The simulation results of antenna ref1 are shown in Figure 3.35. The open slot can be excited to start two operating modes of quarter-wavelength and quarter ter-wavelength, and when the capacitance value of the varactor diode is varied from 0.3 to 2.1 pF, the quarter-wavelength monopole mode can realize scanning for the band of 600–980 MHz, and the three-quarter-wavelength monopole mode can realize scanning for the band of

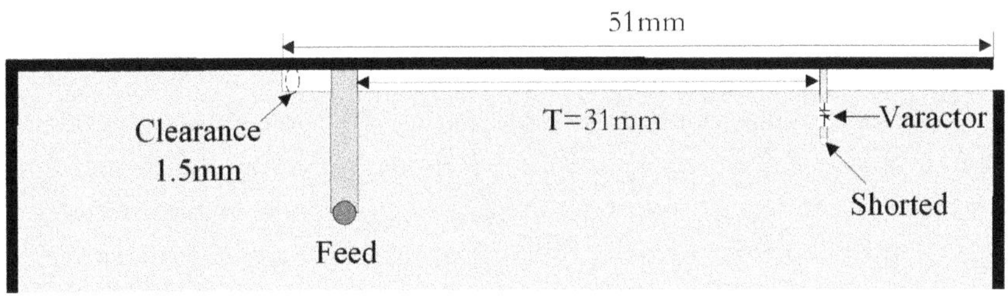

(a) Geometry of Ref1

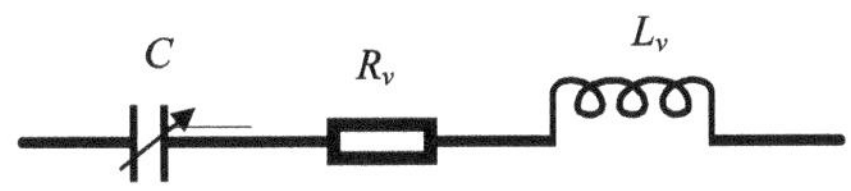

(b) Equivalent circuit of the varactor

FIGURE 3.34 Design of reference antenna ref1: (a) structure of ref1 and (b) equivalent circuit model of varactor diode.

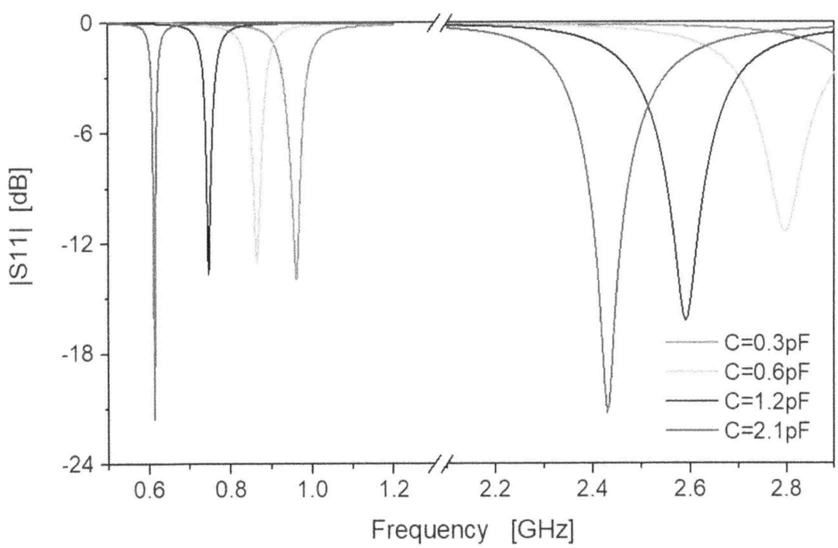

FIGURE 3.35 Simulation results of antenna ref1.

2380 MHz and above. However, the GSM1800, GSM1900, and UMTS bands in the 5G mobile communication band are not fully covered, and the antenna structure needs to be improved so that the dynamic scanning bandwidth of the antenna can cover these middle and high frequency bands.

In antenna design, loading inductance can effectively shorten the current path and thus reduce the monopole branch length. When the design space is insufficient and the antenna metal branch needs to be shortened, an inductor can be loaded at the appropriate position to compensate for the reduction of the electrical length. In general, the larger the inductance, the more obvious the compensation of the electrical length. Using this principle, the electrical length of the three-quarter-wavelength monopole mode is changed by connecting an inductor in series on the short path, and the antenna structure and simulation results after increasing the series inductance are shown in Figure 3.36, with the size of the series inductance of 7.8 nH. It can be seen through the simulation results that, when the capacitance of the varactor diode is changed in size from 0.3 to 2.1 pF, the one-quarter-wavelength monopole mode is changed with the varactor. The variation of the diode realizes the coverage of the 600–970 MHz band, and the variation of the three-quarter wavelength monopole mode enables the antenna to realize the coverage of the middle and high frequency bands.

3.6.2 Analysis of Key Parameters

After design optimization, the final design of the single antenna structure is shown in Figure 3.37, where a DC bias circuit is added to the original structure with a parallel

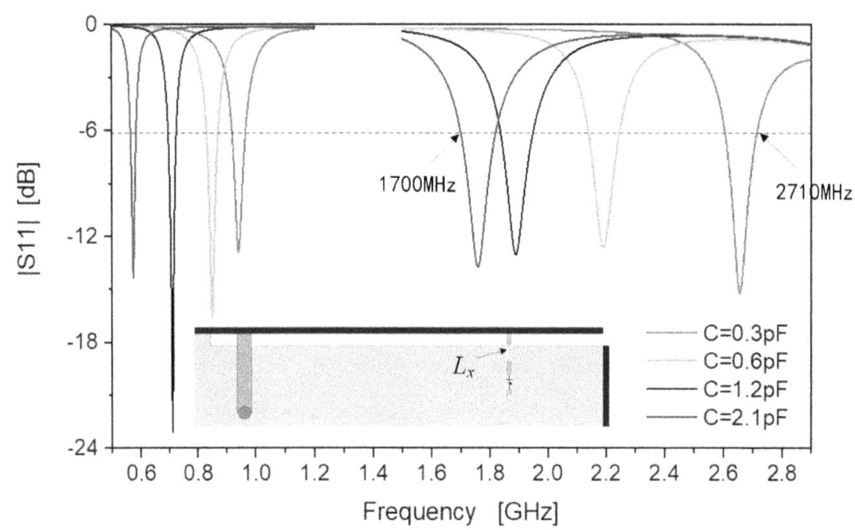

FIGURE 3.36 Antenna structure and simulation results after adding series inductance.

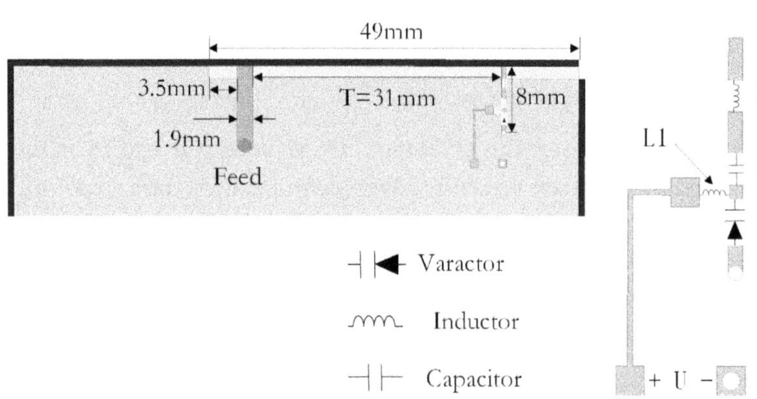

FIGURE 3.37 Single antenna structure.

inductor (L_1= 13 μH) for AC isolation and a series capacitor (C_1=100 nF) for DC isolation, so that the AC signal of the antenna feed and the DC bias of the varactor diode will not affect each other.

Two important parameters of the antenna are scanned and analyzed. First, the series inductance L_x is analyzed. The series inductance designed in this section will change the effective electrical length of the monopole in the high frequency-band, which directly affects the frequency sweep range of the three-quarter wavelength monopole mode. Theoretically, the larger the inductance, the longer the effective electrical length of the monopole, the lower the corresponding center frequency of the mode, and changing the value of the series inductance will affect the frequency coverage of the antenna in the high-frequency band while other parameters and antenna dimensions remain unchanged. From Figure 3.38, it is observed that the resonant frequency in the high-frequency band shifts toward the lower frequency when the inductance becomes larger from 5.8 to 9.8 nH. Although the results are only given in Figure 3.38 for two values of the varactor diode's capacitance, 0.3 and 2.1 pF, this is sufficient to demonstrate that the high-frequency band's frequency shifts to the left overall as the series inductance becomes larger. The inductor is a low-pass component, which generally does not affect the low-frequency band, and the simulation results also show that changing the value of the inductor does not affect the center frequency of the low-frequency band, so the final choice is to adjust the value of the series inductor L_x to 7.8 nH, which can satisfy the antenna's requirement of covering the frequency band of 1710–2690 MHz.

The parameter T is then analyzed. The parameter T is the distance between the short-circuit point and the feedline, and this parameter determines the position of the short-circuit point. Theoretically, changing the size of parameter T does not change the monopole length, i.e., the size of parameter T does not affect the fundamental mode of the antenna, and the resonance of the antenna in the low-frequency band is not affected by the position of the short-circuit point. In the case of keeping other original dimensions unchanged, only

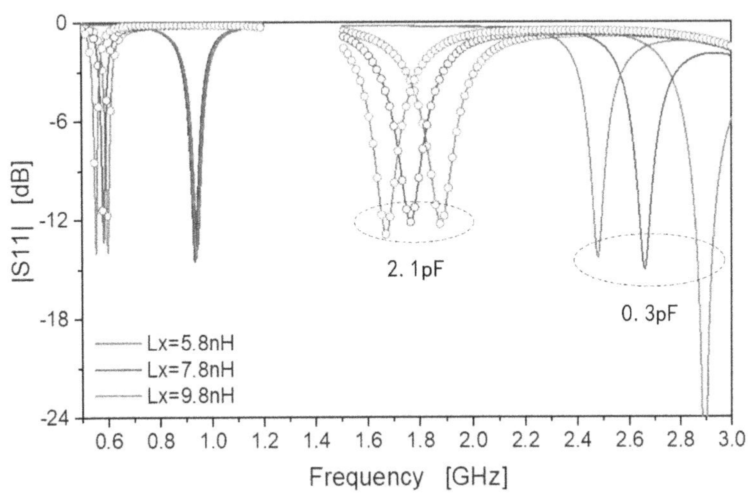

FIGURE 3.38 Series capacitance and inductance scanning results.

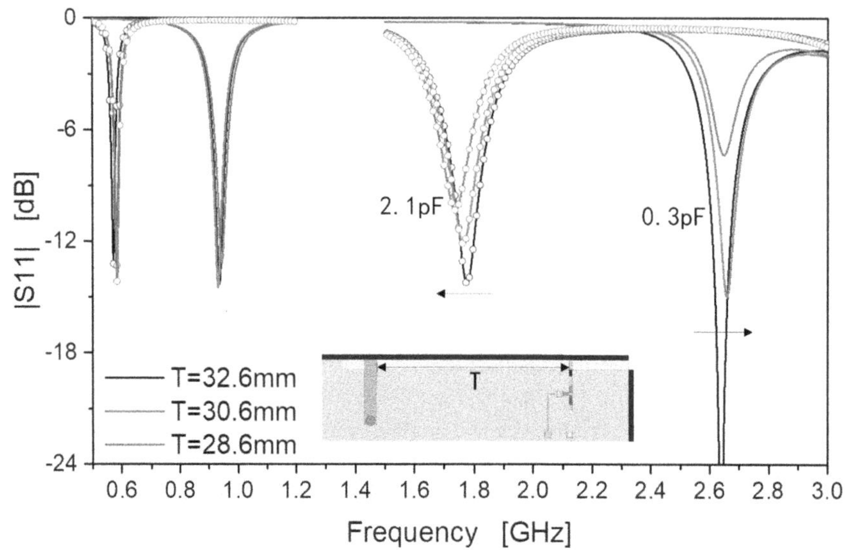

FIGURE 3.39 Scanning results for parameter *T*.

the size of parameter *T is* changed, and parameter *T* is changed from 28.6 to 32.6 mm. From Figure 3.39, it can be seen that the resonant frequency of the quarter-wavelength monopole mode is basically unchanged as the parameter *T* increases, i.e., the resonance in the low-frequency band is not affected by the change of the position of the short-circuit point. However, in the high-frequency band, when the value of *T* becomes smaller, the center frequency becomes lower and the resonance depth becomes shallower when the varactor diode capacitance is 2.1 pF, and the center frequency becomes higher and the resonance depth becomes shallower when the varactor diode is 0.3 pF, which indicates that the closer the short-circuited point is to the feeder, the wider the frequency sweep range of the high-frequency band is, although the resonance strength will be weaker. Considering these two factors, the parameter *T* is set to 30.6 mm, which can satisfy the coverage of the high-frequency band and ensure that the resonance strength will not become weaker.

REFERENCES

1. Wong KL, Chen SC. Printed single-strip monopole using a chip inductor for penta-band WWAN operation in the mobile phone. *IEEE Transactions on Antennas and Propagation*, 2010, 58(3): 1011–1014.
2. Wong KL, Lee CT. Wideband surface-mount chip antenna for eight-band LTE/WWAN slim mobile phone application. *Microwave and Optical Technology Letters*, 2010, 52(11): 2554–2560.
3. Wang P, Cai Z. Planar printed loop antenna with less no-ground space for hepta-band wireless wide area network/long-term evolution mobile handset. *Electronics Letters*, 2016, 52(15): 1284–1286.
4. Kim BN, Park SO, Oh JK, et al. Wideband built-in antenna with new crossed c-shaped coupling feed for future mobile phone application. *IEEE Antennas and Wireless Propagation Letters*, 2010, 9: 572–575.
5. Lee CT, Wong KL. Planar monopole with a coupling feed and an inductive shorting strip for LTE/GSM/UMTS operation in the mobile phone. *IEEE Transactions on Antennas and Propagation*, 2010, 58(7): 2479–2483.

6. Wong KL, Wu TJ, Lin PW. Small-size uniplanar WWAN tablet computer antenna using a parallel-resonant strip for bandwidth enhancement. *IEEE Transactions on Antennas and Propagation*, 2013, 61(1): 492–496.
7. Lin CI, Wong KL. Printed monopole slot antenna for internal multiband mobile phone antenna. *IEEE Transactions on Antennas and Propagation*, 2007, 55(12): 3690–3697.
8. Chi YW, Wong KL. Internal compact dual-band printed loop antenna for mobile phone application. *IEEE Transactions on Antennas and Propagation*, 2007, 55(5): 1457–1462.
9. Deng C, Li Y, Zhang Z, et al. Planar printed multi-resonant antenna for octa-band WWAN/LTE mobile handset. *IEEE Antennas and Wireless Propagation Letters*, 2015, 14: 1734–1737.
10. Chu FH, Wong KL. Planar printed strip monopole with a closely-coupled parasitic shorted strip for eight-band LTE/GSM/UMTS mobile phone. *IEEE Transactions on Antennas and Propagation*, 2010, 58(10): 3426–3431.
11. Wu CH, Wong KL. Ultrawideband PIFA with a capacitive feed for penta-band folder-type mobile phone antenna. *IEEE Transactions on Antennas and Propagation*, 2009, 57(8): 2461–2464.
12. Chang CH, Wong KL. Internal multiband surface-mount monopole slot chip antenna for mobile phone application. *Microwave and Optical Technology Letters*, 2008, 50(5): 1273–1279.
13. Yang CW, Jung YB, Jung CW. Octaband internal antenna for 4G mobile handset. *IEEE Antennas and Wireless Propagation Letters*, 2011, 10: 817–819.
14. Ban YL, Liu CL, Li LW, et al. Small-size coupled-fed antenna with two printed distributed inductors for seven-band WWAN/LTE mobile handset. *IEEE Transactions on Antennas and Propagation*, 2013, 61(11): 5780–5784.
15. Li Y, Zhang Z, Zheng J, et al. A compact hepta-band loop-inverted F reconfigurable antenna for mobile phone. *IEEE Transactions on Antennas and Propagation*, 2012, 60(1): 389–392.
16. Park YK, Sung Y. A reconfigurable antenna for quad-band mobile handset applications. *IEEE Transactions on Antennas and Propagation*, 2012, 60(6): 3003–3006.
17. Zhao W, Wang Y. A reconfigurable antenna with one slit and a small clearance for nona-band metal-bezel mobile phones. *IEEE Transactions on Antennas and Propagation*, 2023, 71(9): 7172–7183.
18. Lee WW, Cho YS. A new frequency-reconfigurable antenna using coupling patterns for mobile handsets. *Microwave and Optical Technology Letters*, 2015, 57(10): 2333–2336.
19. Lee SW, Jung HS, Sung YJ. A reconfigurable antenna for LTE/WWAN mobile handset applications. *IEEE Antennas and Wireless Propagation Letters*, 2014, 14: 48–51.
20. Liu Y, Liu P, Meng Z, et al. A planar printed nona-band loop-monopole reconfigurable antenna for mobile handsets. *IEEE Antennas and Wireless Propagation Letters*, 2018, 17(8): 1575–1579.
21. Deng C, Li Y, Zhang Z, et al. A novel low-profile hepta-band handset antenna using modes controlling method. *IEEE Transactions on Antennas and Propagation*, 2014, 63(2): 799–804.
22. Yang Y, Zhao Z, Yang W, et al. Compact multimode monopole antenna for metal-rimmed mobile phones. *IEEE Transactions on Antennas and Propagation*, 2017, 65(5): 2297–2304.
23. Ban YL, Qiang YF, Chen Z, et al. A dual-loop antenna design for hepta-band WWAN/LTE metal-rimmed smartphone applications. *IEEE Transactions on Antennas and Propagation*, 2014, 63(1): 48–58.
24. Yan Y, Ban YL, Wu G, et al. Dual-loop antenna with band-stop matching circuit for WWAN/LTE full metal-rimmed smartphone application. *IET Microwaves, Antennas & Propagation*, 2016, 10(15): 1715–1720.
25. Aziz RS, Arya AK, Park SO. Multiband full-metal-rimmed antenna design for smartphones. *IEEE Antennas and Wireless Propagation Letters*, 2016, 15: 1987–1990.
26. Liu D. *Research on Multi-antenna for 5G Mobile Terminals for 5G Applications*. Beijing: Beijing Institute of Technology, 2020.

CHAPTER 4

Dual Antenna Design of 5G Mobile Terminals

In the low and medium frequency bands of mobile communication, placing two antennas is a more appropriate design scheme. On the one hand, placing two antennas can achieve better performance than a single antenna; on the other hand, the wavelengths in the low and medium frequency bands are longer, and only two antennas can be placed in general due to the limited space inside the cell phone. In this chapter, several typical dual-antenna design schemes will be proposed for 698–960 MHz (low band) and 1710–2690 MHz (mid-high band), respectively.

4.1 CURRENT STATUS OF DUAL ANTENNA RESEARCH

The coupling between antenna units can be viewed as the effect of the electric field generated by one antenna on the neighboring antennas. In mobile terminals, there are generally four methods used to decouple two antennas. The first method is to adjust the layout of the antennas to pull apart the distance between the antennas, but the method will be limited by the size of the equipment space, and the improvement of this method is limited. The second method is to consider the radiation pattern of the antenna and design two antennas whose patterns are orthogonal to each other. The orthogonality of the patterns can be divided into the orthogonality of the directional beams (i.e., the directional beams pointing to different directions) and the orthogonality of the form of polarization of the radiation field. The orthogonality of the patterns generally achieves lower antenna coupling. The third method is to consider reducing the coupling generated by floor currents between antennas by designing specific decoupling structures, such as resonant slots, to introduce parasitic resonant slots between the antennas and reduce the floor currents flowing to the other antenna. The fourth method is to introduce a new coupling to cancel the coupling between the original antennas, and common methods include designing parasitic cells, designing ground branches, introducing neutralization lines, and introducing decoupling networks.

DOI: 10.1201/9781003770039-4

4.1.1 Dual-Antenna Decoupling in 900 MHz Band

The 900 MHz band corresponds to a wavelength of about 330 mm, while the maximum size of a typical mobile terminal is about 160 mm, which is less than half a wavelength. When integrating two 900 MHz band antennas inside a mobile terminal, there is generally a strong coupling between the antennas, and it is necessary to use a decoupling method to reduce the coupling between the antennas.

Some typical dual antenna designs operating in the 900 MHz band are given in the references [1–5]. For example, two monopole antennas operating in the 710 MHz band are miniaturized by bending, and the two antennas are closely aligned with an edge-to-edge distance of only 8 mm, and by introducing a directional coupler between the two feed ports for decoupling, the isolation between the two ports can be increased from 2 dB to more than 25 dB [1]; by placing the two bent monopole antennas face-to-face, the antennas are operated in the 830 MHz band, the edge-to-edge distance is 15 mm, by building a T-shaped decoupling circuit between the two feed lines, the isolation between the two ports can be increased from 3 dB to more than 20 dB [2]; placing two Planar Inverted-F Antenna (PIFAs) in the two corners of the mobile terminal, there is a T-shaped parasitic branch between the two antennas, by adjusting the length of the branch can effectively reduce the coupling between the antennas, in the 770 MHz band, the isolation between the two ports can be up to 17 dB [3]; two PIFAs are placed orthogonally on the long and short edges of the mobile terminal, and the isolation between the two ports can be up to 20 dB due to the orthogonal polarization of the two antennas, although the port isolation is still affected by the specific structure of the antenna [4]; the eigenmode theory is used for the design of a Multiple-Input Multiple-Output (MIMO) dual antenna, where the floor of the mobile terminal has two orthogonal modes along the long and short edges of the mobile terminal, there are two orthogonal modes, and the isolation between the two ports is more than 10 dB after excitation by a T-shaped monopole and a folded monopole, respectively [5].

4.1.2 Dual-Antenna Decoupling in the 1800 MHz Band

The 1800 MHz band corresponds to a wavelength of about 167 mm, which is comparable to the maximum size of a mobile terminal, and it is easier to realize decoupling compared to lower frequency bands. However, the 1800 MHz band covers the frequency interval of 1710–2690 MHz, with a large relative frequency bandwidth, which requires a wide-band decoupling network to realize the decoupling of the whole band, and the decoupling bandwidth is generally extended by means of a multi-stage cascaded decoupling network.

Some dual antenna designs operating in the 1800 MHz band are given in the references [6–10]. For example, two F-shaped monopole antennas are placed back-to-back with two sets of parasitic branches loaded in the middle for extending the impedance bandwidth and increasing the port isolation to achieve high isolation in the 1710–2690 MHz band [6]; two bent monopole antennas are placed face-to-face with three neutralization lines loaded on the two monopole arms, and a three-stage cascaded decoupling is obtained by controlling the positions and geometrical dimensions of the three microstrip lines network, thus decoupling is achieved over a wide frequency band [7]; two L-shaped open slot antennas are excited by two monopoles to realize the coupling of two modes and thus extend the bandwidth, and the port isolation is maintained above 15 dB in the 1710–2170 MHz band by loading two suspended L-shaped branches [8]; two monopole antennas with different bending morphologies are placed side-by-side, and by inducing two neutralization lines on the two feed lines, two parasitic resonant structures are induced to form a cascade cavity, which can achieve high port isolation, although the bandwidth of this strong resonant structure is relatively small [9]. By applying the eigenmode theory to analyze the eigenmodes of the mobile terminal floor in the 1710–2690 MHz band, a port isolation of 21 dB can be achieved by choosing the appropriate excitation location to place the feed source [10].

4.1.3 Dual-Band Dual-Antenna Decoupling

On the basis of single-band, some researchers try to carry out dual-band dual-antenna design, which arranges two dual-band antennas closely together to improve the integration of the antenna. Obviously, the decoupling of dual-band is more complicated than that of single-band, and the radiating structure and decoupling structure will occupy a larger size.

Four dual antenna designs that can operate simultaneously in the 900 MHz band and 1800 MHz band are given in the references [11–14]. For example, two bent monopole antennas are placed face to face, a Π-shaped decoupling branch is placed in the center region of the two antennas, and the branch is deep in the floor, which can block the current, and the two antennas basically have 10 dB port isolation in the 810–970 MHz and 1360–2800 MHz bands [11]; two capacitively coupled-fed monopole antennas are bent to form a three-dimensional structure, and the two parasitic branches are placed back-to-back in the middle of these two face-to-face antennas, and the port isolation of the 690–960 MHz band and the 1710–2690 MHz band is also on the order of 10 dB on [1–2]; a primary antenna and an auxiliary antenna are integrated in a space of 60 mm × 15 mm × 3 mm, and the blank area between the two antennas is designed as the floor in the USB interface. The design of the floor helps to reduce the coupling between the antennas [13]; four monopole antennas are placed at the four corners of the floor of the mobile terminal, the two antennas on the long-side have greater port isolation due to their distance, and the two antennas on the short-side are decoupled by introducing a ground branch and a collector inductance [1–4].

4.2 EIGENMODE-BASED DUAL ANTENNA DESIGN FOR 900 MHz BAND

Eigenmode theory has received more and more attention in the field of multi-antenna systems for its ability to intuitively analyze the physical properties of antennas. The nature of natural orthogonality between the eigenmodes provides a new solution to realize MIMO antenna design for mobile terminals. Based on the theory of eigenmodes, many MIMO antenna designs for mobile terminals have been proposed. However, most of the designed antennas operate in the middle and high frequency bands. Considering that there are even fewer eigenmodes available in the low-frequency band (generally, only one eigenmode is resonant), it is more difficult to implement MIMO antenna design in the low-frequency band. There are mainly two methods to realize MIMO antenna design in low-frequency band: One method is to optimize the antenna structure and excitation source, and the bandwidth under this method is usually smaller; the other method is to introduce a new structure to provide a new eigenmode, which is more flexible in terms of design, but may increase the size of the antenna. In this design, a new MIMO antenna for mobile terminals is developed based on the theory of eigenmodes. The antenna allows three eigenmodes to work together for the first time, which is able to extend the bandwidth of the low-frequency band as well as realize the MIMO function.

4.2.1 Metal Frame Antenna Modeling

Metal bezel not only has better mechanical properties but also shows a beautiful decoration. This design aims to analyze the eigenmode of the cell phone floor and the metal bezel in the low-frequency band using the eigenmode theory.

Combining a metal border with a floor is a very natural and straightforward idea. A model of the suspended metal border is shown in Figure 4.1. The dimensions of the floor are 120 mm×60 mm, and the metal frame wraps around the floor with a frame size of 130 mm×68 mm×6 mm. The model is discretized using the method of moments, and the computed eigenvalues are shown in Figure 4.2. The first six eigenmodes with eigenvalues closest to 0 are plotted in Figure 4.2. The eigenvalues of modes 1 and 2 are 0 near 800 MHz,

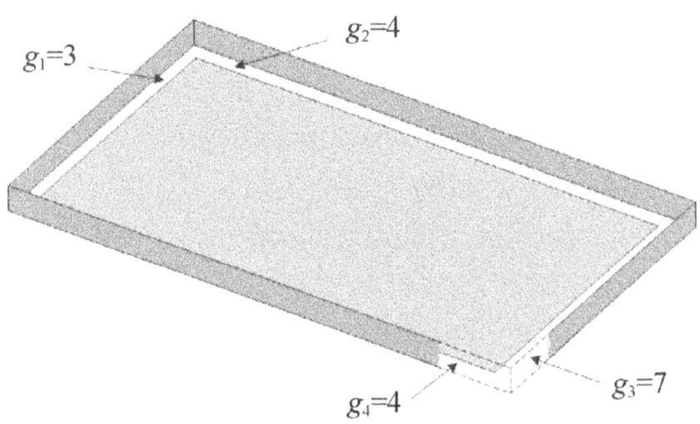

FIGURE 4.1 Suspended metal bezel model (*g* for gap).

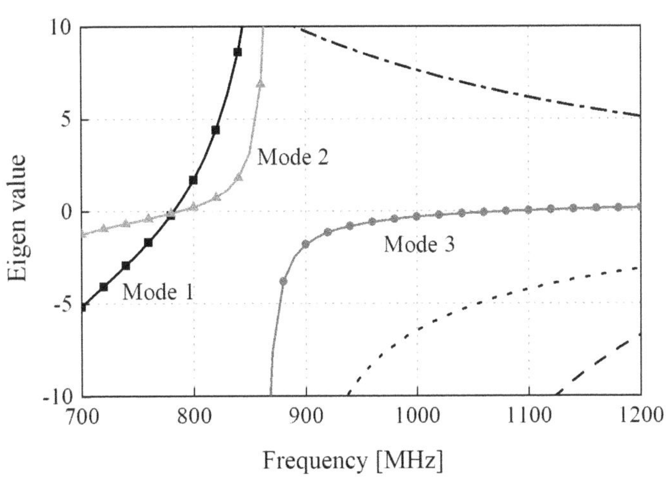

FIGURE 4.2 Eigenvalues of suspended metal border model.

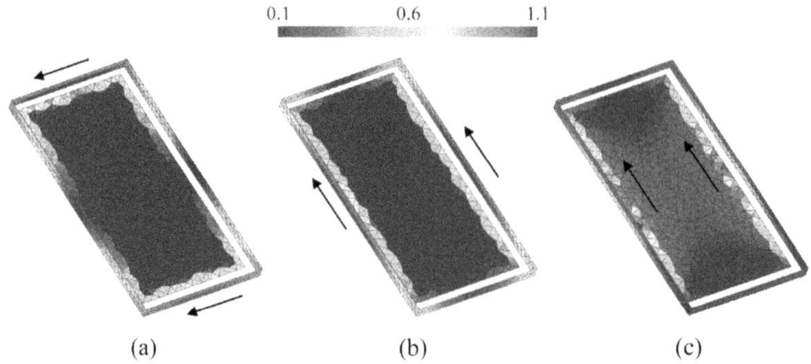

FIGURE 4.3 Intrinsic current distributions of the suspended metal-bezel model (a) Mode 1 at 800 MHz (b) Mode 2 at 800 MHz (c) Mode 3 at 1 GHz.

and the eigenvalue of mode 3 is 0 near 1 GHz. Therefore, these three modes are resonant in the lower frequency band. Comparison of the eigenvalues of this model with the eigenvalues when there is only a floor shows that mode 3 is the eigenmode of the floor and modes 1 and 2 are the eigenmodes of the metal bezel. Therefore, the introduction of the metal bezel produces two new resonances. In addition, it can be seen from Figure 4.2 that the curves corresponding to modes 2 and 3 are flatter, while the curve corresponding to mode 1 is steeper, which indicates that the potential bandwidths of modes 2 and 3 are larger and the potential bandwidth of mode 1 is smaller.

In order to further determine the operating states of these three modes, the intrinsic current distributions of these three modes at their respective resonance points were analyzed, as shown in Figure 4.3. The maximum value of the current of mode 1 is at the midpoint of the two short edges of the metal bezel, and the minimum value is at the midpoint of the two long edges. Similarly, the current maximum value of mode 2 is at the midpoint of the two long edges of the metal bezel and the minimum value is at the midpoint of the two short edges. Thus, both modes are full-wavelength modes of the metal bezel, and the location of the current maximum for one mode corresponds to the location of the current minimum for the other mode. The current of mode 3 is distributed along the two long edges of the metal bezel and corresponds to the one-half wavelength dipole mode of the floor. As can be seen from the current distributions, the locations of the current maxima for modes 2 and 3 are close to each other, as are the locations of the current minima. Based on this, a possible scenario is to simultaneously excite modes 2 and 3, which exhibit similar current distributions, to extend the bandwidth.

4.2.2 Eigenmode Modification

It is well known that the electric field (voltage) at the antenna current maximum is 0. Grounding the antenna at the current maximum will not affect the antenna's modes, which is the concept of a virtual short circuit. In the suspended metal bezel model, the locations of the current maxima for modes 2 and 3 are very close to each other, and if a metal strip is used to connect the floor to the metal bezel at this location, these two modes should not be

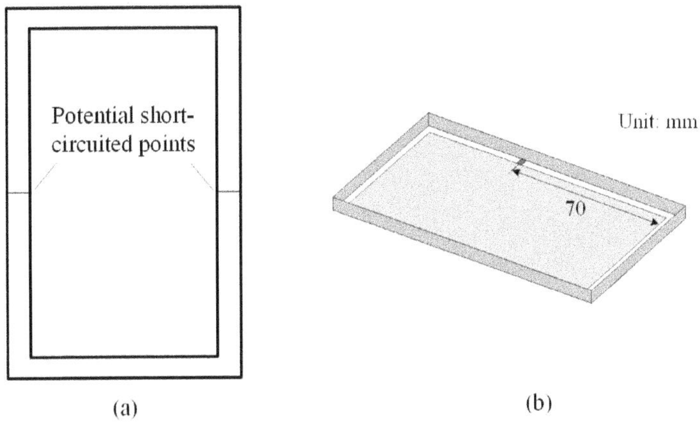

FIGURE 4.4 Model of short-circuiting metal bezel (a) location of possible short circuit points (b) 3D schematic of the model.

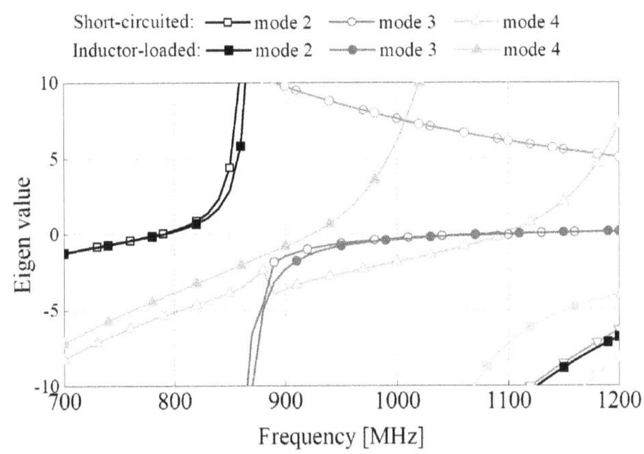

FIGURE 4.5 Effect of inductance loading on the eigenvalues.

affected. The locations of the two possible short-circuit points are labeled in Figure 4.4. For simplicity, we will only consider the case where there is only one short-circuit point. Since the gap between the floor and the metal border is not perfectly symmetrical about the line joining the midpoints of the two long edges, the location of the short-circuit point does not exactly coincide with the midpoint of the long edge of the floor.

After short-circuiting the floor and the metal bezel, as shown in Figure 4.5, modes 2 and 3 are almost unaffected, but mode 1 disappears. This is due to the fact that the short-circuit point is located at the current zero point of mode 1, and the current distribution of this mode is disrupted. However, the introduction of the short-circuit point produces a new resonant mode, mode 4 in Figure 4.5, whose eigenvalue is close to 0 at 1100 MHz. 1100 MHz corresponds to a wavelength of 270 mm, and the circumference of the metal bezel is about one wavelength. Therefore, this mode is a three-fourths perimeter wavelength mode of the metal frame. In summary, by introducing the short-circuit point, the model loses mode 1 and adds mode 4. Considering that the curve corresponding to mode

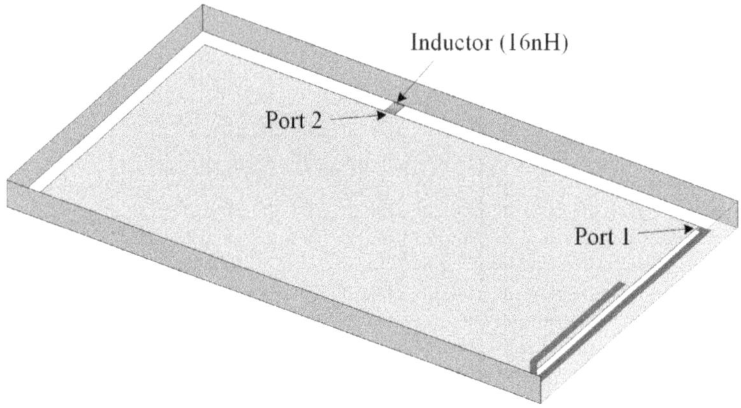

FIGURE 4.6 Antenna model with two external feeds added.

4 is flatter than that corresponding to mode 1, this modification increases the potential bandwidth of the model.

In order to move the operating frequency of mode 4 to the mobile communication band, an inductor was added at the connection between the floor and the metal bezel. Figure 4.5 shows the effect of the loaded inductor (16 nH) on the eigenvalues of the model. The addition of the inductor changes the resonant frequency of mode 4 from 1100 to 920 MHz, while the other modes are little affected. The inductor acts as a matching circuit here, which lengthens the effective current path of mode 4 and thus reduces the resonant frequency. After loading the inductor, the resonant frequencies of modes 3 and 4 are close to each other, and thus, the MIMO function can be attempted using mode 4.

4.2.3 Adding an External Excitation Source

The short-circuit metal-edge model has produced three eigenmodes with resonant frequencies below 1 GHz. Depending on the current distribution, modes 2 and 3 may be excited by the same feed to extend the bandwidth, and mode 4 may form a MIMO channel with mode 3. To verify this conjecture, Figure 4.6 shows the antenna model with the addition of two external feeds. The feed on port 1 is a 3D folded monopole structure, and the feed on port 2 is added between the inductor and the floor. The other parameters of the antenna are the same as those of the short-circuited metal-bezel model.

Figure 4.7 analyzes the surface current distribution of the antenna with different port excitations. Port 1 is designed to excite Mode 2 and Mode 3, while port 2 is designed to excite Mode 4. Mode 2 has a resonant frequency of 800 MHz and its coupling to Mode 4 is small due to the frequency separation. The resonant frequencies of modes 3 and 4 are close to each other and the orthogonality of these two modes needs to be focused on. When port 1 is excited, the current on the floor is larger and distributed along the long side of the floor, which indicates that mode 3 is excited. When port 2 is excited, the current is concentrated on the metal rim and there are three clear current zeros on the rim, which indicates that mode 4 is excited. In addition, the currents on both sides of the short-circuit point have different distributions under the two excitations, i.e., the port 1 excitation exhibits isotropic

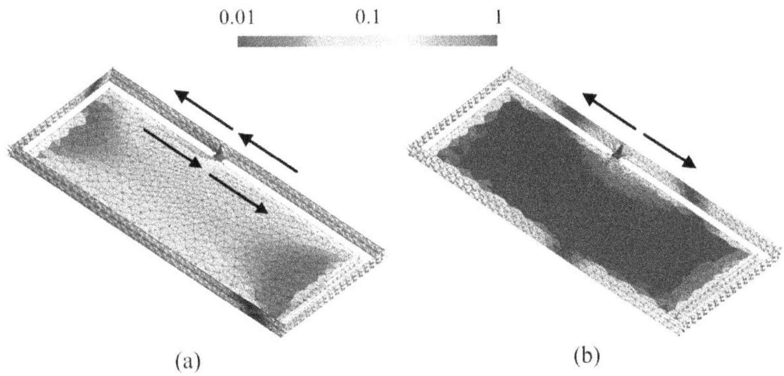

FIGURE 4.7 Surface current distribution of the antenna under different port excitations (a) port 1 excitation (b) port 2 excitation.

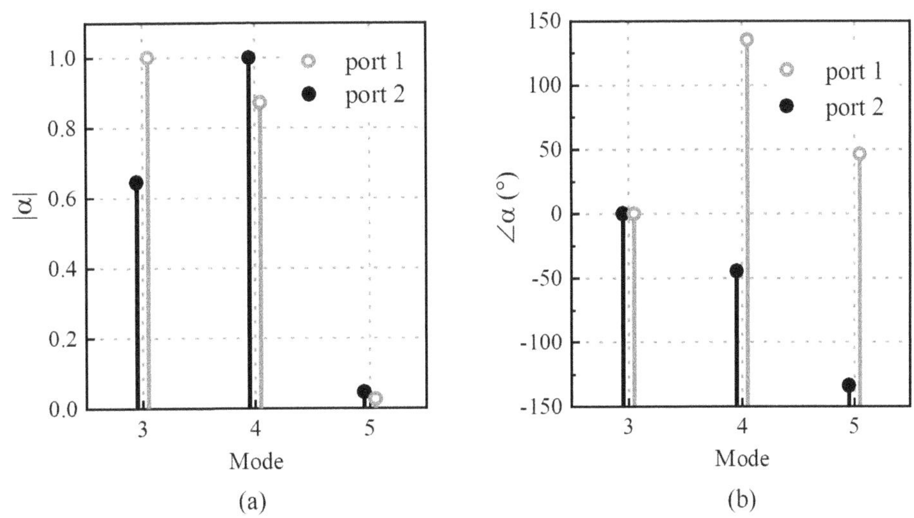

FIGURE 4.8 Normalized excitation coefficients for each mode (mode 3–mode 5) for different port feeds (a) amplitude (b) phase.

currents and the port 2 excitation exhibits reverse currents. This difference is important to increase the port isolation.

The normalized excitation coefficients for each mode (modes 3–5) at different port feeds are shown in Figure 4.8. Since the amplitude of the normalized excitation coefficients of the other modes at 920 MHz is very small, the calculated values of the normalized excitation coefficients for mode 3–mode 5 are given in Figure 4.8. Under both feeding configurations, modes 3 and 4 exhibit significantly larger normalized excitation coefficient amplitudes, indicating that these two modes serve as the dominant resonant modes regardless of whether port 1 or port 2 is excited.

These two modes are the dominant resonant modes. However, the phase performance of the normalized excitation coefficients of these two modes is quite different. This is reflected in the fact that the phase of the normalized excitation coefficients of mode 3

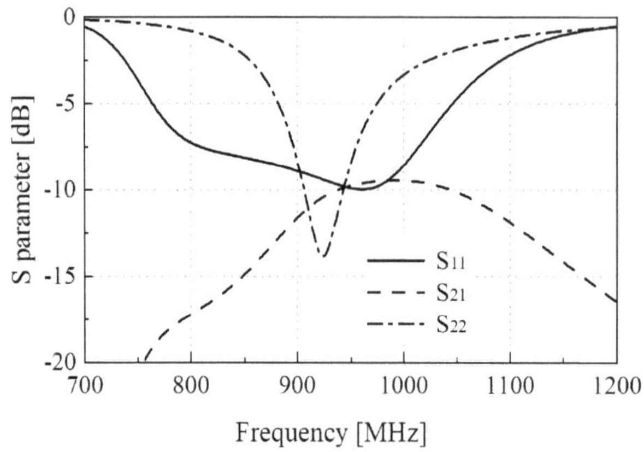

FIGURE 4.9 S-parameters of MIMO antenna simulation.

remains constant under the excitation of these two feed structures, and the phase of the normalized excitation coefficients of mode 4 under the excitation of the two feed structures differs by 180°. The isolation between the two ports can be measured by the inner product of the excitation coefficients. Based on the data shown in Figure 4.8, the inner product of the excitation coefficients is 0.0207, indicating a small coupling between the two ports.

Based on the eigenmode analysis, HFSS is used to further validate the feasibility of the model. Figure 4.9 shows *the* S-parameters of the MIMO antenna simulation. The reflection coefficient curve of port 1 has two distinct resonance points. The −6 dB bandwidth is 254 MHz (778–1032 MHz), covering the GSM850 and GSM900 bands. The reflection coefficient curve for port 2 has only one resonance point. The −6 dB bandwidth is 77 MHz (888–965 MHz), essentially covering the GSM900 band. The isolation between the ports is over 10 dB in the bandwidth where the two ports overlap.

4.3 EIGENMODE-BASED DUAL ANTENNA DESIGN FOR 1800 MHz BAND

In the past decades, various methods have been proposed to reduce the coupling between antennas. Among them, the introduction of bandstop structures or coupling-controllable structures, which can block or cancel the original coupling, is very popular in dual antenna design. There are various types of decoupling structures such as gaps, defective ground structures, parasitic branches, and neutralization lines. Although these designs can achieve high isolation, the introduced decoupling structures take up additional antenna space. To solve this problem, some scholars have designed decoupling circuits that use lumped elements to build decoupling networks. However, this method adds extra cost. In recent years, the eigenmode theory has been applied to the design of mobile terminal antennas. The natural orthogonality of the eigenmodes of the metal floor can be used to achieve high port isolation. Many MIMO antenna designs based on eigenmode theory have been reported for mobile terminals. However, the limited number of eigenmodes available in mobile terminals and the small bandwidth of individual modes make it difficult to realize wide-band designs in MIMO antennas.

Reasonable arrangement of antenna positions is another way to realize low coupling. The related reference can be categorized into three main types. The first type utilizes different antenna positions and/or asymmetric feed port positions [15,16]. For example, reference [16] proposes a dual antenna system consisting of an Inverted-F Antenna and a monopole with a port isolation of 20 dB. The second type places two antennas in orthogonal positions at the floor edge [17], where the coupling between the two antennas is low due to the orthogonality of the polarizations. The third type is to place the two antennas at the top and bottom of the floor [18–21], where the spacing between the two antennas is made larger, allowing the coupling to decrease. In the reference [20], two inverted L-shaped antennas are placed diagonally across the floor, achieving high port isolation in the 1710–2690 MHz band. However, these designs generally require large ground clearance dimensions.

In order to realize a small ground clearance design, this section presents a compact planar dual antenna design operating in the 1710–2690 MHz band. Placing the feed ports at the corners of the floor and placing the antennas at the top and bottom of the floor allows for both large bandwidth and high port isolation, which can be further extended by combining the bandwidth of multiple modes. The ground clearance for each antenna is only 5 mm × 30 mm.

4.3.1 Eigenmode of the Cell Phone Floor

The eigenmodes of the cellular floor are first analyzed by applying the eigenmode theory. The eigenvalues can be calculated by the following equations:

$$X(J_n) = \lambda_n R(J_n) \tag{4.1}$$

where X and R are the imaginary and real parts, respectively, of the impedance matrix $\boldsymbol{Z}$ obtained from the integral equation of the electric field. J_n and λ_n are the intrinsic current and intrinsic value of the nth mode, respectively. If λ_n is equal to 0, then mode n is resonant in that frequency band.

Based on the eigenmode theory, the eigenvalues of the cell phone floor with a size of 120 mm × 65 mm are calculated using a self-compiled code, as shown in Figure 4.10.

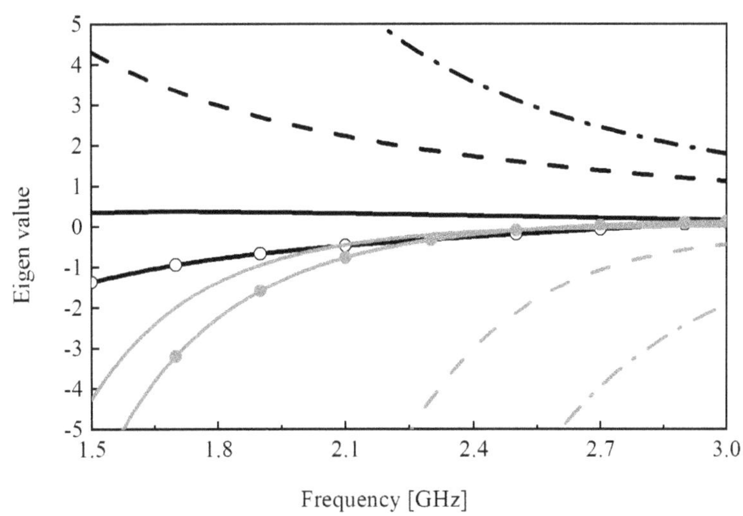

FIGURE 4.10 Eigenvalues of cell phone floor with dimensions 120 mm × 65 mm.

From Figure 4.10, it can be observed that there are multiple eigenmodes at 1500–3000 MHz. Among them, there are four modes whose eigenvalue curves cross the zero point, indicating that these four modes are resonant. Figure 4.11 shows the eigencurrents of these four resonant modes at 2.2 GHz. These modes can be clearly identified by the current distribution. In all the modes, the current intensity at the four corners of the floor is weak. This means that if a feed port is placed at one of the corners, multiple modes will be excited. Therefore, the corners of the cell phone floor are the best locations to introduce feed ports in a single antenna design.

To quantitatively assess the effect of feed port location on bandwidth, two typical feed port locations were compared. One feed port location is at the corner of the floor (edge feed) and the other is in the middle of the short side of the floor (interior feed). In the 1710–2690 MHz band, 2200 MHz is considered the center frequency. An L-shaped monopole is used as the feed source. Figure 4.12 shows the geometry of the edge-fed and internally fed L-shaped monopoles. In both cases, the structure of the monopole is the same, but the length needs to be adjusted so that both monopole antennas resonate at the same frequency. Figure 4.13 shows the simulated reflection coefficient curves for both edge-fed and internally-fed. Both monopole antennas resonate at 2200 MHz, but the edge-fed has better resonance depth and impedance bandwidth parameters. The edge-fed not only outperforms the internally-fed in terms of bandwidth performance, but also in terms of structural miniaturization.

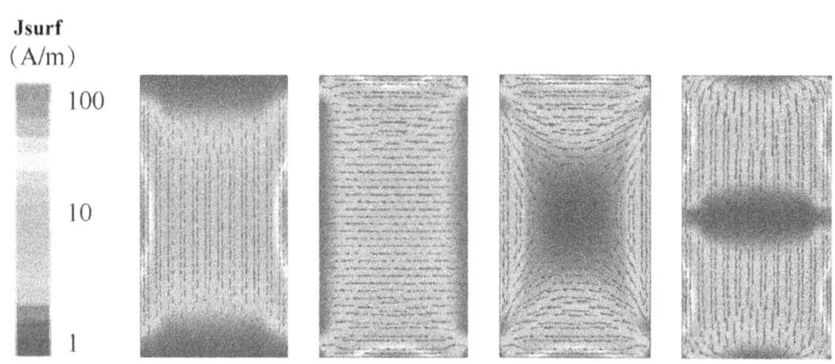

FIGURE 4.11 Eigencurrents of four resonant modes at 2.2 GHz.

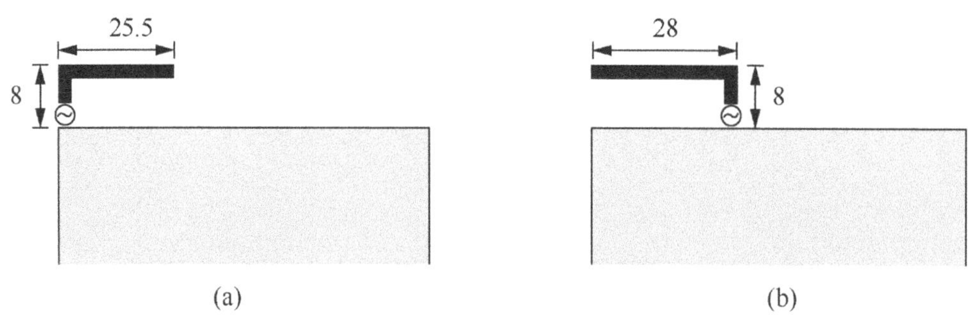

FIGURE 4.12 Geometries of edge-fed and internally fed L-shaped monopoles (in mm) (a) edge-fed (b) internally fed.

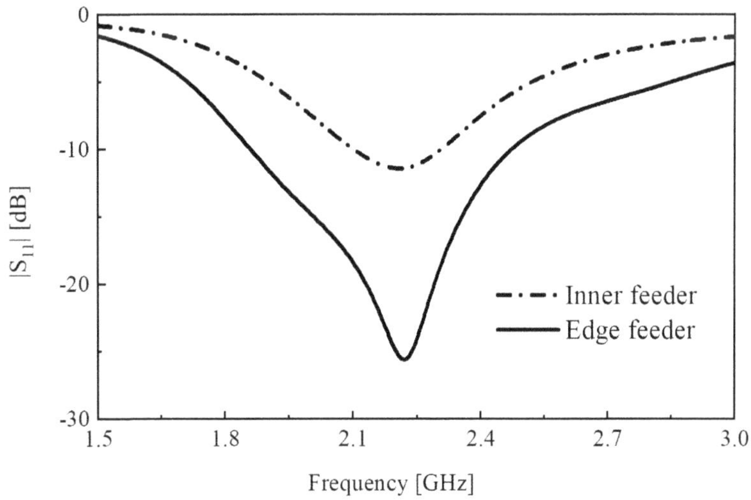

FIGURE 4.13 Simulated reflection coefficient curves for edge-fed and internally fed.

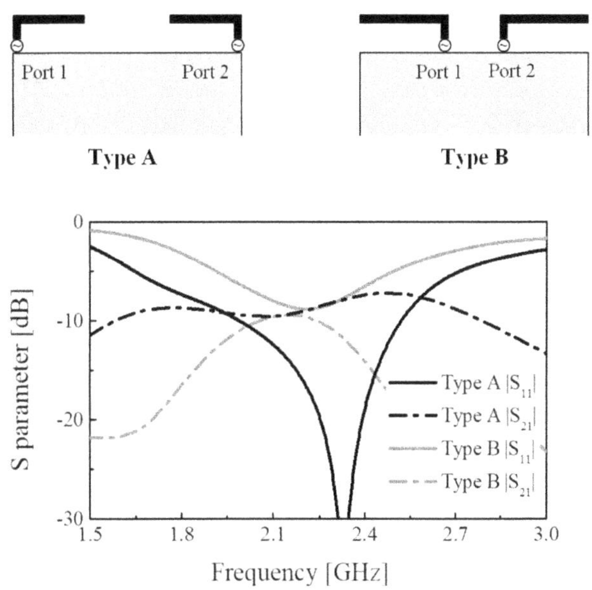

FIGURE 4.14 Two configurations and reflection coefficient curves when both antennas are placed on the same side of the floor.

4.3.2 Dual Antenna Configuration

Next, the performance of the dual antenna configuration is analyzed for both edge-fed and internally-fed feeds. It is known that edge-feeding has an advantage over internal-feeding in single-antenna design, but whether this advantage is still applicable when forming a MIMO antenna needs further investigation. In order to find the optimal arrangement of dual antenna configurations, various dual antenna models are compared.

Two configurations of a dual antenna system based on an L-shaped monopole are shown in Figure 4.14. In both configurations, the two antennas (i.e., the two ports) are placed on the same side of the floor. Type A and Type B are edge-fed and internally-fed, respectively. The simulated reflection coefficient results show that the resonant frequency of the dual antenna is similar to that of the single antenna. However, the port isolation is poor for both configurations, with a peak port isolation of 9.5 dB for Type B and only 7.2 dB for Type A. Edge-feeding is not suitable for MIMO antenna design if it is required to place both antennas on the same side of the floor.

To analyze the contribution of each mode to the total radiated field, the total current is decomposed. The total current J excited by the feed source can be expressed as a linear superposition of the individual eigencurrents:

$$J=\sum \alpha_n J_n \tag{4.2}$$

where α_n is the weighting factor for the nth mode, i.e., the normalized excitation factor. The radiated power contributed by all modes is normalized using the following equation:

$$\sum_n |\alpha_n|^2 = 1 \tag{4.3}$$

Figure 4.15 shows α_n at 2.2 GHz under different port excitations. The amplitude of each mode is the same regardless of which port is excited, but the phase of each mode is not always the same. Some modes have a constant phase when port 1 or port 2 is excited, while some modes have a 180° phase difference. This phenomenon will be used in high isolation designs.

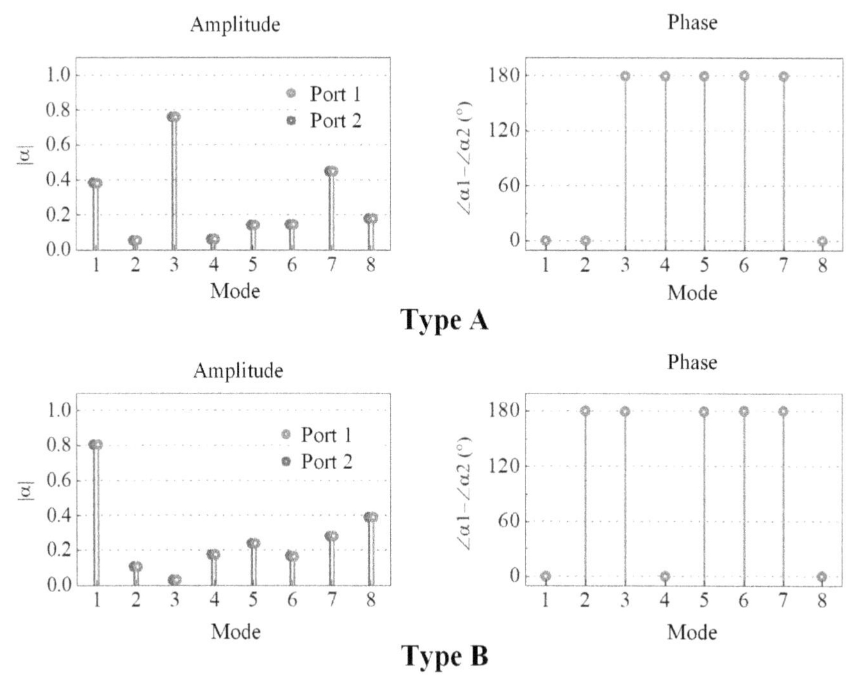

FIGURE 4.15 α_n at 2.2 GHz for different port excitations.

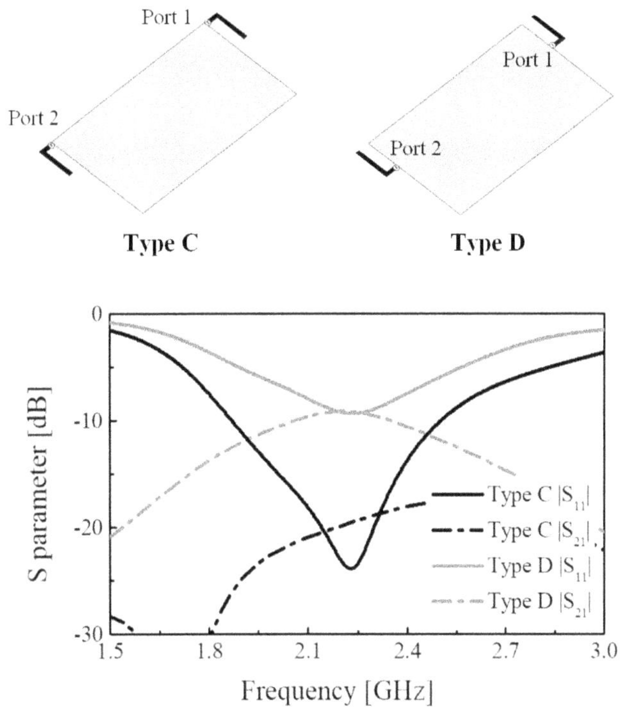

FIGURE 4.16 Two configurations and reflection coefficient curves when two antennas are placed at the top and bottom of the floor.

To take advantage of the bandwidth of the edge feed, two other configurations are given in Figure 4.16. In both configurations, the two ports are placed at the top and bottom of the floor (i.e., on either side of the long side of the floor). The results show that the resonance depth is much deeper for Type C. The peak port isolation is 17.5 dB for Type C and 9.1 dB for Type D. Type C provides better performance in terms of bandwidth and also provides better performance in terms of isolation. Comparison of Type B and Type D reveals no significant improvement in isolation performance for Type D despite the increased antenna spacing. This is a good example that can be used to refute the claim that coupling can be reduced by increasing the distance.

Figure 4.17 illustrates α_n for two configurations with different port excitations. The amplitude of α_n does not change when port 1 or port 2 is excited, but for some modes, the phase of α_n may flip 180°.

Finally, the model with two antennas located on either side of the long edge of the floor and on either side of the diagonal was analyzed, as shown in Figure 4.18. The two antennas in Type E have the largest spacing. However, the port isolation for Type E is smaller than that of Type C. This further indicates that port isolation does not always increase with antenna distance.

The phase of the normalized excitation coefficients is an internal factor that determines the port isolation. In order to evaluate the port isolation degree from a pattern perspective, the orthogonality between the two current distributions excited by port 1 and port 2

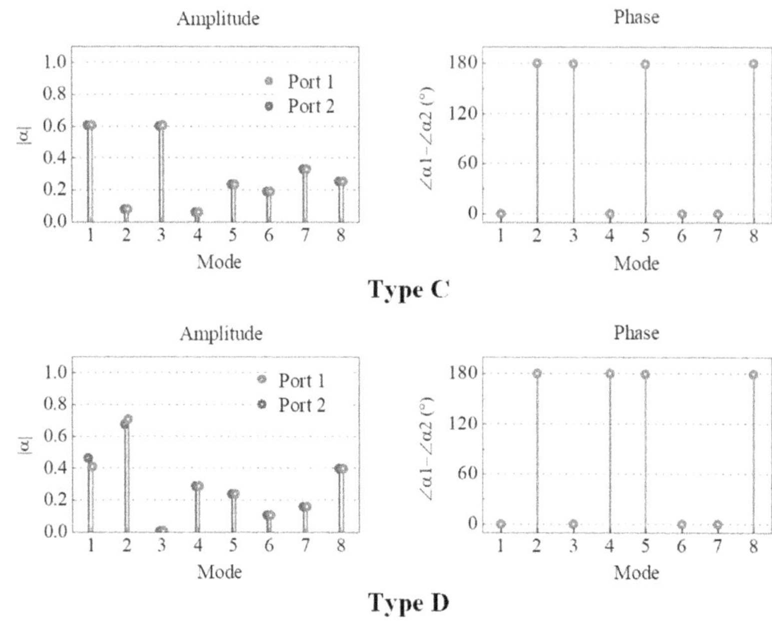

FIGURE 4.17 α_n for two configurations with different port excitations.

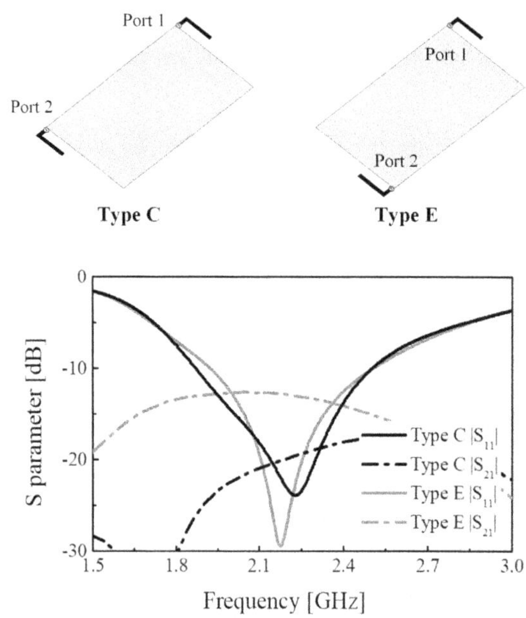

FIGURE 4.18 Configuration and reflection coefficient curves for two antennas when they are located on both sides of the long side of the floor and on both sides of the diagonal, respectively.

is calculated. Let J_{port1} and J_{port2} be the currents on the antenna when port 1 and port 2 are excited, respectively, and the inner product of these two currents is expressed as follows:

$$J_{\text{port1}} \cdot J_{\text{port2}} = \left(\sum \alpha_{n,1}\right) \cdot \left(\sum \alpha_{n,2}^{*}\right) = \sum \alpha_{n,1} \cdot \alpha_{n,2}^{*} \tag{4.4}$$

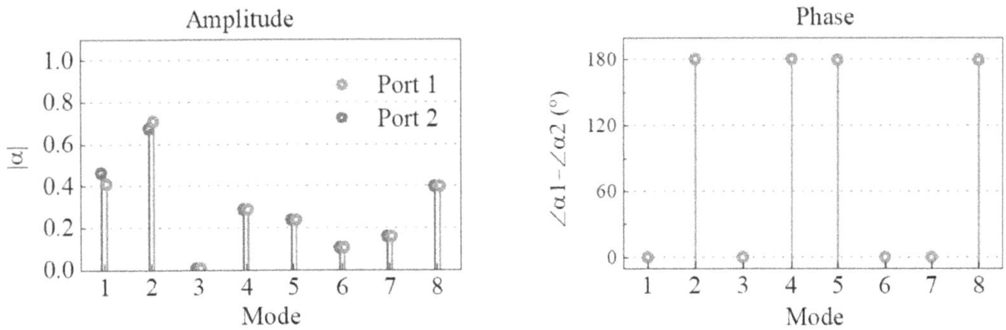

FIGURE 4.19 Excitation coefficient of each mode for different port excitation.

TABLE 4.1 Calculated Values of Inner Product at 2.2 GHz for Different Types of Configurations

Type	Internal Product
A	0.655
B	0.641
C	0.028
D	0.547
E	0.437

where $\alpha_{n,1}$ denotes the normalized excitation coefficient for the nth mode when port 1 is excited and $\alpha_{n,2}$ denotes the normalized excitation coefficient for the nth mode when port 2 is excited. Figure 4.19 shows each mode for different port excitations. Considering that the excitation amplitudes of the two port excitation modes are the same and the phase difference is 0° or 180°, the result of Eq. (4.4) is a real number.

Table 4.1 lists the calculated inner product values for different types of configurations at 2.2 GHz. The calculated value of inner product for type C is very small, which indicates that the individual currents of this configuration are highly uncorrelated with each other, a good indication of its ability to achieve high port isolation.

From the above analysis, it is clear that type C is the most promising candidate for a dual antenna design, which uses edge-feeding techniques and places the antennas on both sides of the long edge of the floor, achieving both large bandwidth and high port isolation. MIMO antenna designs covering the entire 1710–2690 MHz band are highlighted next.

4.3.3 Wide-Band MIMO Antenna Design

Since only one resonant mode of the L-shaped monopole is excited, its bandwidth is insufficient to cover the middle and high frequency bands in a limited space. Extending the bandwidth by exciting multiple resonant modes is described below.

The structure of the designed wide-band MIMO antenna is shown in Figure 4.20. The two antennas are located at the upper right and lower right corners of the cell phone floor, and the FR4 dielectric substrate (ε_r=4.4, tanδ=0.02) used has dimensions of 130 mm×65 mm×1 mm. the ground clearance is 5 mm×30 mm and consists of three branches (one drive monopole branch and two parasitic branches). The L-shaped drive monopole branch is fed by a 50

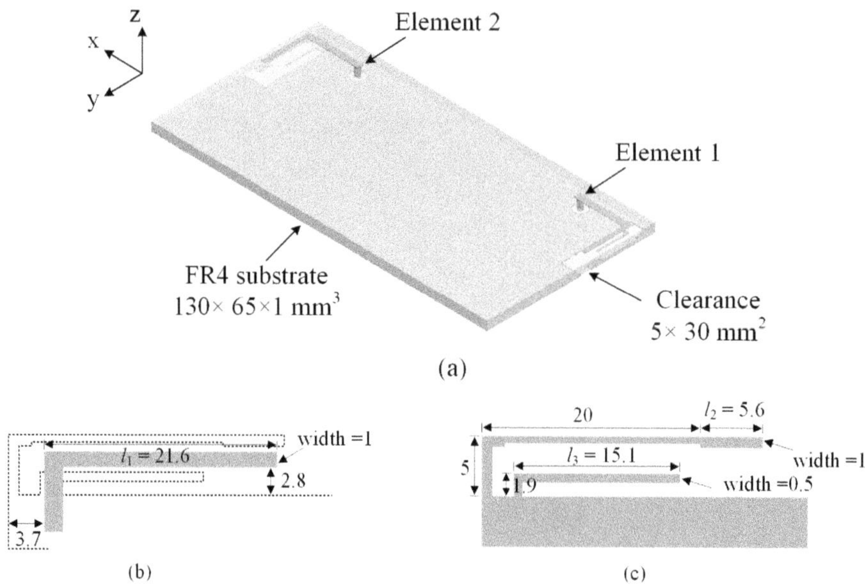

FIGURE 4.20 Structure of the wide band MIMO antenna (in mm) (a) three-dimensional structure (b) driven monopole branch (c) parasitic branches.

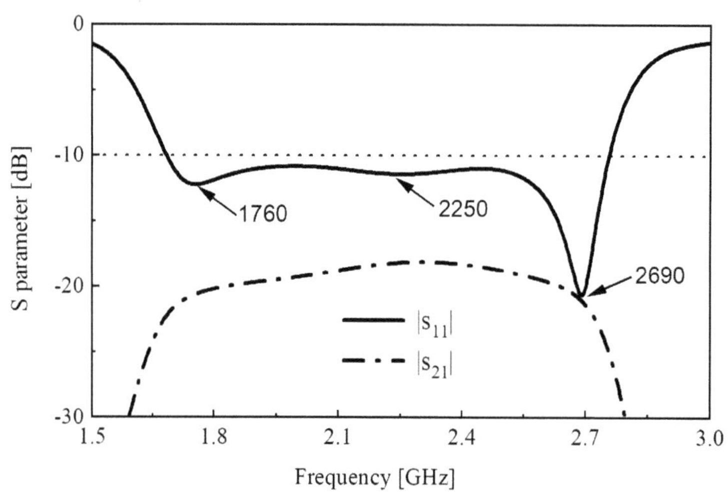

FIGURE 4.21 Simulated S-parameters of the MIMO antenna.

Ω microstrip line, and the two parasitic branches are of unequal lengths, with the shorter branch having a line width of 0.5 mm. A metallic region with a size of 0.4 mm × 16 mm is etched on the longer branch. All parameters are optimized by HFSS software.

The simulated S-parameters of the MIMO antenna are shown in Figure 4.21. The −10 dB bandwidth is 1075 MHz covering 1685–2760 MHz. There are three resonance points at 1760, 2250 and 2690 MHz. Port isolation is higher than 18 dB across the band.

In order to determine the radiation structure corresponding to the three resonance points, the current distribution at the three resonance points was observed. As shown in Figure 4.22, the currents are concentrated on different branches in different frequency bands. At 1760 MHz,

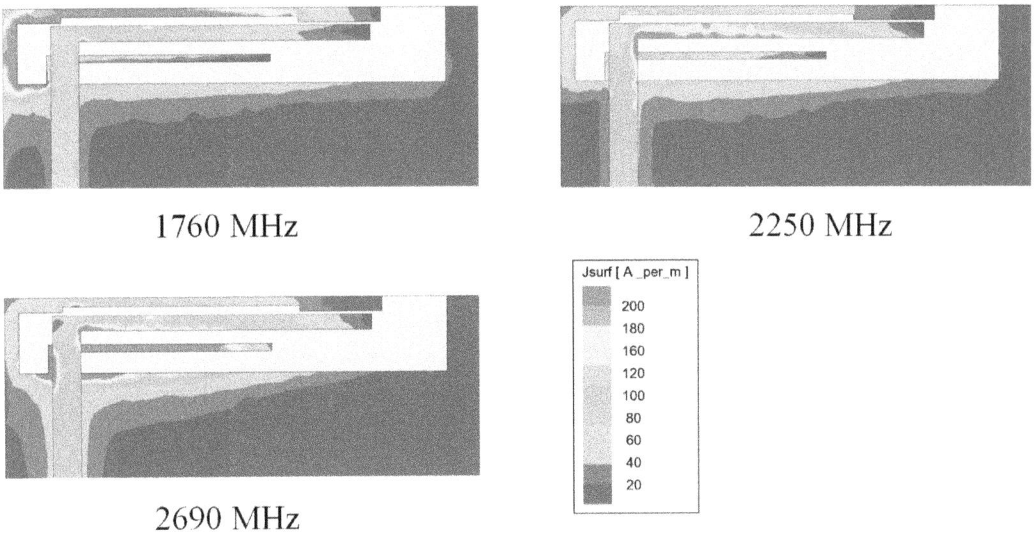

FIGURE 4.22 Current distribution at the three resonant frequencies (a) 1760 MHz (b) 2250 MHz (c) 2690 MHz.

the current signals on the longer parasitic branch are stronger, while at 2250 and 2690 MHz, the current signals on the driving monopole branch and the shorter parasitic branch are stronger. Therefore, the main radiation structure corresponding to these three resonance points can be determined from the strength of the current signals on the three branches. In addition, from the current distribution of the main radiation structure, it can be seen that the modes corresponding to these three resonance points are all quarter-wavelength modes.

All the parameters are studied to find the key parameters that can tune the resonant frequencies of these 3 modes. According to the current distribution, the longer parasitic branch, the driving monopole branch and the shorter parasitic branch are the main radiation structures corresponding to the three resonance points of 1760, 2250, and 2690 MHz. Therefore, the lengths of the three branches should be able to effectively tune the resonant frequencies of the modes. Figure 4.23 shows the trend of the reflection coefficient with respect to the 3 parameters. The port isolation curves are not shown in Figure 4.23 for simplicity since the port isolation is sufficiently high. As can be seen from the variation of the curves, the increase in l_2 (the length of the longer parasitic branch) mainly reduces the frequency of the first resonance point, while the other two resonance points are hardly affected. As l_1 increases, i.e., the length of the driving monopole branch becomes longer, the frequency of the second resonance point gradually decreases. Similarly, as l_3 changes, i.e., the length of the shorter parasitic branch changes, the frequency of the third resonance point also changes. The parameter scanning results show that the resonant frequencies of these three modes can be tuned independently.

4.3.4 Physical Processing Validation

To validate the performance of the model, the performance of the designed MIMO antenna is measured. The physical object of the machined antenna is shown in Figure 4.24. The antenna is printed on a flat circuit board with two antennas distributed at the two corners

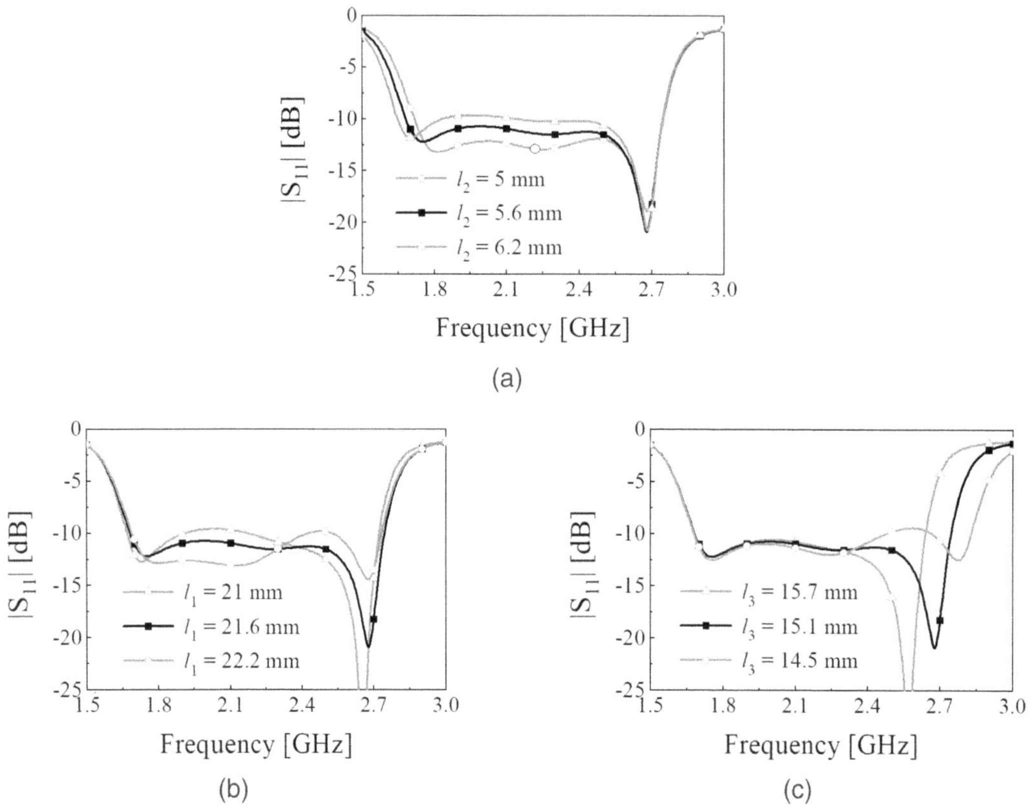

FIGURE 4.23 Trend of reflection coefficient with 3 parameters (continued) (a) l_2 (b) l_1 (c) l_3.

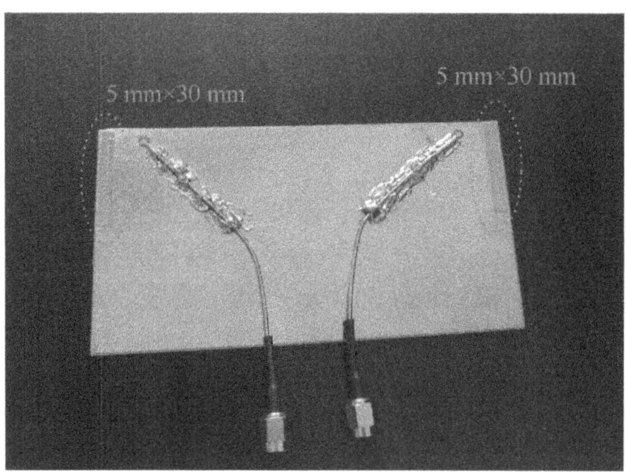

FIGURE 4.24 Processed antenna model.

of the dielectric substrate, fed by two coaxial cables. The ground clearance dimensions of both the top and bottom are 5 mm × 30 mm.

The simulated and measured S-parameters are shown in Figure 4.25, where three resonance points can be clearly observed (the S_{22} curve is similar to the S_{11} curve, which is not

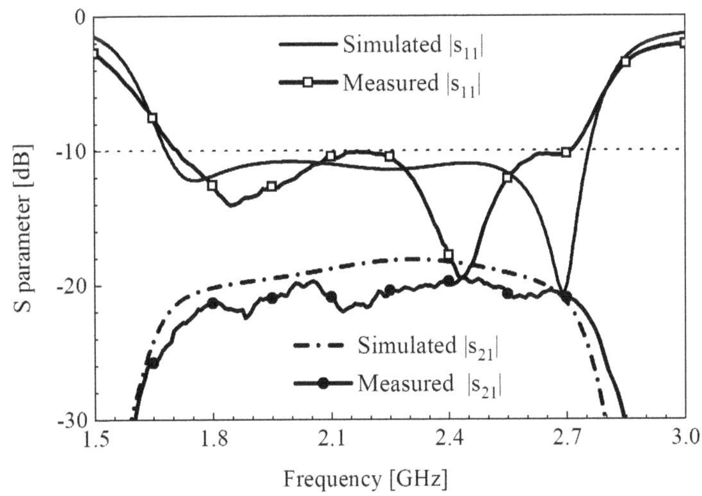

FIGURE 4.25 Simulated and measured S-parameters.

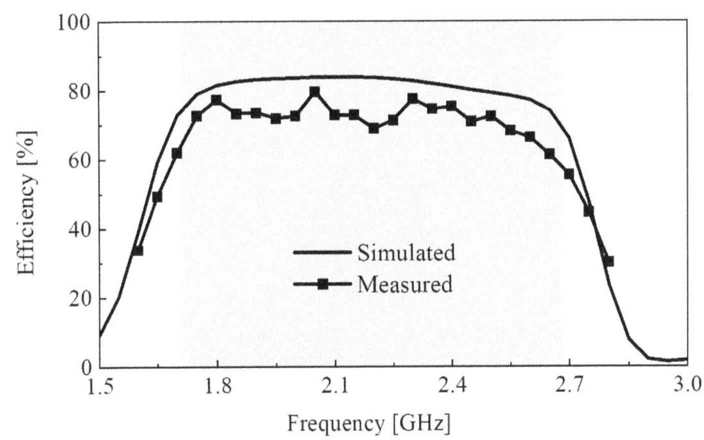

FIGURE 4.26 Simulated and measured radiation efficiency.

shown here). There are some errors between the simulated and measured values. These errors may be caused by fluctuations in the dielectric constant of the FR4 dielectric substrate and processing errors. The measured –10 dB bandwidths of port 1 and port 2 are 1006 MHz (1708–2714 MHz) and 1071 MHz (1704–2775 MHz), respectively. Both ports cover the 1710–2690 MHz band. The measured port isolation is greater than 19 dB in the overlapping bands of both ports.

The radiation performance of the MIMO antenna is measured in a microwave chamber. Considering that the two ports of the MIMO antenna are identical and symmetric about *the y*-axis, only the results when port 1 is excited are given. When port 1 is excited, port 2 is connected with a 50 Ω matched load. The simulated and measured radiation efficiencies are shown in Figure 4.26. It can be seen that the measured radiation efficiency is 60%–80% in the 1710–2690 MHz band.

TABLE 4.2 Performance Comparison of the Present Design with MIMO Antennas from other References

Design	Antenna Position	Antenna Size (mm×mm)	Total Ground Clearance (mm×mm)	−10 dB Bandwidth (Frequency Range, MHz)	Isolation (dB)	Radiation Efficiency (%)
Reference [6]	Floor long side single side	17×60	17×60 (× 1)	1670–2750	15	79–89
Reference [22]	Floor long side single side	15×60	15×60 (× 1)	1660–2840	15	79–81
Reference [23]	Floor long side single side	15×60	15×60 (× 1)	1680–2700	15	57–86
Reference [18]	Both sides of the long side of the floor	>9×39	9×39 (× 2)	1700–2850	12	84–86
Reference [20]	Both sides of the long side of the floor	>14.5×65	14.5×65 (× 2)	1700–2900	16	83–89
This design	Both sides of the long side of the floor	5×25.6	5×30 (× 2)	1690–2770	19	60–80

Note: (× 1) in column 4 indicates 1 antenna and (× 2) indicates 2 antennas.

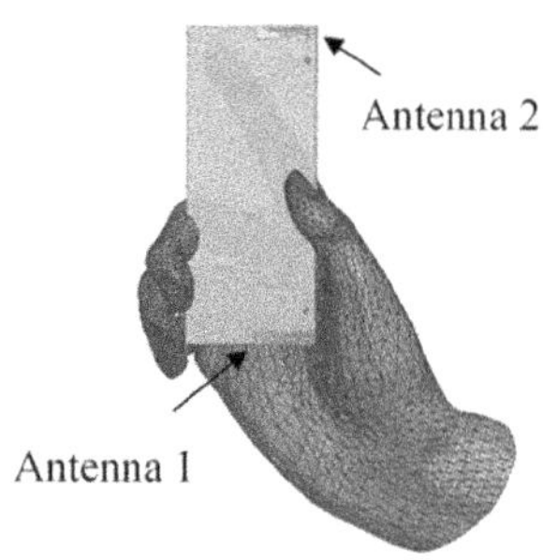

FIGURE 4.27 Position of the antenna in relation to the user hand model.

Table 4.2 shows the performance comparison of this design with the MIMO antennas in other references. All designs can cover the 1710–2690 MHz band with port isolation greater than 10 dB. Compared to other antennas, this design significantly reduces the antenna and total ground clearance size and improves the port isolation.

4.3.5 Influence of User Behavior

This section investigates the impact of the user model on the performance of the designed MIMO antenna. Figure 4.27 shows the positional relationship between the antenna and the human hand model, where the human hand directly grasps the base plate of the cell phone, and the part of the human hand model that overlaps with the antenna is subtracted. The simulated S-parameters and radiation efficiency of the antenna with the human hand model are shown in Figure 4.28. It can be seen that *the* S-parameters are almost unaffected by the hand grip because the hand does not directly touch the two antennas. However, the radiation performance is quite different when multiple ports are excited. The radiation

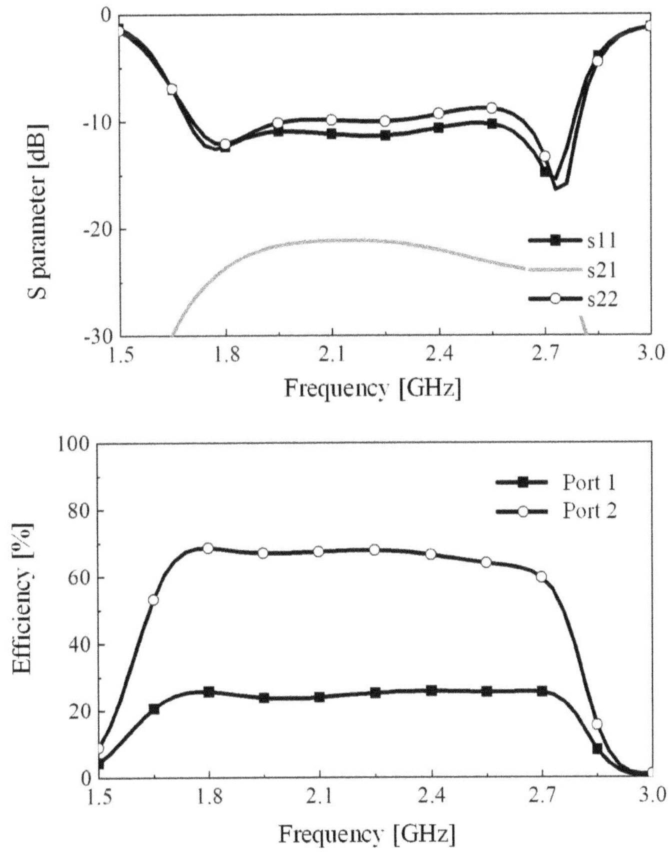

FIGURE 4.28 Simulated S-parameters and radiation efficiency of the antenna with a user hand model (a) S-parameters (b) radiation efficiency.

efficiency of port 2 is almost unaffected by the hand-holding because it is placed on the top of the phone, while the radiation efficiency of port 1 is only about 30% of the unshielded one because it is directly shielded by the hand-holding, which is mainly due to the fact that the model of the hand-holding with loss absorbs most of the energy radiated by antenna 1. This indicates that the MIMO antenna design can effectively eliminate the effect of the human hand.

REFERENCES

1. Bhatti RA, Yi S, Park SO. Compact antenna array with port decoupling for LTE-standardized mobile phones. *IEEE Antennas and Wireless Propagation Letters*, 2009, 8: 1430–1433.
2. Cheor WL, Al-Hadi AA, Soh PJ, et al. A decoupling network for resonant and non-resonant sub-1 GHz MIMO mobile terminal antennas with improved compactness and efficiency. *IEEE Access*, 2021, 9: 59475–59485.
3. Lee B, Harackiewicz FJ, Wi H. Closely mounted mobile handset MIMO antenna for LTE 13 band application. *IEEE Antennas and Wireless Propagation Letters*, 2014, 13: 411–414.
4. Lee WW, Rhee BH. Characterization of performance of a mobile MIMO antenna in free space. *IEEE Antennas and Wireless Propagation Letters*, 2013, 12: 1153–1156.

5. Li H, Miers ZT, Lau BK. Design of orthogonal MIMO handset antennas based on characteristic mode manipulation at frequency bands below 1 GHz. *IEEE Transactions on Antennas and Propagation*, 2014, 62(5): 2756–2766.
6. Wang Y, Du Z. A printed dual-antenna system operating in the GSM1800/GSM1900/UMTS/LTE2300/LTE2500/2.4-GHz WLAN bands for mobile terminals. *IEEE Antennas and Wireless Propagation Letters*, 2014, 13: 233–236.
7. Wang Y, Du Z. A wideband printed dual-antenna with three neutralization lines for mobile terminals. *IEEE Transactions on Antennas and Propagation*, 2013, 62(3): 1495–1500.
8. Kang G, Du Z, Gong K. Compact broadband printed slot-monopole-hybrid diversity antenna for mobile terminals. *IEEE Antennas and Wireless Propagation Letters*, 2011, 10: 159–162.
9. Zhao L, Yeung LK, Wu KL. A coupled resonator decoupling network for two-element compact antenna arrays in mobile terminals. *IEEE Transactions on Antennas and Propagation*, 2014, 62(5): 2767–2776.
10. Kishor KK, Hum SV. A two-port chassis-mode MIMO antenna. *IEEE Antennas and Wireless Propagation Letters*, 2013, 12: 690–693.
11. Zhu J, Feng B, Peng B, et al. Multiband printed mobile MIMO antenna for WWAN and LTE applications. *Microwave and Optical Technology Letters*, 2017, 59(6): 1446–1450.
12. Wang S, Du Z. A dual-antenna system for LTE/WWAN/WLAN/WiMAX smartphone applications. *IEEE Antennas and Wireless Propagation Letters*, 2015, 14: 1443–1446.
13. Wong KL, Kang TW, Tu MF. Internal mobile phone antenna array for LTE/WWAN and LTE MIMO operations. *Microwave and Optical Technology Letters*, 2011, 53(7): 1569–1573.
14. Zhang S, Zhao K, Ying Z, et al. Adaptive quad-element multi-wideband antenna array for user-effective LTE MIMO mobile terminals. *IEEE Transactions on Antennas and Propagation*, 2013, 61(8): 4275–4283.
15. Zhou X, Li RL, Tentzeris MM. A compact broadband MIMO antenna for mobile handset applications. *Microwave and Optical Technology Letters*, 2011, 53(12): 2773–2776.
16. Lee WW, Jang B. 2×2 MIMO antenna system with different antenna types for mobile terminals. *Microwave and Optical Technology Letters*, 2016, 58(6): 1337–1340.
17. Zhao K, Zhang S, Ying Z, et al. SAR study of different MIMO antenna designs for LTE application in smart mobile handsets. *IEEE Transactions on Antennas and Propagation*, 2013, 61(6): 3270–3279.
18. Aljarafreh SS, Huang Y, Xing L, et al. A low-profile and wideband PILA-based antenna for handset diversity applications. *IEEE Antennas and Wireless Propagation Letters*, 2015, 14: 923–926.
19. Zhang S, Zhao K, Ying Z, et al. Investigation of diagonal antenna-chassis mode in mobile terminal LTE MIMO antennas for bandwidth enhancement. *IEEE Antennas and Propagation Magazine*, 2015, 57(2): 217–228.
20. Shoaib S, Shoaib I, Shoaib N, et al. MIMO antennas for mobile handsets. *IEEE Antennas and Wireless Propagation Letters*, 2015, 14: 799–802.
21. Vasilev I, Plicanic V, Lau BK. Impact of antenna design on MIMO performance for compact terminals with adaptive impedance matching. *IEEE Transactions on Antennas and Propagation*, 2016, 64(4): 1454–1465.
22. Wang Y, Du Z. A wideband printed dual-antenna system with a novel neutralization line for mobile terminals. *IEEE Antennas and Wireless Propagation Letters*, 2013, 12: 1428–1431.
23. Huang HF, Wu JF. Decoupled dual-antenna with three slots and a connecting line for mobile terminals. *IEEE Antennas and Wireless Propagation Letters*, 2015, 14: 1730–1733.

CHAPTER 5

Multi-Antenna Design for 5G Mobile Terminals

With the full implementation of the 5G standard, the 5G Mobile Terminals antennas have been upgraded in some ways for the 5G system, mainly in two aspects. On the one hand, it is the support of new frequency bands. In order to meet the demand for higher wireless transmission rates, 5G adds a new Sub-6 GHz band. In 2015, the World Radiocommunication Conference (WRC-15) divided the 3300–3600 MHz band as the new spectrum for 5G. In 2017, the Ministry of Industry and Information Technology (MIIT) issued a relevant document, which determines the working frequency band of China's 5G [1]. This requires terminal antennas to cover the new Sub-6 GHz band on the basis of compatibility with the 4G band. On the other hand, it is supported by high-order MIMO technology. MIMO technology can exponentially improve spectrum utilization by arranging multiple independent antennas at the same time at the base station and the mobile terminal, which is an important means to enhance the wireless transmission rate. In 4G mobile terminals, the number of MIMO antennas in the same frequency band is generally two, and the multiplier of transmission rate improvement is relatively limited. Integrating more multi-antenna into 5G Mobile Terminals has become the development trend of mobile terminal antenna design. In order to meet the new requirements of 5G transmissions, it is generally necessary to integrate 6–10 antennas operating in the new Sub-6 GHz band within 5G Mobile Terminals [2].

In this chapter, we investigate the miniaturization technology of MIMO antennas operating in the 3300–6000 MHz band in restricted space for 5G Mobile Terminals as an application platform to provide new solutions for 5G Mobile Terminals antenna design.

5.1 CURRENT STATUS OF MIMO MULTI-ANTENNA RESEARCH

The 3300–6000 MHz frequency band corresponds to a shorter wavelength, and multiple antennas can be integrated within the mobile terminal to realize the high-order MIMO function. Common antenna forms include line antennas and surface antennas, where line

DOI: 10.1201/9781003770039-5

antennas are usually mounted on both sides of the long edge of the cell phone floor and surface antennas are usually mounted inside the back cover of the cell phone, and a large number of designs have been proposed for these two antenna forms.

5.1.1 Wire Antenna

The design of wire antenna in 3300–6000 MHz band is similar to the antenna design in 900 and 1800 MHz bands. The main antenna form is monopole antenna, the available space is the area on both sides of the long side of the cell phone floor, and the antenna is integrated in the metal bezel. The development of full-screen requires that the ground clearance of the antenna is as small as possible, and the size occupied by the MIMO antenna module is as small as possible.

Several dual antenna designs operating in the 3600 MHz band are given in the references [3–7]. For example, two L-shaped monopoles are placed face-to-face with an edge-to-edge spacing of only 1 mm, and the open ends of the two antennas are connected through an aggregate inductor, which acts as a decoupling structure, and the port isolation can be increased from 5 dB to more than 15 dB by optimizing the value of the inductance [3]; the two face-to-face monopoles are treated as a slot antenna, and the horizontal electric field is excited by port 1, which is coupled through the two monopoles to feed the electric field directly, and the vertical electric field is excited by port 2, which is coupled through the L-shaped monopoles to feed the electric field directly. The horizontal electric field is excited by port 1 and fed directly through the two monopoles, and the vertical electric field is excited by port 2 and coupled and fed through the L-shaped monopoles, which achieves a port isolation of 16 dB [4]; the two bent monopoles undergo a mirror-flip operation and are placed on both sides of the long edge of the cell phone floor, and due to the mirroring principle, the mirrored current paths of antenna 1 are centrally symmetric to those of antenna 2, so that a port isolation of 10 dB can be achieved without using a decoupling structure [4]; two faces placed as a slot antenna, the horizontal electric field is excited by port 1 and fed directly through the two monopoles, achieving high isolation [5]; a T-shaped monopole and a U-shaped dipole are placed on the top and bottom sides of the cell phone floor, common mode excitation can be realized through the monopole feed, and differential mode excitation can be realized through the dipole capacitively coupled feed to achieve a good degree of isolation [6]; by using common-mode/differential-mode theory of decoupling [7], the two feed ports share the same radiator, and a power divider circuit with a 180° phase difference is used to provide a good degree of isolation [8][1]. Phase difference power divider circuit is used to provide the phase required for differential mode, and the port isolation is higher than 16.5 dB [7].

On the basis of single-band MIMO antenna designs, some dual-band MIMO antenna designs have been proposed. Four typical dual-band MIMO antenna designs are given in the references [8–11]. For example, the mirror-flip technique is applied to dual-band MIMO antenna design, where antenna 1 adopts a capacitively coupled feed in the form of a dual-band monopole, which can operate in the 3400–3600 MHz and 5725–5875 MHz bands, and mirror-flip decoupling is still applicable to dual-band antenna design with port isolation of more than 12 dB [9]; the antenna consists of a T-shape open slit and a dual-branch monopole, which can operate in the 2450–2690 MHz and 3400–3800 MHz bands, and the two antennas are decoupled by a wide-band neutralization line with a port isolation of more than 10 dB [8]; the two back-to-back capacitively coupled monopoles are arranged in a close proximity, and each monopole has two branches distributed on the two sides of the cell phone floor, and the antenna can be used in the 3400–3600 MHz frequency bands. By adding a parasitic branch between the two monopoles, the antenna can operate in the 3400–3600 MHz and 4800–5000 MHz frequency bands, achieving good resonance and a port isolation of more than 17.5 dB [10]; the common mode/differential mode theory is used for dual-band antenna design, in which both the T-shaped open slit and the T-shaped monopole are modified into dual-band units by loading them with lumped components, and the antenna can operate in the 3400–3600 MHz and 4800–4900 MHz bands and the port isolation is above 11.8 dB [11].

In addition to dual-band design, wide-band design has also stimulated the research interest of many researchers. The fundamentals of wide-band design are similar to those of dual-band design, with the focus on designing wide-band antenna units and wide-band decoupling structures. Some typical wide-band MIMO antenna designs are given in the references [12–15]. For example, an antenna unit that can cover the 3300–6000 MHz band is designed by the multi-antenna resonance technique, and the cell coupling is reduced by maintaining a certain cell spacing between the eight antenna units [12]; the two monopoles are placed back-to-back, and the driver monopole and the parasitic monopole are fed through capacitive coupling, and the parasitic monopole is directly connected to the cell phone floor, so that the inter-antenna coupling can be effectively reduced. The port isolation in the 3300–6000 MHz band is higher than 10 dB [13]; two back-to-back monopoles are placed vertically above the cell phone floor, and the parasitic branch and the T-shaped defective ground branch in the middle of the two monopoles not only extend the bandwidth but also reduce the inter-antenna coupling, and the isolation of the antenna in the 3300–6000 MHz band is higher than 15 dB [14];two monopoles with different path lengths are fed through capacitive coupling; two monopoles with different path lengths are closely arranged together, by controlling the length and bending degree of the two paths can make the multiple coupling paths at the feed ports canceled, and the isolation of the ports in the 3300–5000 MHz band is kept more than 20 dB [15].

5.1.2 Surface Antenna

The mounting location of the face antenna is typically located inside the back cover of the cell phone, which requires the face antenna to have as low a profile as possible; in addition, the antenna ground clearance needs to be as small as possible considering the demands of MIMO multi-antenna. Numerous face antenna structures have been proposed, including single antenna, dual antenna, quad antenna, and integration of more multi-antenna units.

Dual-band or wide-band dual antenna structures are given in the references [16–19]. For example, shorting posts were used to control the magnetic coupling between the parasitic and driving cavities, and open slits were used to generate dual-band resonances, and the antenna covered the 3400–3600 MHz and 4800–5000 MHz bands at a profile height of 1 mm [16]; shorting posts and open slits were used to modify the performance of the microstrip patch antenna by exciting multiple modes to extend the bandwidth, thereby extending its bandwidth to cover the 3300–4200 MHz frequency range while maintaining a low profile height of 1.7 mm [17]; the common mode/differential mode of the same microstrip patch antenna is excited to realize a dual antenna design, with two shorting posts loaded on the centerline of the microstrip patch antenna, and a U-shaped slit on each side is used to adjust the resonance frequency of the common mode/differential mode, and the structure covers the 4800–4900 MHz band at 3 mm profile height, achieving port isolation greater than 25 dB across the operational band [18]; four Planar InvertedF Antenna (PIFA) units form two pairs of magnetic dipoles, and by adjusting the number and position of the shorting posts, the antenna covers the 4300–5500 MHz band with port isolation above 22 dB at a profile height of 1.2 mm [19].

Integrated designs with four or eight antennas are given in the references [20–25]. For example, four microstrip patch antennas are rotationally placed in sequence, each antenna is internally loaded with a number of shorting pillars, and shorting pillars are also loaded in the gaps between the connected antennas, and the antennas cover the 3300–5000 MHz band at a profile height of 2.5 mm with an isolation degree of 12 dB or more between the four antennas [20]; four magnetic antennas are rotationally placed in sequence, and since the radiating port face of each magnetic antenna is in the middle region of the antenna, the isolation between the antennas is high, and the antennas cover the 3400–3600 MHz and 4800–5000 MHz bands at a profile height of 0.5 mm [21]; four antennas are placed in sequential rotation, the microstrip patch antenna is engraved with four I-shaped slits and one ring-shaped slit, and there is a cross-shaped shorting post in the center region, and the antennas cover the 4200 MHz and 4800–5000 MHz bands at a profile height of 2 mm. The antenna covers the 4250–5130 MHz band at a profile height of 2 mm [22]. Similarly, different slits are etched on the square microstrip patch antenna to improve the antenna performance [23,24]. Splitting the loop-shaped microstrip patch antenna into more sectorial cells, the antenna performance is improved by adding shorting posts, and the eight multi-antennas covers the 5900–7200 MHz band at a profile height of 1.4 mm and the port isolation is higher than 10 dB [25].

5.2 HIGH ISOLATION ANTENNA PAIR BASED ON PARITY MODE THEORY

In this section, a highly isolated antenna pair based on odd/even mode (or common mode/differential mode) theory is designed to construct a radiation pattern diversity by stimulating the odd and even modes of two antenna units placed face to face, thus realizing high isolation of the antenna.

5.2.1 Pattern Diversity of Two Face-to-Face Antenna Cells

The excitation method of two face-to-face antenna cells is analyzed below. The antenna cells are in the form of planar monopoles, with the advantages of simple fabrication and low cost. The L-shaped monopoles are further folded to reduce the cell size. The geometry and dimensions of a two-cell MIMO antenna in a cell phone are shown in Figure 5.1. Two face-to-face units are placed along the edges of the long side of the cell phone floor and are symmetric with respect to the center of the long side. The antenna units and the floor are

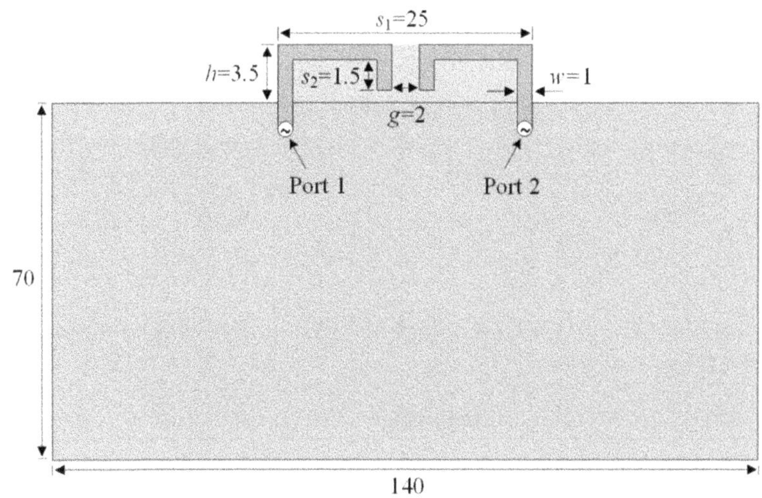

FIGURE 5.1 Geometry and dimensions of the two-cell MIMO antenna in cell phone (in mm).

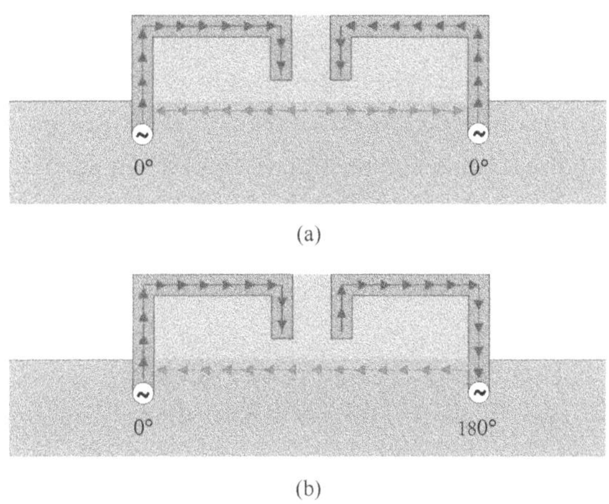

FIGURE 5.2 Current distributions of two cells for in-phase and 180° inverted excitation: (a) in-phase, (b) 180° in-phase.

printed on a 0.8 mm thick FR4 dielectric substrate ($\varepsilon_r = 4.4$, $\tan\delta = 0.02$). The two antenna units are printed on the upper surface of the dielectric substrate and occupy a space of 25 mm × 3.5 mm. The edge-to-edge distance of the two antenna units is 2 mm. The floor is printed on the lower surface of the dielectric substrate with a length and width of 140 and 70 mm, respectively. The center frequency of the MIMO antenna is set to 3500 MHz (3.5 GHz). All parameters are optimized using HFSS software.

The current distribution of the two cells during in-phase and 180° inverse excitation is shown in Figure 5.2. Since the antenna is close to the floor, the direction of the mirror current on the floor is opposite to the direction of the horizontal current on the monopole, which results in poor radiation capability of the horizontal section of the monopole, and thus, the radiation pattern is mainly provided by the vertical section. In in-phase excitation, the currents on both monopoles are in the same direction, and the vertical segment

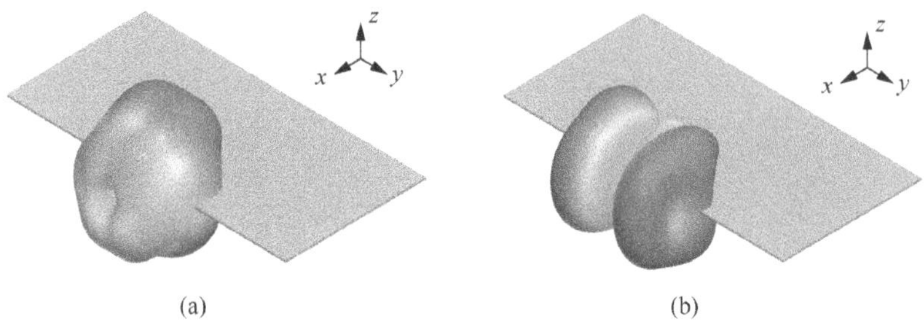

FIGURE 5.3 3D radiation patterns for in-phase and 180° in-phase excitation: (a) in-phase excitation, (b) 180° in-phase excitation.

TABLE 5.1 ECC of Radiation Pattern for 3.5 GHz Band

Excitation of Radiation Pattern 1	Excitation of Radiation Pattern 2	ECC
P_1	P_2	0.162
$P_1 0° + P_2 0°$	$p_1 0° + p_2 180°$	0.013

can be regarded as a two-cell array with in-phase excitation that has maximum gain in the zenith direction. Figure 5.3a shows the corresponding three-dimensional radiation pattern with the main beam pointing to *the* z-axis. In addition, when two units are fed with inverted signals, the currents on the two monopoles are in opposite directions, and the vertical section can be regarded as a two-unit array with 180° inverted excitation, with the main beam splitting into two parts. As a result, the two radiation patterns generated by the two-unit array with in-phase and antiphase signal excitation can be realized as radiation pattern diversity.

The three-dimensional radiation pattern shown in Figure 5.3 simply demonstrates the complementarity of the two excitation modes (i.e., radiation modes), but this assessment is only qualitative. In order to quantitatively assess the orthogonality of the two radiation modes, the envelope correlation coefficient based on the far-field radiation modes is calculated. For the detailed calculation procedure, refer to the reference [26].

Table 5.1 shows the ECCs of the radiation patterns in the 3.5 GHz band. Radiation pattern 1 and radiation pattern 2 are determined by the corresponding excitation signals. P_1 and P_2 are the excitation amplitudes of port 1 and port 2. To simplify the calculation, it is assumed that P_1 and P_2 are equal. It can be seen that the ECC for in-phase and inverted-signal excitation is much smaller than that for separate excitations. This indicates that the two radiation patterns for in-phase and inverted-signal excitation are highly uncorrelated.

5.2.2 Compact Two-Port Feed Network Design

In order to provide both in-phase and 180° inverted-phase excitation, a compact two-port feed network design is proposed in this section. It is well known that the squirrel cage structure is the traditional four-port network for generating in-phase and inverted-phase feeds. However, there is strong coupling between the two antenna units in this design; the two output ports do not fulfill the requirement of mutual isolation of the output ports in the squirrel-cage structure, and the size of the squirrel-cage structure is large.

The designed two-port feed network with antennas is shown in Figure 5.4. The size of the two-port feed network is much smaller compared to the squirrel cage structure. The two feed lines of the feed network are directly connected to the monopole radiator, and the two feed ports lead from the center and edges of the feed network through a 50 Ω microstrip line, which is easy to integrate with the circuit plane on the floor.

The mechanism of operation of in-phase and inverted excitation is discussed first. When port 1 is excited, port 1 provides in-phase excitation since the port is located on the axis of symmetry of the feed network and is equidistant from the two current paths on the two monopoles. When port 2 is excited, the difference between the distance of this port and the two current paths on the two monopoles is one-half wavelength (the dielectric wavelength corresponding to 3.5 GHz), which means that port 2 provides in-phase excitation. The current distribution of the MIMO antenna with different port excitations is shown in Figure 5.5.

The port isolation is then analyzed. In Figure 5.5a, port 1 is located at the minimum value of the current and port 2 is located at the maximum value of the current. In Figure 5.5b, port 1 is located at the maximum value of the current and port 2 is located at the minimum value of the current. These observations indicate that high isolation is achieved between the two ports.

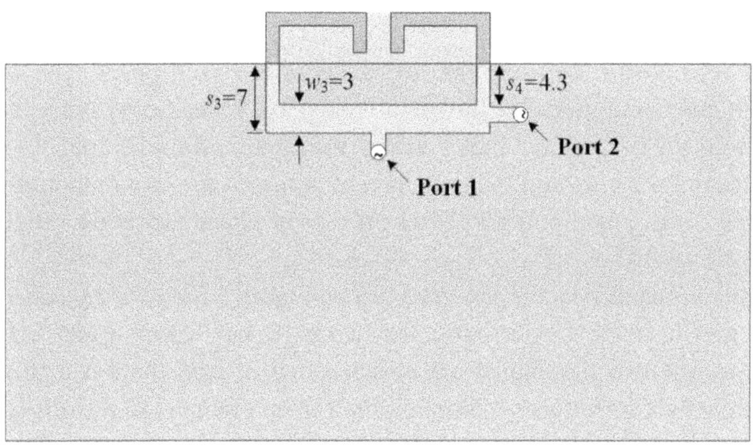

FIGURE 5.4 The designed two-port feed network and antenna (in mm).

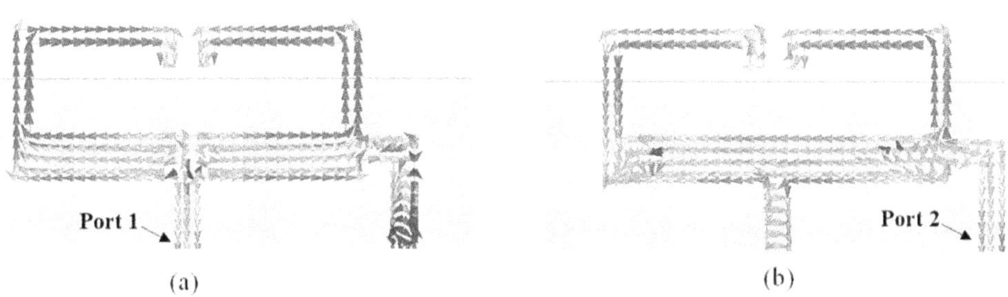

FIGURE 5.5 Current distribution of MIMO antenna with different port excitations: (a) in-phase excitation, (b) inverse excitation.

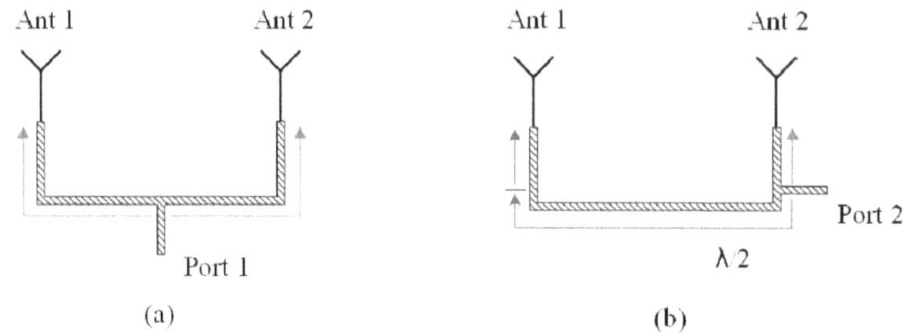

FIGURE 5.6 Current paths for different port excitations : (a) port 1, (b) port 2.

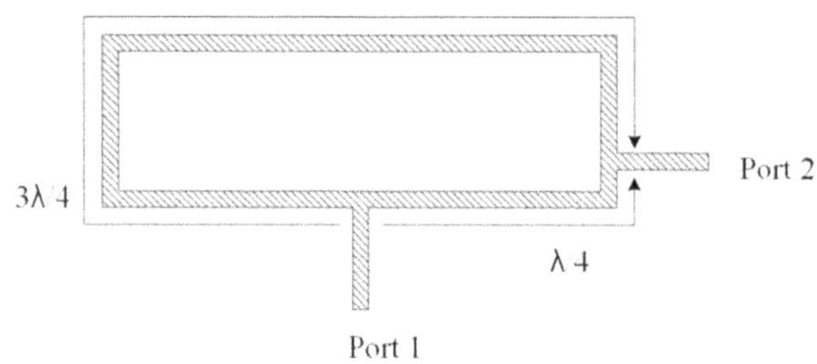

FIGURE 5.7 Equivalent circuit of a feeder network.

The phase relationship between the two ports is analyzed below. The current paths under different port excitations are shown in Figure 5.6. When port 1 is excited, the port is equidistant from the two current paths on the two monopoles; when port 2 is excited, the difference between the port and the two current paths on the two monopoles is one-half the wavelength. Thus, port 1 and port 2 can provide in-phase excitation and inverse-phase excitation, respectively.

The isolation between ports is analyzed below. Figure 5.7 shows the equivalent circuit of the feeder network. On the one hand, the two ports have directly connected paths; on the other hand, the two monopoles are closely lined up and there is a strong coupling between them, which introduces a coupling path from one port to the other. The current phase difference and amplitude ratio of the two paths can be controlled by changing the length of the coupling path and the coupling strength. The currents on the two paths after the optimization process have a 180° phase difference and the same amplitude, allowing the currents from the two paths to cancel out, resulting in high port isolation. As shown in Figure 5.7, a direct path length of one-quarter wavelength and a coupled path length of three-quarters wavelength produce the desired 180° phase based on the difference in their lengths. Mutual coupling between antennas is beneficial in a compact two-port feed network, and this effect is very different from that of a squirrel-cage structure.

The impedance matching of the antenna is analyzed below. Figure 5.8 shows the impedance transformation relationship for different port excitations. For port 1, there are two quarter-wavelength impedance converters for impedance matching. For port 2, two quarter-wavelength impedance converters are connected in series to form a one-half-wavelength impedance converter, which has almost no effect on the impedance of port 2.

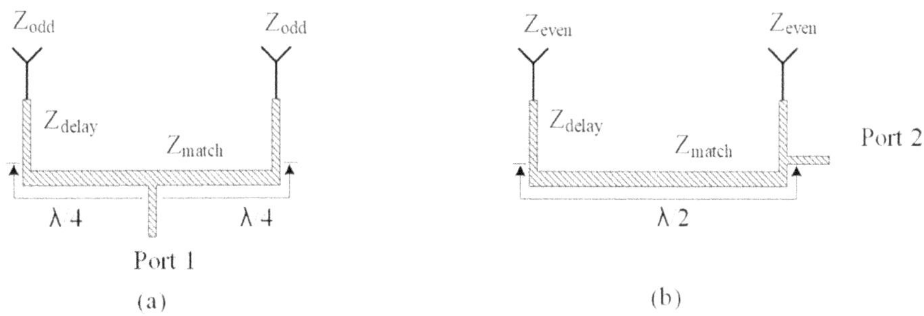

FIGURE 5.8 Impedance transformation relations under different port excitations: (a) port 1 has two quarter-wavelength impedance converters, (b) port 2 has a one-half-wavelength impedance converter.

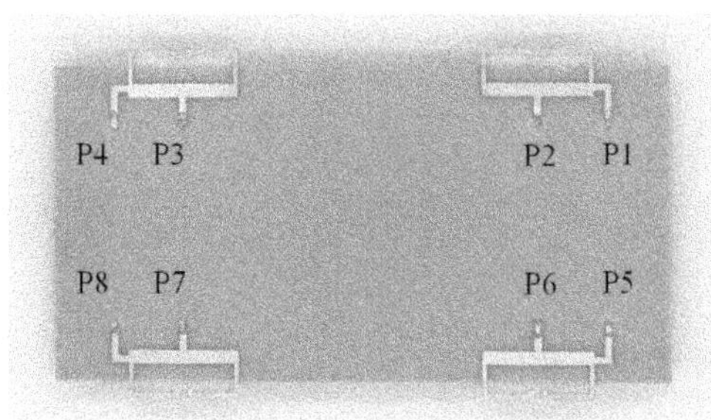

FIGURE 5.9 Eight-cell MIMO antenna design.

Based on this observation, the impedance matching of the two feed ports can be explained by a two-step impedance transformation: first, optimize the impedance of port 2 by changing the antenna impedance to achieve the impedance matching of port 2; and then optimize the impedance of port 1 by changing the linewidth of the one-quarter-wavelength impedance converter to achieve the impedance matching of port 1.

5.2.3 Physical Processing Verification

An eight-cell MIMO antenna design is proposed based on the two-cell MIMO antenna design, as shown in Figure 5.9. The antenna consists of four antenna pairs, each with a distance of 40 mm from the center of the long side of the floor. Eight Sub-Miniature version A (SMA) coaxial connectors feed the signal from the bottom of the floor. Since the monopole is the main radiator, the positional offset of the antenna pairs has little effect on the antenna performance.

The simulated and measured S-parameters based on the first antenna pair in the MIMO antenna are shown in Figure 5.10. The simulation results are in good agreement with the measured results. The −6 dB bandwidths of port 1 and port 2 are 750 MHz (3060–3810 MHz) and 340 MHz (3330–3670 MHz), respectively. The overlapping bandwidth of 340 MHz is sufficient to cover the 3400–3600 MHz band with port isolation greater than 14 dB.

The radiation performance of the antenna was measured in a microwave chamber room. When one port was excited, all other ports were connected to a 50 Ω matched load.

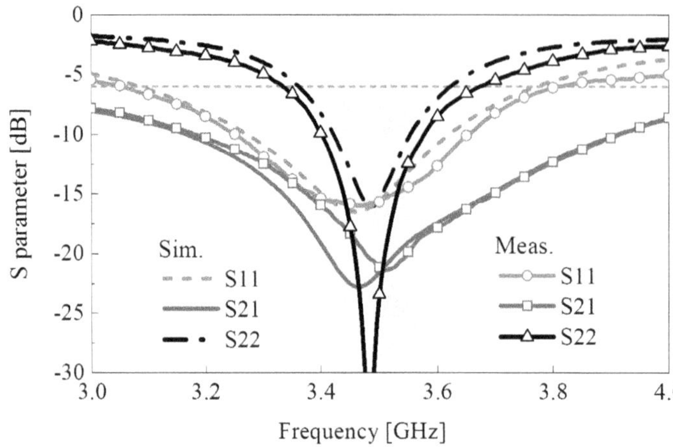

FIGURE 5.10 Simulated and measured S-parameters based on the first antenna pair in the MIMO antenna.

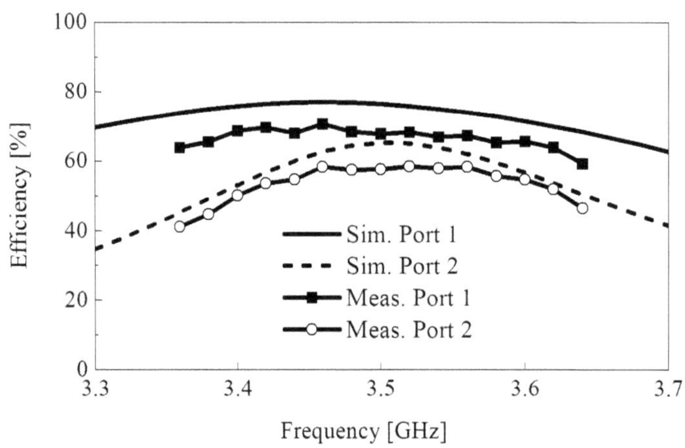

FIGURE 5.11 Simulated and measured radiation efficiency based on the first two ports.

The simulated and measured radiation efficiency based on the first two ports is shown in Figure 5.11. The measured radiation efficiency is slightly lower than the simulated radiation efficiency, which may be caused by processing and measurement errors. In the 3400–3600 MHz band, the measured radiation efficiencies of port 1 and port 2 are higher than 65% and 52%, respectively.

Table 5.2 shows the performance comparison of this design with the antenna designs in other references. The results show that this design has the advantages of small antenna size and wide operating band.

5.2.4 Common Mode/Differential Mode Method Extension

In fact, the common-mode/differential-mode excitation feed method is highly versatile and only requires mirror symmetry of the two units within the antenna pair.

TABLE 5.2 Performance Comparison of this Design with Antenna Designs from Other References

Design	Antenna Size (mm × mm)	Overlap Bandwidth (MHz)	Isolation (dB)
Reference [26]	19 × 3	220	10
Reference [27]	>52 × 6.5	270	29
Reference [6]	12 × 7	220	17
Reference [28]	28.5 × 5	280	15.7
Reference [5]	10 × 7	180	10
Reference [29]	20 × 7	320	17
This design	25 × 3.5	340	14

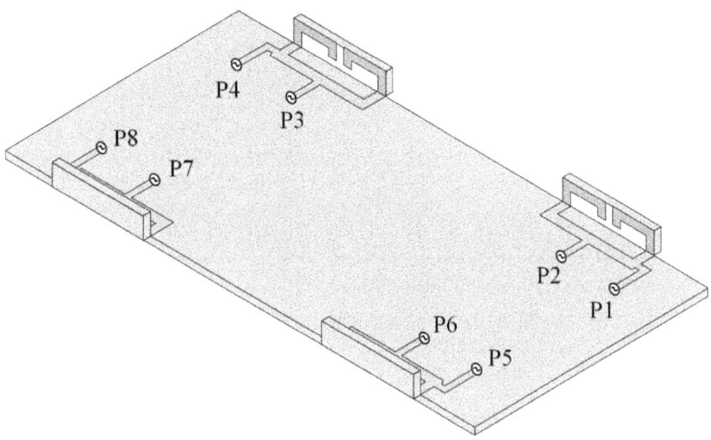

FIGURE 5.12 Model extension: a zero-clearance MIMO antenna design.

A zero-clearance MIMO antenna designed based on the common-mode/differential-mode excitation feed method is shown in Figure 5.12. The monopole is vertically folded 90° along the edge of the cell phone floor and the radiating antenna is perpendicular to the cell phone floor. All parameters remain unchanged except for fine-tuning the length of the feed line. This modification reduces the antenna clearance from the floor to 0, realizing a zero ground clearance design. The performance of the vertically placed MIMO antenna is comparable to that of the planar structure.

This common mode/differential mode excitation feed method is also applicable to other types of radiating structures. Another zero-clearance MIMO antenna design using a similar configuration is shown in Figure 5.13. The radiating cell is in the form of a PIFA, with a vertical feed strip connecting the cell to the feed port and a vertical shorting strip shorting the patch to the ground plane. By optimizing the parameters, this design also has good MIMO performance.

The above analysis shows that the common-mode/differential-mode excitation feed method is general for many two-cell MIMO antenna designs, i.e., if the two antenna cells are in a face-to-face configuration and the proposed feed network is used to provide both in-phase and 180° inverted-phase excitation feeds, high isolation at the ports and good directional map diversity can be achieved.

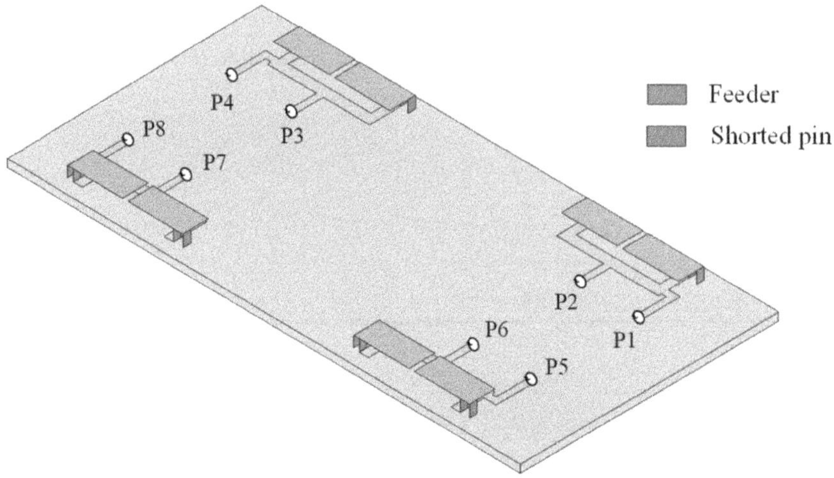

FIGURE 5.13 Model extension: another zero ground clearance MIMO antenna design.

5.3 ANTENNA PAIR COVERING THE 3300–6000 MHz BAND

This section describes a MIMO antenna pair with wide-band coverage capability, which extends the bandwidth by multi-antenna branching on the basis of a narrow-band structure.

5.3.1 Antenna Structure

The structure of the designed MIMO antenna is shown in Figure 5.14 and consists of eight vertically placed antenna units, where every two units are combined to form an antenna pair. The antenna pair consists of two driving monopoles and a parasitic branch. Because of sharing the same parasitic branch, the edge-to-edge distance between the two cells is 0. The driving monopoles consist of two metal strips, i.e., a longer L-shaped strip and a shorter linear strip. The parasitic branch includes a T-shaped ground strip and a linear suspension strip. There is a 1 mm gap between the vertical part of the antenna and the ground plane. The dimensions of the ground plane are 140 mm × 68 mm, and the antenna unit is 16 mm away from the top corner of the cell phone floor and is fed through a 50 Ω microstrip line. The MIMO antenna and ground plane are printed on a 0.8 mm thick FR4 dielectric substrate ($\varepsilon_r = 4.4$, $\tan\delta = 0.02$).

5.3.2 Principle of Operation

The simulated S-parameters of the MIMO antenna are shown in Figure 5.15. It can be observed that the reflection coefficient profile has three resonance points at 3.5, 5, and 6 GHz. The −6 dB bandwidth is 3.26 GHz. Mutual coupling between the antenna pairs is below −10 dB across the band.

In order to study the working principle of MIMO antenna, its design process is discussed below. As shown in Figure 5.16, the whole design process includes five steps. First, two folded monopoles are placed face-to-face, and the corresponding S-parameters are shown in Figure 5.17. The resonance is observed near 5.5 GHz, which is generated by the quarter-wavelength mode of the monopole. The addition of a shorter metal strip produces

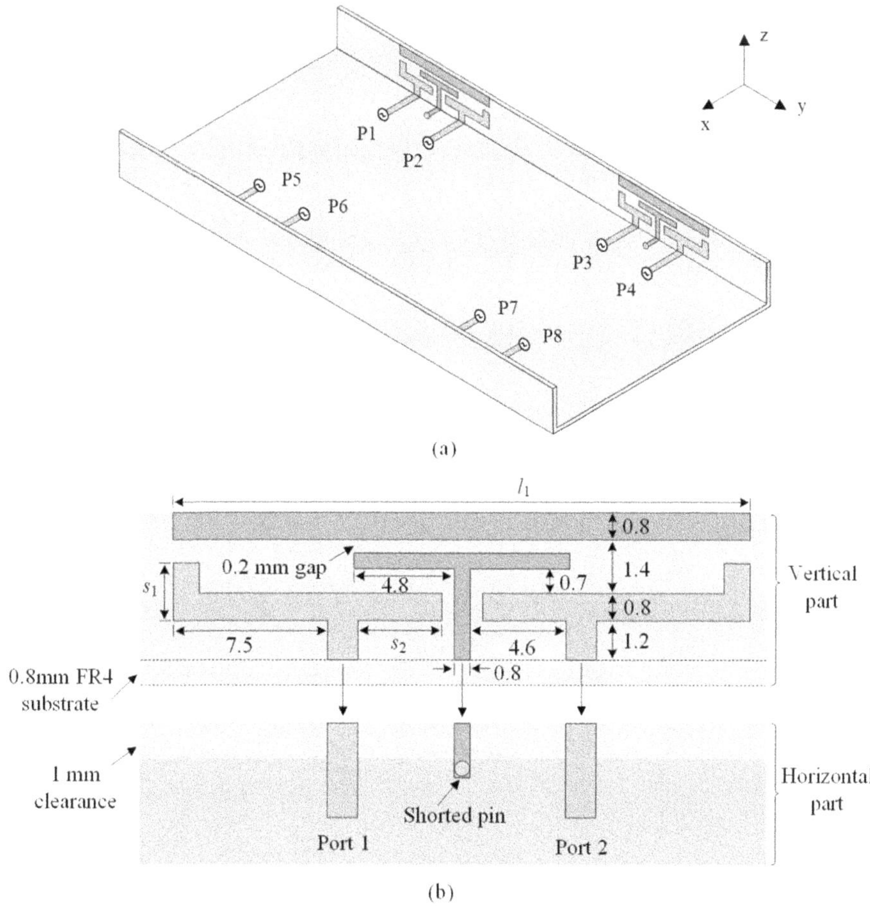

FIGURE 5.14 Structure of the ultra-wideband eight-element MIMO antenna (in mm): (a) 3D view, (b) dual cell size.

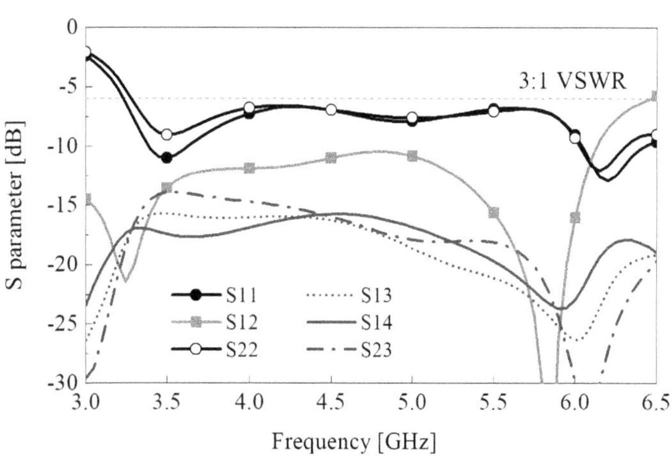

FIGURE 5.15 Simulated S-parameters of MIMO Antenna.

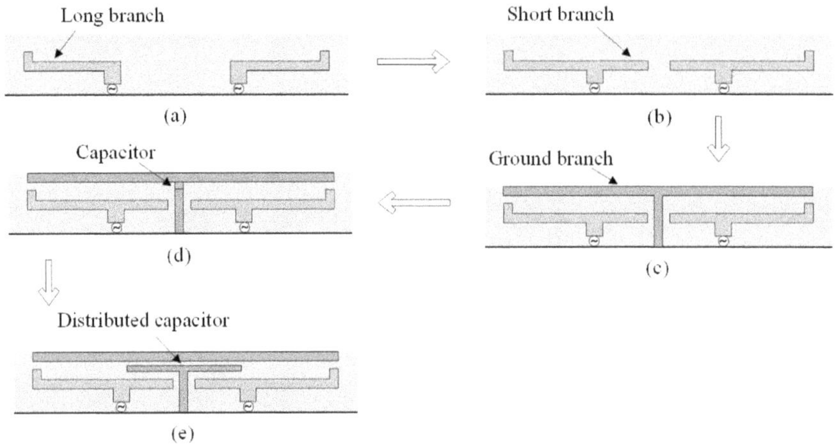

FIGURE 5.16 Antenna design process in side view: (a) Type I, (b) Type II, (c) Type III, (d) Type IV (e) this design.

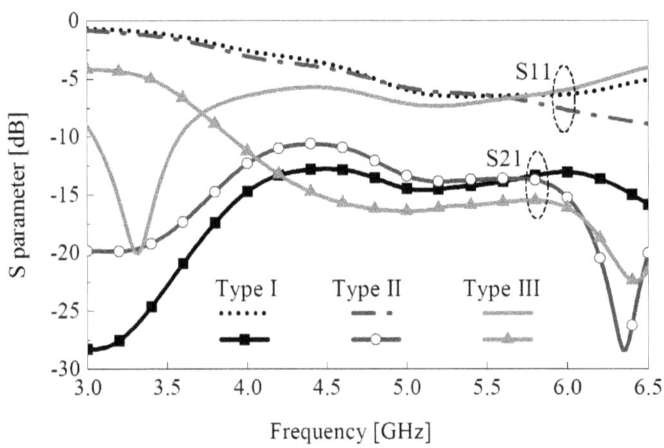

FIGURE 5.17 S-parameters for Type I–Type III.

another resonance at about 6.5 GHz, extending the bandwidth by combining the two resonances. Next, a T-shaped parasitic branch is placed in the center of the two cells, and an additional resonance is observed at 3.3 GHz, but the mutual coupling of the antenna cells is strong (about −4 dB). To minimize the coupling, a collector capacitor with a capacitance value of 0.8 pF was added to the grounded branch. As shown in Figure 5.18, the peak coupling decreases while S_{11} is hardly affected. Considering the uncertainty in the parameter magnitude of the lumped component (e.g., chip capacitor) at high frequencies, a distribution capacitor is used instead of the collector capacitance. The grounding and suspension bars form the distributed capacitance, and the coupling strength can be controlled by a gap.

In order to find the main radiating structure corresponding to the three resonance points, the current distribution on the antenna and ground plane is observed when the left cell is excited at 3.5, 5, and 6 GHz. As shown in Figure 5.19, at 3.5 GHz, strong currents are

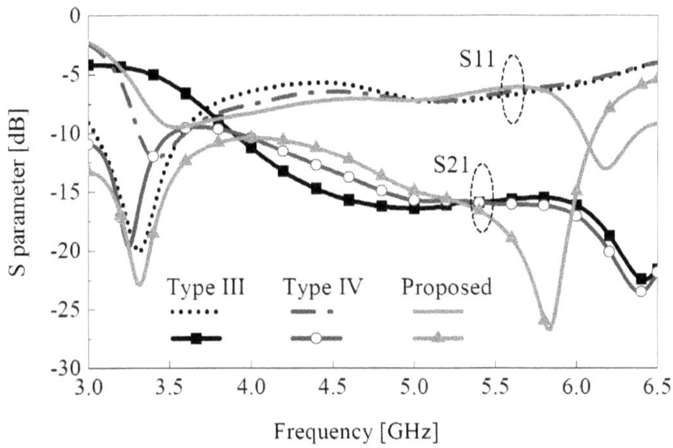

FIGURE 5.18 S-parameters for Type III, Type IV and this design.

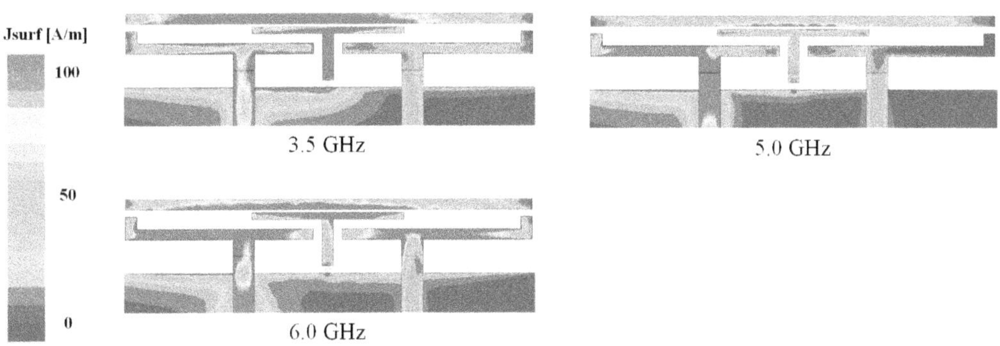

FIGURE 5.19 Current distribution on the antenna and ground plane when the left cell is excited.

observed on the T-shaped parasitic branch; at 5 GHz, the currents are strong on the long strip driving the monopole; and at 6 GHz, there are strong currents on both the long and short strips driving the monopole. These observations indicate that the T-shaped parasitic branch and the long strip of the driver monopole are the main radiation structures at the 3.5 and 5 GHz resonances, respectively, while both the long and short strips of the driver monopole are the main radiation structures at the 6 GHz resonance.

5.3.3 Experimental Results

The designed MIMO antenna was physically processed and experimentally measured. The fabricated MIMO antenna prototype is shown in Figure 5.20, where two thin FR4 dielectric substrate strips are placed vertically along the long side of the cell phone floor, and eight SMA coaxial connectors are soldered at the bottom of the floor to feed the signal through a 50 Ω microstrip line.

The simulated and measured S-parameters are shown in Figure 5.21, and the measured curves are slightly shifted toward higher frequencies compared to the simulated curves.

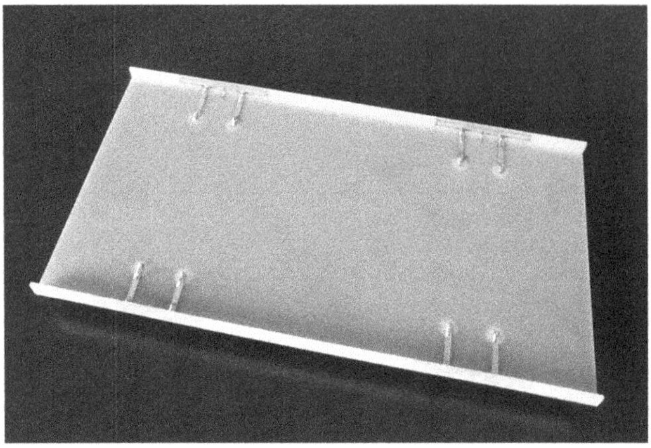

FIGURE 5.20 Fabricated MIMO antenna prototype.

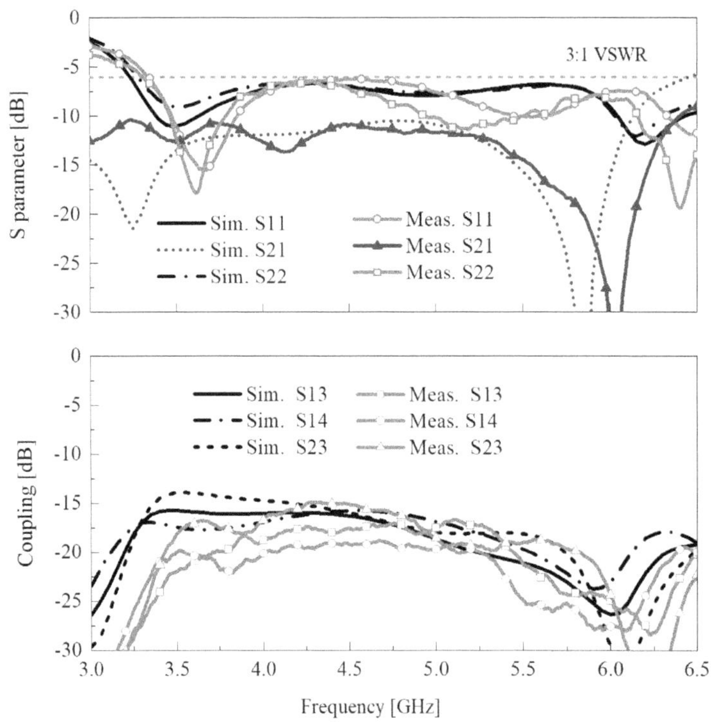

FIGURE 5.21 Simulated and measured S-parameters.

The difference between the simulated and measured curves is caused by processing errors and fluctuations in the dielectric constant of the FR4 dielectric substrate. For the measured curves, three resonance points are observed at about 3.6, 5.3, and 6.4 GHz. The overlapping

−6 dB bandwidth of port 1 and port 2 measurements is 3.02 GHz (3.34–6.36 GHz), and the port isolation of the antenna pairs is higher than 10.6 dB across the band.

The radiation efficiency of the MIMO antenna is measured in a microwave chamber room. When one port is excited, all other ports are connected to a 50 Ω impedance load. The simulated and measured radiation efficiency of the first antenna pair is shown in Figure 5.22. The results show that the measured radiation efficiency of port 1 and port 2 is higher than 45% in the 3.3–6 GHz band.

Table 5.3 shows the performance comparison of this design with other dual-band or wide-band MIMO antenna designs in the references. The antenna pair size of this design is relatively small compared to the existing work. In addition, the present design has advantages in terms of large bandwidth.

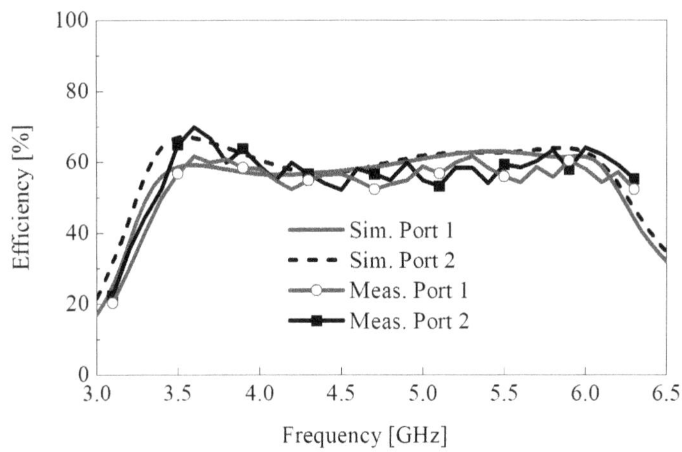

FIGURE 5.22 Radiation efficiency of the first antenna pair simulated and measured.

TABLE 5.3 Performance Comparison of the Present MIMO Antenna Design with Dual-Band or Wide-Band MIMO Antenna Designs in Other References

Design	Antenna Pair Size (mm × mm)	Frequency Bandwidth (Frequency Range, GHz)	Port Isolation (dB)
Reference [8]	15 × 7	3.4–3.6, 5.2–6	10
Reference [30]	40 × 7	3.4–3.6, 4.8–5.0	11.5
Reference [9]	22 × 7	3.4–3.6, 4.8–5.0	17.5
Reference [31]	51.6 × 4.3	3.4–3.8, 4.8–5.0	15.5
Reference [32]	32.2 × 8	2.3–2.6, 3.7–6.0	9
Reference [33]	52.2 × 5.5	3.3–4.2, 4.8–5.0	11.5
Reference [12]	35 × 4.2	3.3–6.0	10
Reference [34]	20 × 6	3.3–4.2	12
Reference [35]	53 × 6	3.3–5.0	13.5
Reference [11]	44.4 × 6.2	3.3–6.0	10
Reference [36]	40 × 7.5	3.3–5.0	12
This design	28 × 4.2	3.3–6.0	10.6

5.4 MULTI-ANTENNA MODULE WITH FOUR-UNIT AGGREGATION

In this section, a four-cell aggregated multi-antenna module is proposed to reduce the coupling between the antenna cells by introducing lumped components such as inductors and capacitors.

5.4.1 Decoupling of Two Closely Spaced Units

5.4.1.1 Cell Design

The monopole is one of the most popular forms of MIMO antenna design and can be bent into an L-shape to reduce the antenna profile height. When the monopole is close to the floor, a parallel stub acting as an inductor is usually added to match the input impedance, and this modified structure is known as an inverted-F antenna (IFA).

Figure 5.23 shows the design of an IFA cell operating in the 3.5 GHz band. The IFA is placed along the long edge of the cell phone floor. The distance between the open-circuit interceptor of the IFA and the short edge of the floor is 35 mm. The ground plane and the IFA are printed on a 0.8 mm thick FR4 dielectric substrate ($\varepsilon_r = 4.4$, $\tan\delta = 0.02$). The profile height of the unit was 3 mm and the dimensions of the ground plane were 140 mm × 70 mm.

The impedance performance of the IFA and the L-shaped monopole (IFA without parallel short cut-off line) is shown in Figure 5.24. It can be seen that the impedance matching performance of the IFA is better than that of the L-shaped monopole, and the −6 dB bandwidth of the IFA is about 420 MHz.

In a two-cell MIMO antenna design, the two cells have three configurations, i.e., face-to-face, back-to-back, and face-to-back configurations. Figure 5.25 shows three different dual cell designs based on the proposed IFA cell. In all designs, the dimensions of the units are the same. The offset distance of the cell from the floor corners is fixed at 35 mm. The antenna coupling in the first two configurations will be discussed next, while the third configuration is structurally asymmetric and will not be analyzed in depth here.

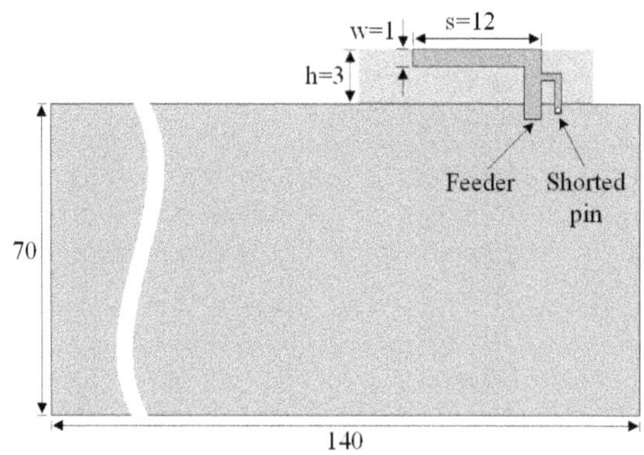

FIGURE 5.23 IFA cell design operating in the 3.5 GHz band (unit: mm).

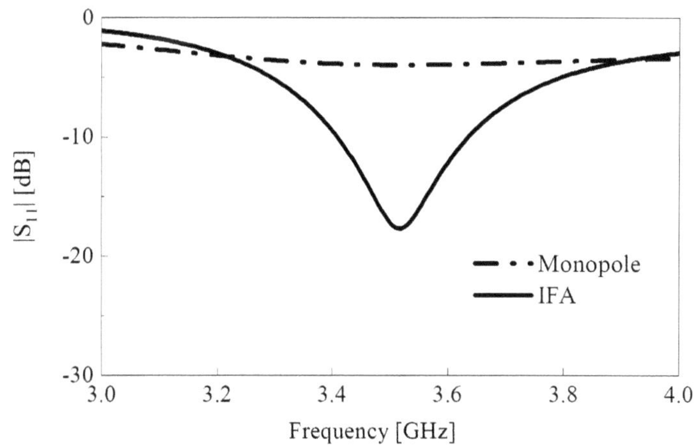

FIGURE 5.24 Impedance performance of IFA and L-shaped monopole.

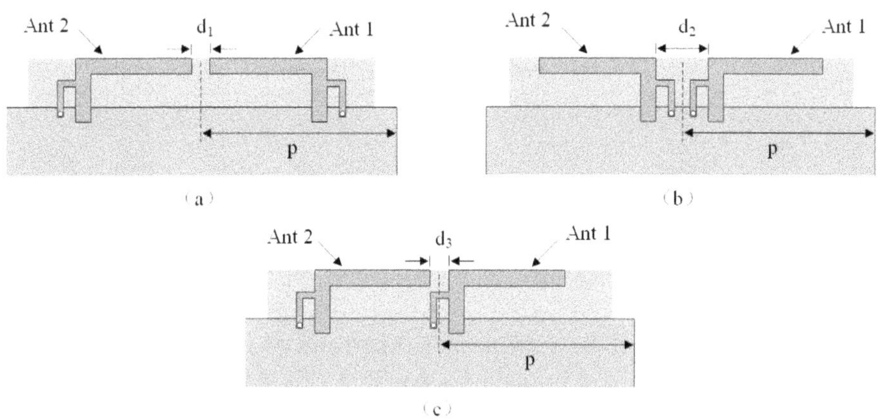

FIGURE 5.25 Three different dual cell designs based on the proposed IFA cell: (a) face-to-face, (b) back-to-back, (c) face-to-back.

5.4.1.2 Face-to-Face Configuration

The dual cell design in face-to-face configuration is studied first. When the cell spacing is 5 mm, the mutual coupling between the cells is strong (about −5 dB). As the cell spacing increases, the port isolation can be improved. However, even with a cell spacing of 20 mm, the isolation is still only 10.5 dB, and continuing to increase the cell spacing has limited effect.

A collector inductor is used on top of this to reduce mutual coupling. As shown in Figure 5.26, a decoupling inductor is added at the end of the two parallel stubs of the IFA (current minimum position). The edge-to-edge distance between the two cells is only 1 mm, corresponding to 0.01 λ_0 at 3.5 GHz, with λ_0 being the free-space wavelength.

The decoupling principle of the inductor is analyzed next. The IFA unit operates in the quarter-wavelength mode, a mode in which the current is maximum at the feed ports and minimum at the end of the stub. When two IFAs are placed face to face, the current

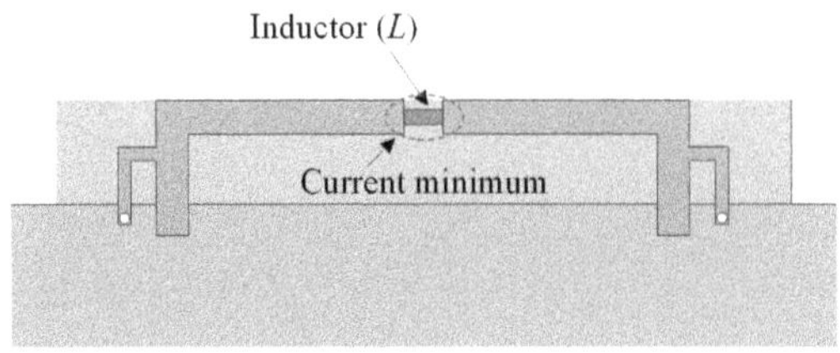

FIGURE 5.26 Introduction of decoupling inductor at current minimum position.

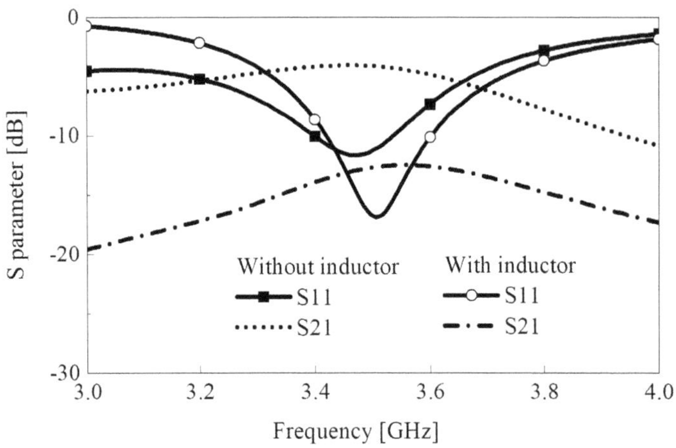

FIGURE 5.27 Effect of presence or absence of inductance on S-parameters.

minima of these two units will be close to each other, and an inductor can be added at the current minima. Numerous studies have shown that this modification ensures that the pattern of the cells is not affected. The introduction of an inductor provides an additional current path that can cancel out the existing coupling. The strength and phase of the introduced coupling is controlled by the inductance value (*L*).

The effect of the presence or absence of inductance on the S-parameters is shown in Figure 5.27. In order to resonate at 3.5 GHz, the length of the IFA was extended from 12 to 13.6 mm, and the inductance value was adjusted to 39 nH. It can be observed that the inductance has a small effect on the resonance frequency, but can significantly change the port isolation. Therefore, the use of decoupling inductors is an effective way to improve port isolation.

The current distribution at 3.5 GHz is shown in Figure 5.28. In both cases, with and without inductance, the excitation unit on the right side operates in quarter-wavelength mode, but there is a clear difference in the performance of the parasitic unit on the left side. As shown in Figure 5.28a, there is a strong current on the parasitic unit, and its quarter-wavelength mode is similarly excited, with more energy coupled to the left port. As shown in Figure 5.28b, the current density on the parasitic cell is much less and there is almost no coupled energy at the left port. This comparison directly characterizes the decoupling effect of the inductor.

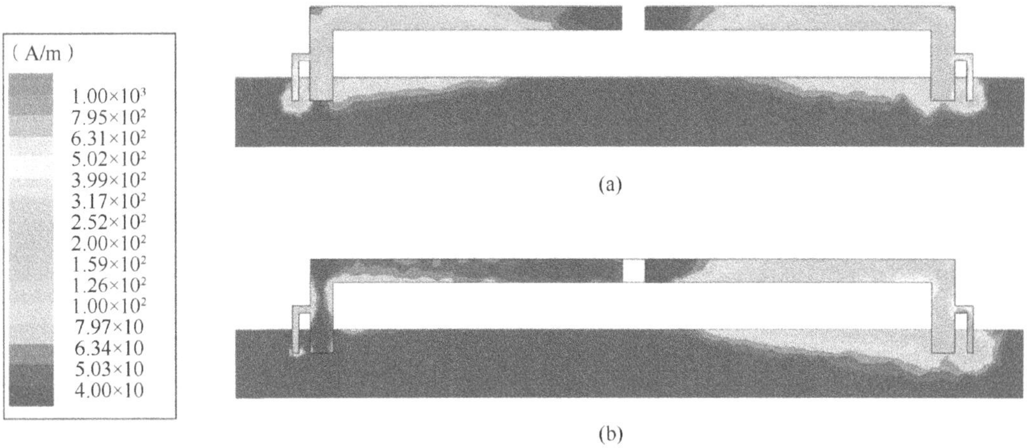

FIGURE 5.28 Current distribution at 3.5 GHz: (a) without inductance, (b) with inductance.

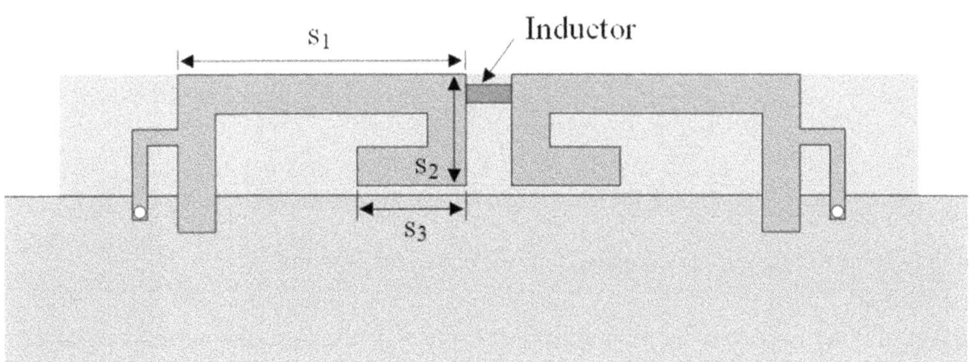

FIGURE 5.29 Dual-cell model using folded stubs.

The model was further modified to reduce the size of the IFA cell. A two-cell model using folded stubs is shown in Figure 5.29. The stubs of each IFA were folded into a U-shape, and the total length of the stubs was approximately one-quarter wavelength. Optimization is performed using HFSS software and the optimized values are as follows: l = 24 nH, s_1 = 9 mm, s_2 = 2.8 mm, and s_3 = 3.5 mm. The total dimensions of the antenna are 20.6 mm × 3 mm. Figure 5.30 shows the S-parameters of the simulation. It is observed that −6 dB bandwidth is still available for 225 MHz. Port isolation can be optimized by adjusting the size of the inductance value so that the port isolation is higher than 14 dB.

5.4.1.3 Back-to-Back Configuration

The back-to-back configuration, where the two feed ports are close to each other and the openings of the cells are far away from each other, is investigated here. When the cell spacing is 5 mm, the mutual coupling between the cells is about −6.2 dB, and as the cell spacing increases, the mutual coupling between the cells is still very strong, and at a spacing of 20 mm, the mutual coupling is still as high as −9.3 dB. Therefore, back-to-back configurations require the addition of decoupling structures to reduce the mutual coupling.

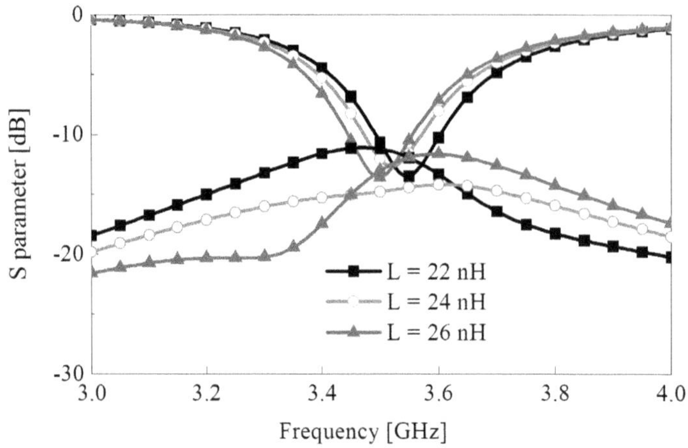

FIGURE 5.30 S-parameters for simulation (solid and dashed lines indicate S_{11} and S_{21} respectively).

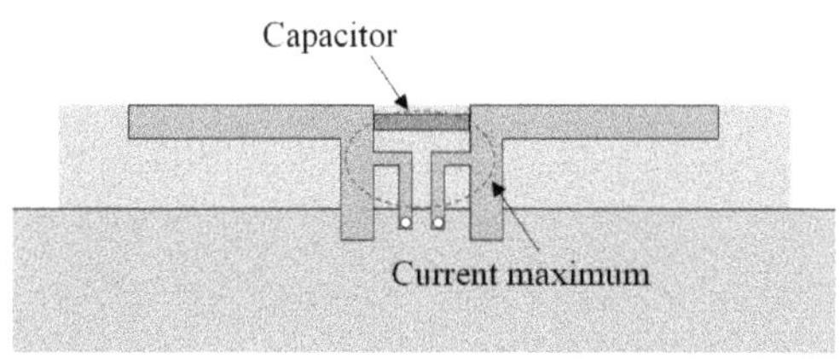

FIGURE 5.31 Introducing capacitance at current maximum location.

It has already been verified that adding inductors at the current-minimizing locations of two face-to-face cells is an effective way to improve port isolation. Based on the pairwise principle, a possible conjecture is that adding a capacitor at the current-maximum position of the two back-to-back cells might be an effective decoupling method in a back-to-back configuration. Based on this conjecture, the design of the back-to-back antenna with capacitors is shown in Figure 5.31. 0.2 pF of capacitance is placed near the current-maximum locations of the two IFAs, and the antenna cells are separated from each other by an edge-to-edge distance of 2.5 mm.

The S-parameters for the back-to-back configuration are shown in Figure 5.32. The results show that the capacitance significantly changes the resonant frequency, but the isolation is only improved by 1 dB. This indicates that the introduction of capacitance does not show the expected decoupling effect, and the coupling introduced by the capacitance does not cancel the initial coupling well; and further improvement of the decoupling structure is needed.

As we all know, inductive feed and capacitive feed are two methods to stimulate the antenna. Theoretically, placing the inductive feed in the current maximum position is equivalent to placing the capacitive feed in the current minimum position, as shown in Figure 5.33a. However, it is not clear whether these two feeding methods have the same performance in a two-cell MIMO antenna design.

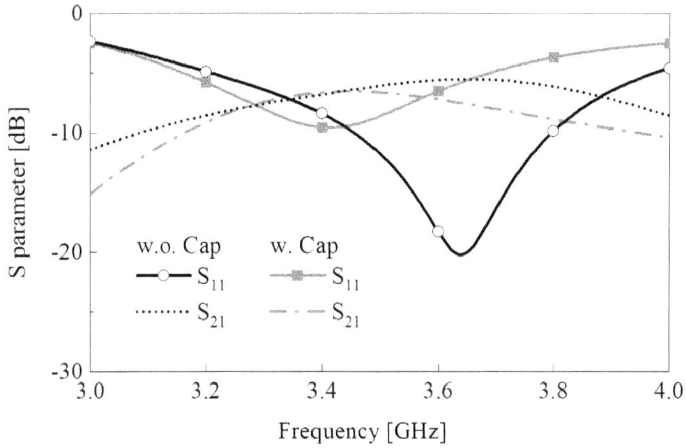

FIGURE 5.32 S-parameters for back-to-back configuration.

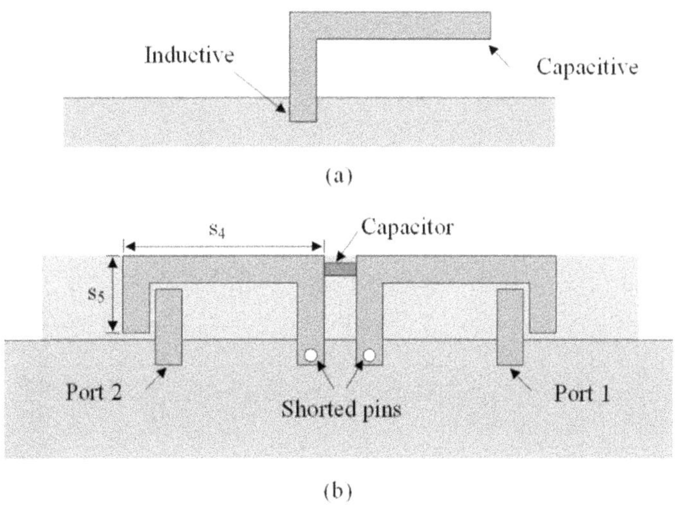

FIGURE 5.33 Improved antenna design using capacitive feeding: (a) two feeding techniques, (b) dual cell MIMO antenna design.

To analyze the effect of capacitive feeds, the back-to-back model was modified as shown in Figure 5.33b. The inductive feed ports of each cell are replaced by shorting pins to keep the quarter-wavelength mode unchanged. The parallel shorting line and the shorting pin were combined into a single pin and connected to ground through the shorting point. In addition, a short cut-off wire is introduced as a capacitive feed structure to feed the signal. The gap between the capacitive feed structure and the folded branch is 0.2 mm. The edge-to-edge distance of the antenna unit is 1 mm. The capacitance value (C) is 0.5 pF. The parameters are the same as the configuration of a single unit, except that $s_4 = 8.4$ mm and $s_5 = 2.6$ mm. The antenna unit is connected to ground through the shorting point. The antenna unit is connected to ground through the shorting point.

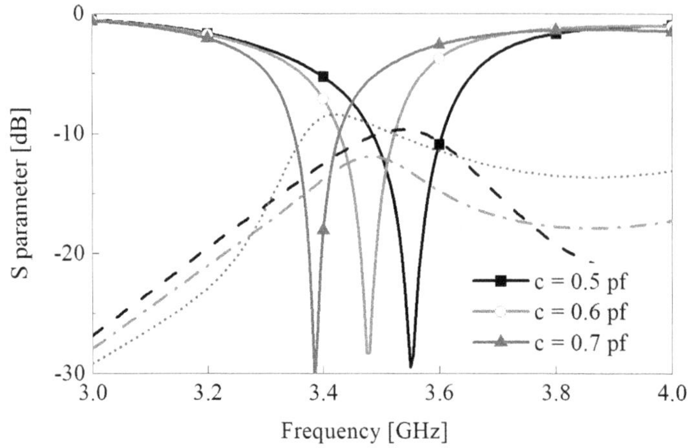

FIGURE 5.34 Effect of capacitance on S-parameters (solid and dashed lines denote S_{11} and S_{21} respectively).

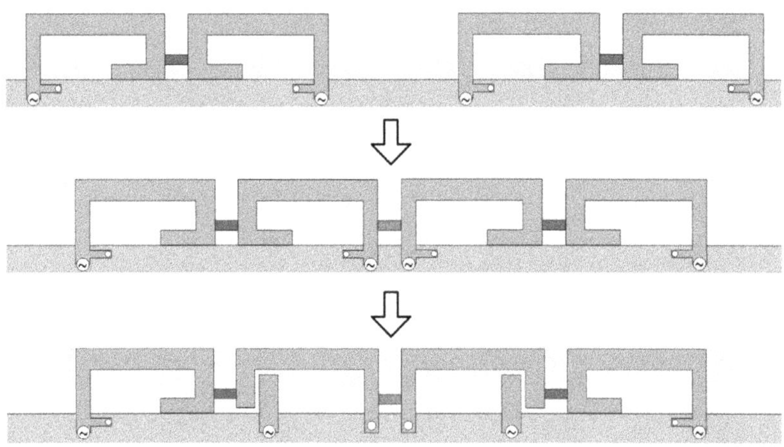

FIGURE 5.35 Evolution of multi-antenna module design with four-cell aggregation.

The effect of capacitance on S-parameters is shown in Figure 5.34. The results show that the variation of capacitance value has an effect on both resonant frequency and port isolation. The resonant frequency and peak isolation move to lower frequencies as the capacitance value increases. The performance in terms of impedance bandwidth and isolation is better when the capacitance value is 0.6 pF.

5.4.2 Multi-Antenna Module Design

Based on the previous two types of highly isolated two-cell designs, the four-cell MIMO antenna is analyzed below. The evolution of the multi-antenna module design for four-cell aggregation is shown in Figure 5.35. First consider two face-to-face two-cell structures with a decoupling inductor between each antenna pair. Then the two antenna pairs are brought closer together so that the two antenna pairs have an edge-to-edge distance of

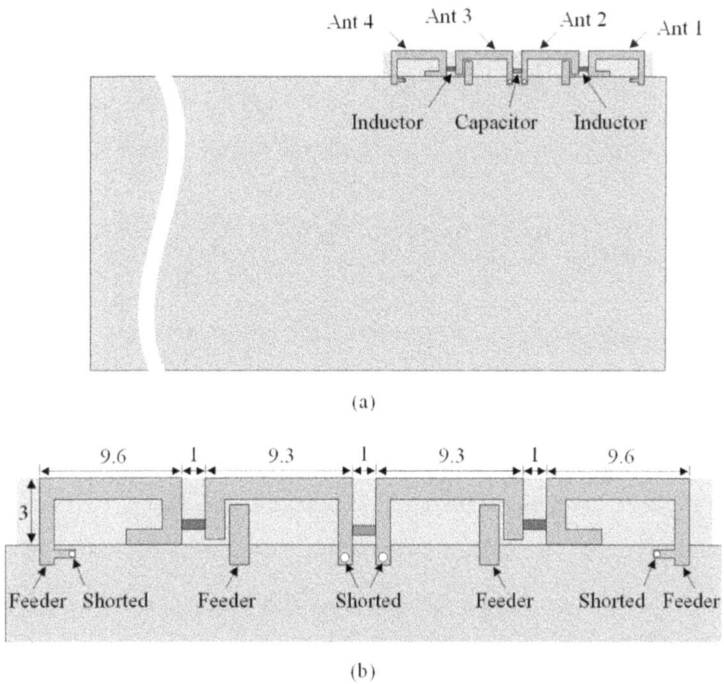

FIGURE 5.36 Geometry and dimensions of the four-cell MIMO antenna design (in mm): (a) top view, (b) structural details of the four-cell MIMO antenna.

only 1 mm and a capacitor is added between them. Finally, the feeding method of the two cells inside the quad cell is modified from inductive to capacitive feeding. With these three steps, the four cells can be tightly aggregated into a miniaturized multi-antenna module.

The four-cell design can be decomposed into two sets of two-cell designs. Inductive decoupling analysis is used for face-to-face configuration and capacitive decoupling analysis is used for back-to-back configuration. The geometry and dimensions of the proposed four-cell MIMO antenna design are shown in Figure 5.36. The four cells are closely connected with inductance or capacitance between two neighboring cells. The dimensions of the MIMO antenna are 40.8 mm × 3 mm. The edge-to-edge distance of the neighboring cells is 1 mm. The symmetry axis of the MIMO antenna is offset from the floor corners by a distance of 45 mm. The values of both decoupling inductances are 27 nH and the values of decoupling capacitances are 0.5 pF.

The S-parameters of the four-cell MIMO antenna simulation are shown in Figure 5.37. It can be observed that the curves of S_{22} and S_{33} match well, and the curves of S_{11} and S_{44} almost overlap. The overlapping bandwidth of −6 dB for the four cells is 165 MHz. The port isolation of any two cells is higher than 13.5 dB.

Next, the current distribution at 3.5 GHz is analyzed. The current distribution at 3.5 GHz simulated with different port excitations is shown in Figure 5.38. The current distribution on the excitation unit indicates that the unit operates in quarter-wavelength mode. For the remaining three parasitic units, although strong currents can be observed on some

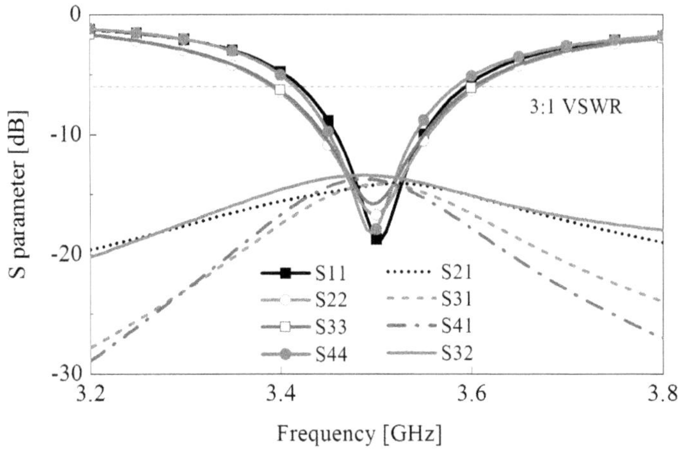

FIGURE 5.37 S-parameters of the four cell MIMO antenna simulation.

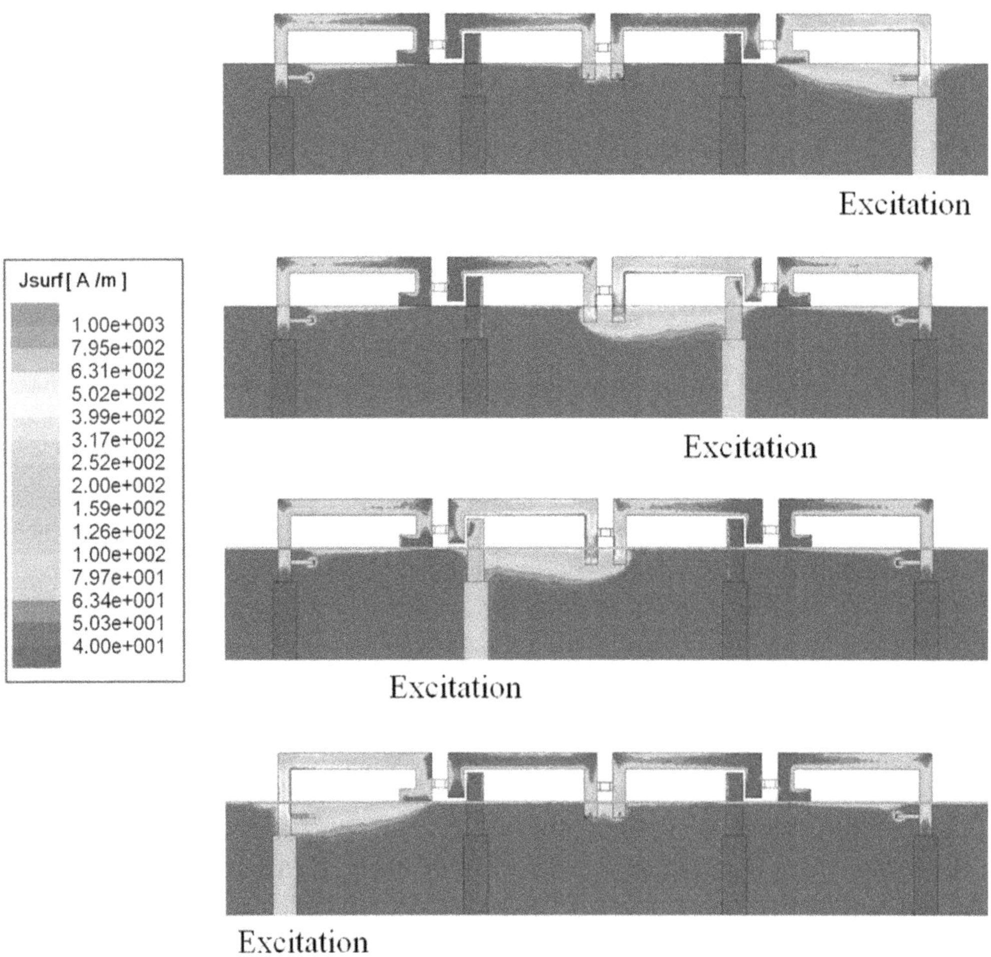

FIGURE 5.38 Current distribution at 3.5 GHz simulated with different port excitations.

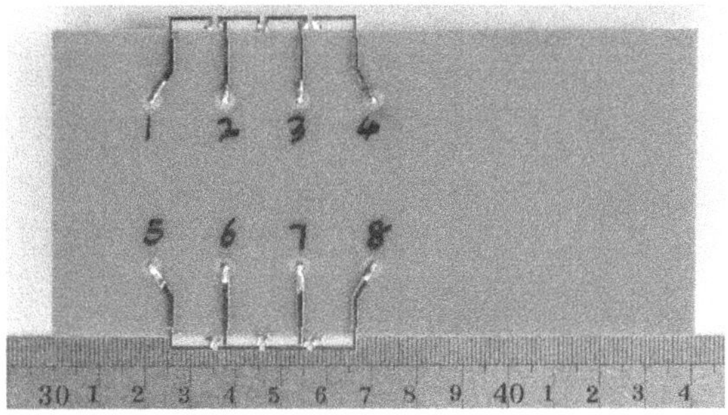

FIGURE 5.39 Eight-cell MIMO antenna in kind.

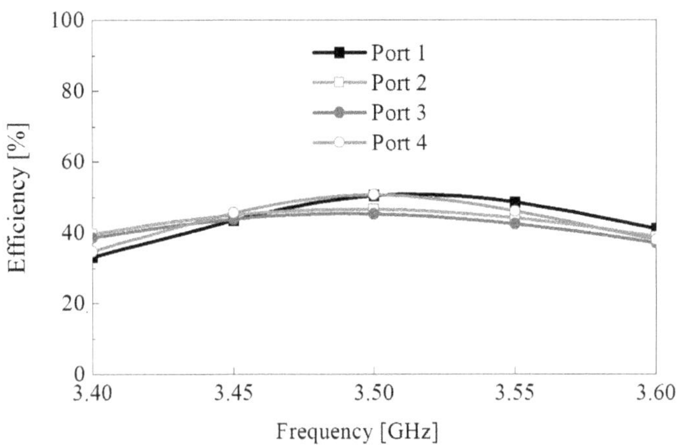

FIGURE 5.40 Radiation efficiency of the antenna when the four ports are excited separately.

of the radiating structures, the currents on the feed ports of the parasitic units are weak, indicating that high isolation is achieved between the ports.

Figure 5.39 shows the actual eight-cell MIMO antenna designed, containing two four-cell aggregation modules. Figure 5.40 shows the radiation efficiency of the antenna when each of the four ports is excited. When port 1 or port 4 is excited, the peak efficiency is about 51%; when port 2 or port 3 is excited, the peak efficiency is about 46%. There are two main reasons for the low radiation efficiency: one is that the reduced cell size reduces the radiation efficiency, and the other is that there are four closely spaced cells within each antenna module, and the coupling between all cells should be considered. Compared with the traditional two-unit module, there are more coupling paths between the four units, and the combined effect of multi-antenna coupling paths reduces the radiation efficiency, so it is necessary to consider these factors comprehensively in the antenna design.

5.4.3 Model Extension

In the planar MIMO antenna design, the profile height of the antenna cell is 3 mm. This means that the gap between the phone bezel and the ground plane is at least 3 mm. This dimension may not be small enough for some narrow bezel phones. Figure 5.41 shows

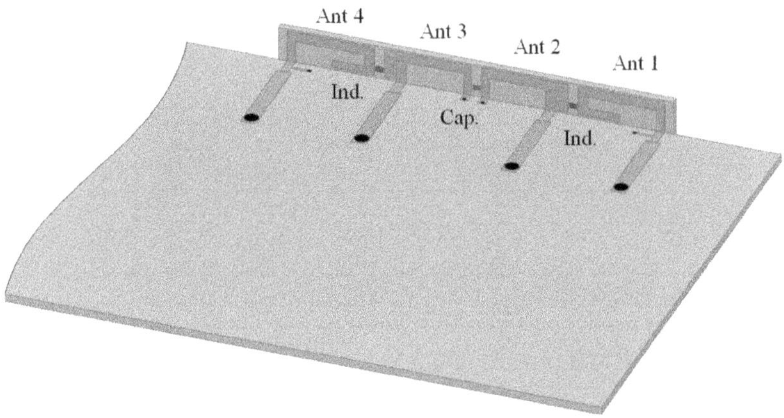

FIGURE 5.41 Geometry of the improved MIMO antenna design.

the geometry of the modified MIMO antenna design, based on a planar four-cell antenna with four cells folded vertically by 90° along the long side of the ground plane and the MIMO antenna placed vertically at the edge of the floor. Because of this, the gap between the floor and the antenna is 0. The area of each cell is 8.8 mm × 3.2 mm, and the size of the MIMO antenna is 38.2 mm × 3.2 mm. The other parameters are the same as those of the planar design.

The working principle of vertically placed MIMO antenna design is similar to the working principle of planar four-cell antenna design. Inductance is used to reduce the mutual coupling between two face-to-face units and capacitance is used to reduce the mutual coupling between two back-to-back units. The antenna performance obtained from simulation is similar and is not described here.

5.5 ANTENNA PAIR COVERING 3.5/4.9 GHz DUAL BAND

In this section, the design of the dual-band dual-cell MIMO antenna is discussed. First, the antenna pair design for the 4.9 GHz band is analyzed to explain the decoupling principle based on the common mode/differential mode theory; then the antenna pair design for the 3.5 GHz band is discussed to replicate the decoupling structure of the higher band for the 3.5 GHz band; and finally, the performance of the MIMO antenna is tested completely.

5.5.1 Single-Band Antenna Pair

Figure 5.42a shows the geometry of a two-unit antenna pair operating in the 4.9 GHz band. Two L-shaped monopoles are closely arranged face to face. The edge-to-edge distance between the two cells is 1 mm. A 14.5 nH collector inductance is loaded at the openings of the two cells. The characteristic impedance of the feedline is 50 Ω. A short cut for short-circuiting is connected to the feedline to match the input impedance. The antenna module is printed on a FR4 dielectric substrate ($\varepsilon_r = 4.4$, $\tan\delta = 0.02$) with a thickness of 0.8 mm. The floor plane measures 140 mm × 70 mm and the antenna modules are placed on either side of the midpoint of the long side of the floor.

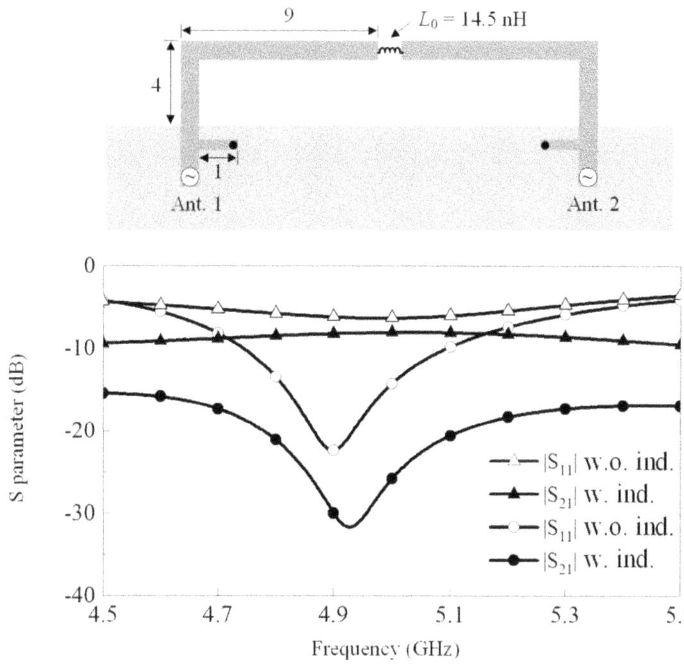

FIGURE 5.42 Geometric structure (in mm) of a two-cell antenna pair operating in the 4.9 GHz band and the effect of the presence or absence of inductance on the S-parameters: (a) geometric structure, (b) S-parameters.

Inductance plays an important role in reducing cell coupling. Figure 5.42 compares the S-parameters of the antenna pair with and without inductance. When an inductor is added, the crater near 4.9 GHz can be observed from the S_{21} curve, and the port isolation near 4.9 GHz is improved from 8 dB to more than 30 dB when comparing with the no inductor result.

In addition to being explained in terms of current cancellation theory, the workings of inductive decoupling can also be analyzed in terms of common mode/differential mode theory. This theory provides a simple explanation for the decoupling mechanism between two symmetrically placed cells.

According to this theory, *the* S-parameter (S_{cc11}/S_{dd11}) of the common mode/differential mode can be calculated using the following equation:

$$S_{cc11} = (S_{11} + S_{12} + S_{21} + S_{22})/2 \tag{5.1}$$

$$S_{dd11} = (S_{11} - S_{12} - S_{21} + S_{22})/2 \tag{5.2}$$

Considering that the antenna pair is a symmetric and reciprocal two-port network, it can be shown that $S_{11} = S_{22}$, $S_{12} = S_{21}$. The common mode/differential mode equation can be simplified as:

$$S_{cc11} = S_{11} + S_{21} \tag{5.3}$$

$$S_{dd11} = S_{11} - S_{21} \tag{5.4}$$

The expression for the port mutual coupling S_{21} is as follows:

$$S_{21} = (S_{cc11} - S_{dd11})/2 \tag{5.5}$$

Equation (5.5) shows that the port isolation can be calculated based on the difference between *the* S-parameters of the common mode/differential mode; more specifically, the smaller the difference between the reflection coefficients of the common mode/differential mode, the higher the isolation of the two feed ports.

Because the antenna pair uses a face-to-face configuration, the two units are symmetrical with respect to the centerline. As shown in Figure 5.43a, if the common mode is excited, a virtual perfect magnetic conductor boundary is created. As shown in Figure 5.43b, if the differential mode is excited, a virtual perfect electric conductor (PEC) boundary will result. These two boundaries can be used to explain the mechanism of port isolation, and they behave very differently for inductive loading. From the antenna theory, it can be seen that inductive loading has no effect on the perfect magnetic conductor boundary and has a significant effect on the perfect electric conductor boundary. Therefore, when the inductance value is changed, the S_{11} of the differential mode changes significantly, while the S_{11} of the common mode is almost unchanged. By choosing a suitable inductance value, the reflection coefficient curve of the differential mode can be adjusted so that the difference

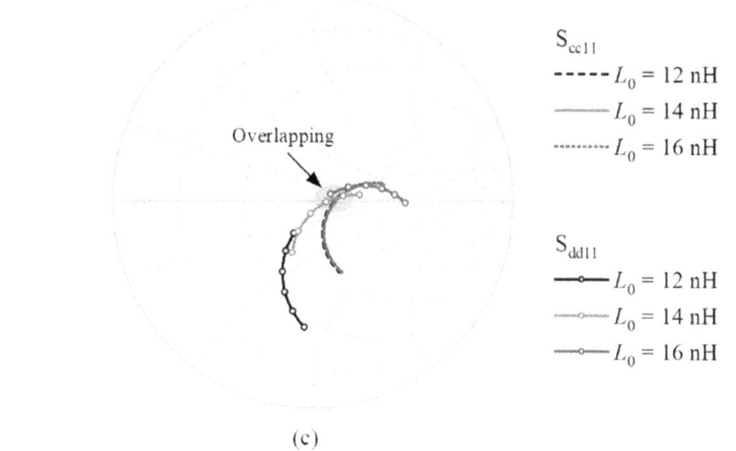

FIGURE 5.43 Principle explanation of inductive decoupling: (a) common mode excitation, (b) differential mode excitation, (c) effect of inductive decoupling on common mode/differential mode.

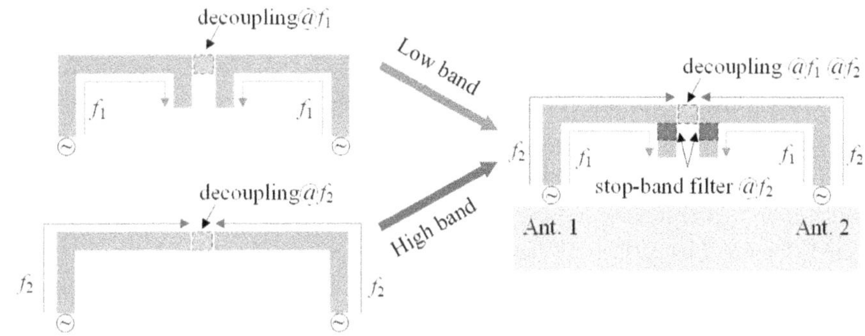

FIGURE 5.44 Dual-band MIMO antenna design.

between the reflection coefficients of the common mode and the differential mode is reduced, thus achieving high port isolation. After optimization, when the inductance value is 14 nH, the reflection coefficient curves of common mode/differential mode are both close to the center point of Smith Chart diagram, indicating that the two impedances are well matched and the impedance difference is small. Therefore, inductive decoupling is an effective way to decouple two closely spaced units.

5.5.2 Using the Same Inductor Decoupling for Dual Frequency Bands

The proposed dual-band MIMO antenna design based on the single band is shown in Figure 5.44. First, the U-shaped monopole is used to generate resonance in the low band f_1, while the L-shaped monopole is used to generate resonance in the high band f_2. These two frequency bands can be decoupled separately using inductors. A bandstop filter is then added to the radiating unit such that the unit behaves like a U-shaped monopole in the lower frequency band but like an L-shaped monopole in the higher frequency band. The challenge here is to design a dual-band decoupling structure that is effective in both the low and high-frequency bands. Decoupled inductor is popular in 5G single-band MIMO antenna design. However, it has not been used in dual-band MIMO antenna design.

Theoretically, it is always possible to find inductors (L_1 and L_2) to decouple the low- and high-frequency bands, respectively, i.e., these two bands satisfy the following two equations:

$$a(f_1)+b(f_1,L_1)=0 \tag{5.6}$$

$$a(f_2)+b(f_2,L_2)=0 \tag{5.7}$$

where $a()$ denotes the original coupling path signal between the two cells, and $b()$ denotes the coupling path signal introduced by the inductor. The two path signals cancel each other, resulting in a synthesized signal of 0.

For the proposed structure, due to the different radiation structures in the low and high-frequency bands, a suitable inductor (L) can be found which decouples both bands as shown in Eq. (5.8):

$$a(f_1, f_2) + b(f_1, f_2, L) = 0 \tag{5.8}$$

Details of how to find the inductance L, how to design the band-stop filter, and how to design the dual-band antenna unit are described next.

Figure 5.45a shows an L-shaped monopole in a face-to-face configuration, which is printed on a 0.8 mm thick FR4 dielectric substrate ($\varepsilon_r = 4.4$, $\tan\delta = 0.02$). The dimensions of the floor are 140 mm × 70 mm and the antenna pair is placed near the midpoint of the long side of the floor. The line widths of the feedline and the monopole are 1.5 and 1 mm, respectively. The edge-to-edge distance between the two cells is only 1 mm. A decoupling inductor of 14.5 nH is loaded at the open end of the two cells. From the S-parameter curves, it can be observed that a good decoupling effect is presented in the 4.9 GHz band.

An L-shaped monopole operating in the 3.5 GHz band is shown in Figure 5.45b. The length of the L-shaped monopole is extended from 9 to 12.2 mm, and the inductance value is increased from 14.5 to 28.5 nH, which achieves good decoupling. Comparing the models in the 3.5 and 4.9 GHz bands, it can be concluded that the inductance value increases with decreasing frequency.

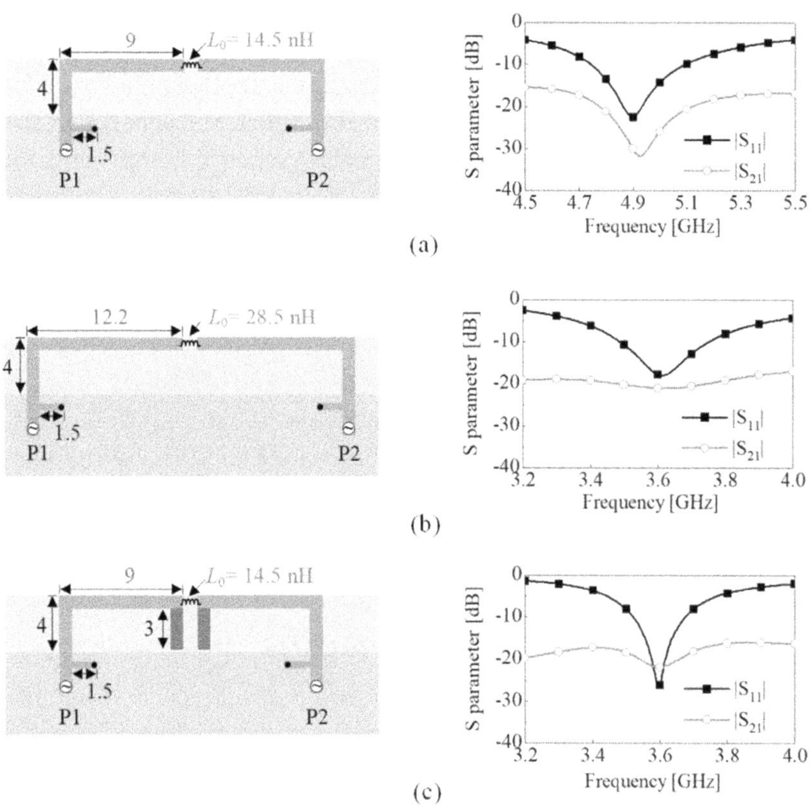

FIGURE 5.45 Geometry (in mm) and S-parameter curves of single-band antenna pair: (a) L-shaped monopole operating at 4.9 GHz band, (b) L-shaped monopole operating at 3.5 GHz band, (c) U-shaped monopole operating in the 3.5 GHz band.

In order to share the same decoupling inductance in both bands, either the inductance value in the 3.5 GHz band is reduced or the inductance value in the 4.9 GHz band is increased. In practical engineering, the self-resonant frequency of commercial inductors decreases with increasing inductance value. Therefore, to avoid self-resonance of the inductor, reducing the inductance value from 28.5 to 14.5 nH is a better choice.

A modified model operating in the 3.5 GHz band is shown in Figure 5.45c, in which the L-shaped monopole is folded into a U-shaped structure. It is found that increasing the length of the folded part decreases the inductance value. The inductance value decreases from 28.5 to 14.5 nH when the folded length is 3 mm. Thus, the model for the 3.5 GHz band has the same size and inductance value as the model for the 4.9 GHz band.

5.5.3 Using a Single Branch to Generate a Dual-Band Band

The following describes the design of a single-branch monopole structure covering a dual-band by using a band-stop filter.

According to microwave theory, a capacitor in parallel with an inductor can form a bandstop filter. Figure 5.46 shows the design and S-parameters of a band-stop filter operating in the 4.9 GHz band. A capacitor with $C_1 = 0.6$ pF and an inductor with $L_1 = 0.8$ nH are placed side-by-side somewhere on the 50 Ω microstrip line. A capacitor with $C_1 = 0.6$ pF and an inductor with $L_1 = 0.8$ nH are placed side-by-side. In the 4.9 GHz band, S_{21} is below −10 dB and S_{11} is above −0.4 dB. The S-parameter curves show that the structure has a stop band filtering function at 4.9 GHz. In the 3.5 GHz band, S_{21} is above −1 dB, indicating that the structure has passband capability at 3.5 GHz. It should be noted that adding another series capacitor in the 3.5 GHz band can improve the bandpass performance.

Based on the proposed band-stop filter, a dual-band single-branch monopole cell geometry is designed as shown in Figure 5.47. A band-stop filter is added to the folded branch at the end of the monopole, and the position of the filter can be optimized according to the performance of the dual-band, and the closer to the end, the lower the operating frequency of the higher frequency band. The S-parameters of the antenna unit with and without the bandstop filter are shown in Figure 5.48. The current can pass through the filter in the lower band, so

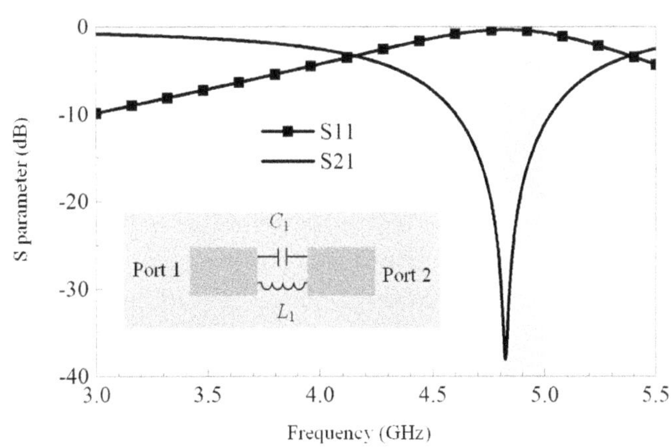

FIGURE 5.46 Bandstop filter design and its S-parameters operating at 4.9 GHz band.

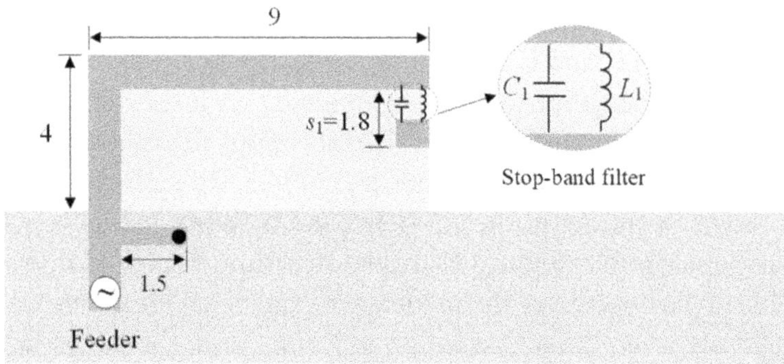

FIGURE 5.47 Geometry of the dual-band single-branch monopole unit (in mm).

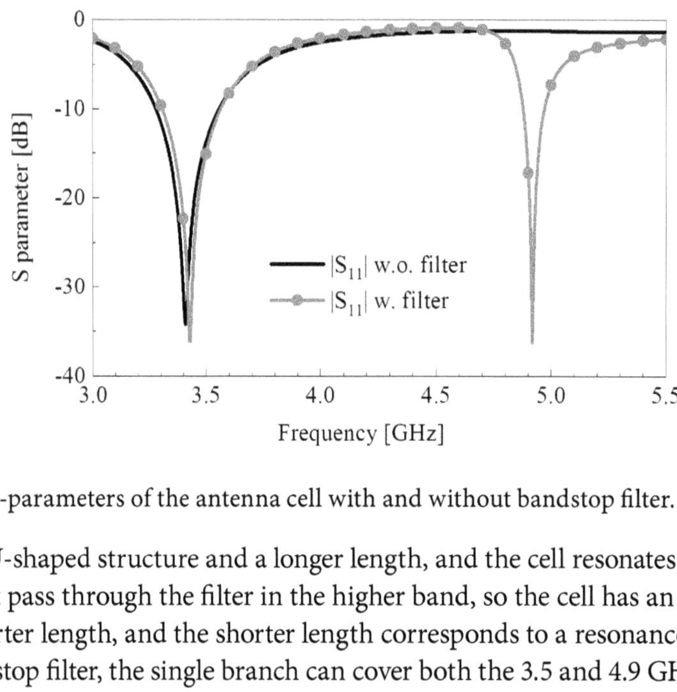

FIGURE 5.48 S-parameters of the antenna cell with and without bandstop filter.

the cell has a U-shaped structure and a longer length, and the cell resonates at 3.5 GHz. The current cannot pass through the filter in the higher band, so the cell has an L-shaped structure and a shorter length, and the shorter length corresponds to a resonance at 4.9 GHz. By adding a bandstop filter, the single branch can cover both the 3.5 and 4.9 GHz bands.

The key parameters of the cell are scanned to study the resonance of the two frequency bands. The frequency tuning of different bands is shown in Figure 5.49. As shown in Figure 5.49a, the resonance point of the higher band can be tuned by changing the parameters of the filter, while the lower band is almost unaffected. As shown in Figure 5.49b, the change of the folded arm (s_1) changes the resonance point of the low frequency band only. The parameter scanning results show that the two resonance points of the two frequency bands can be tuned independently, and also verify that the filter can reflect the electromagnetic waves in the high-frequency band and pass the electromagnetic waves in the low frequency band.

5.5.4 Dual-Band MIMO Antenna Design

A novel dual-band MIMO antenna design is proposed based on decoupled inductors and dual-band single-branch monopoles. The geometry of the designed antenna pair with dimensions of 19 mm × 4 mm is shown in Figure 5.50, where two U-shaped monopoles are

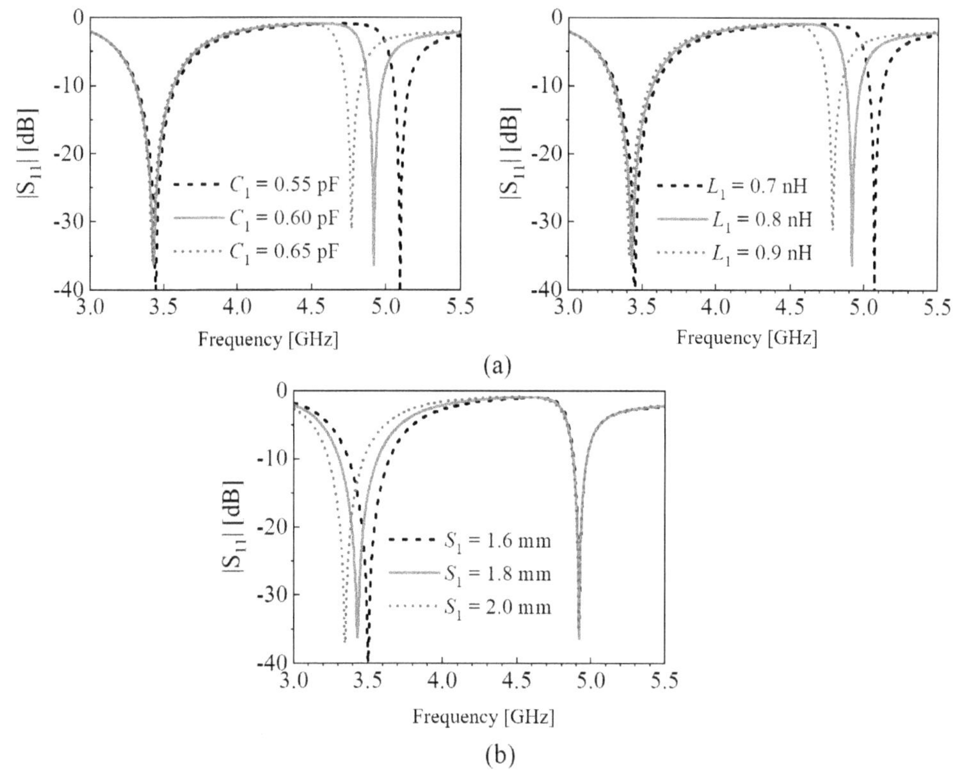

FIGURE 5.49 Frequency tuning of different frequency bands: (a) frequency tuning in high-frequency band, (b) frequency tuning in low frequency band.

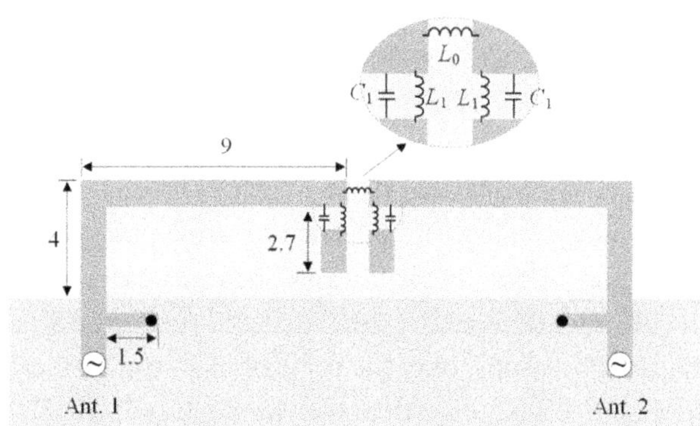

FIGURE 5.50 Dual-band antenna pair geometry (in mm).

configured face-to-face. An inductor of $L_0 = 13$ nH is added between the two monopoles. Bandstop filters are placed on the folded branches of each monopole with $L_1 = 0.8$ nH and $C_1 = 0.6$ pF.

The simulated S-parameters of the antenna pair are shown in Figure 5.51. If the decoupling inductor is removed, the coupling in both bands is very strong, and the port isolation is minimized to 5 dB. In contrast, the port isolation in both bands improves very

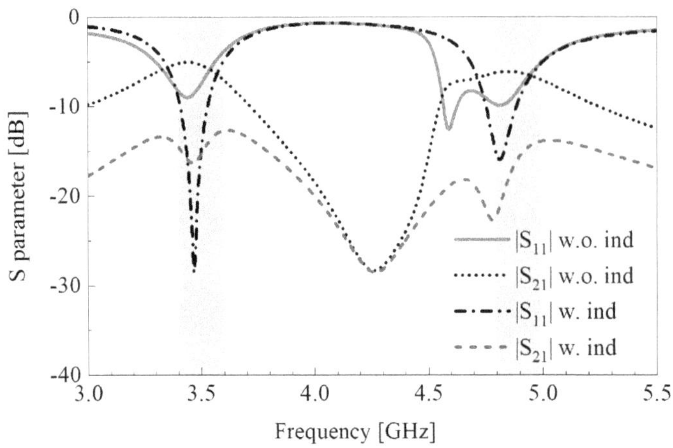

FIGURE 5.51 Simulated S-parameters of the antenna pair.

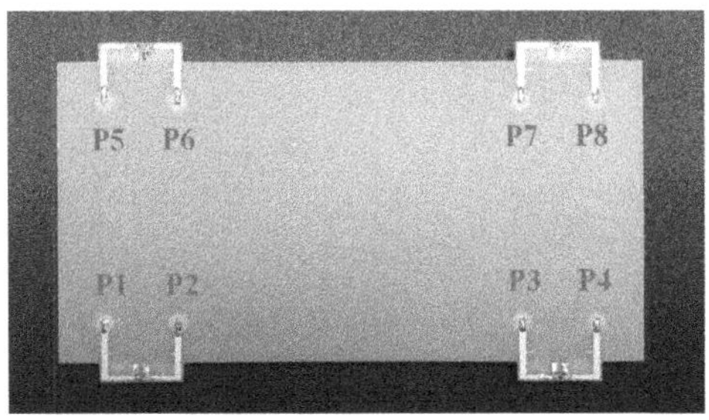

FIGURE 5.52 Eight-cell dual-band MIMO antenna prototype.

significantly after adding the inductor, which is above 12 dB. Observation of *the* S_{21} curve shows that there are two troughs at 3.5 and 4.9 GHz, which implies that the decoupling inductor is effective in both low and high-frequency bands.

To verify this design, an eight-cell dual-band MIMO antenna prototype was fabricated, and the physical object is shown in Figure 5.52. Each antenna pair is offset from the corner of the floor by 20 mm. It should be noted that the antenna pairs can also be folded vertically to achieve a zero ground clearance design. The simulated and measured S-parameters are shown in Figure 5.53. For symmetry reasons, only the test results of the first antenna pair are given. The measured overlapping −6 dB bandwidth of the two ports is 260 MHz in the low band and 220 MHz in the high band, essentially covering the 3.4–3.6 GHz and 4.8–5.0 GHz bands. The measured port isolation in these two bands is higher than 12 and 15 dB, respectively.

The radiation performance of the MIMO antenna was measured in a microwave chamber room. The radiation efficiencies in the low and high-frequency bands are shown in Figure 5.54. In the 3.4–3.6 GHz band, the measured efficiencies of port 1 and port 2 are higher than 44% and 41%, respectively. In the 4.8–5.0 GHz band, the measured efficiencies range from 38% to 62%, which is lower than the simulated efficiency. This difference is mainly attributed to the systematic error of the microwave chamber room and the larger loss of the FR4 dielectric substrate in the high-frequency band.

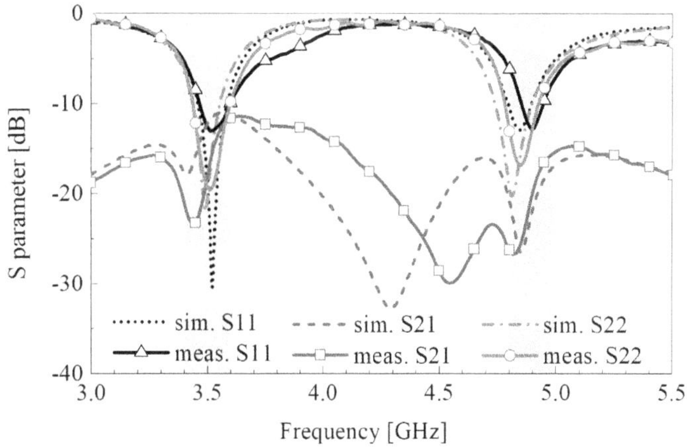

FIGURE 5.53 Simulated and measured S-parameters.

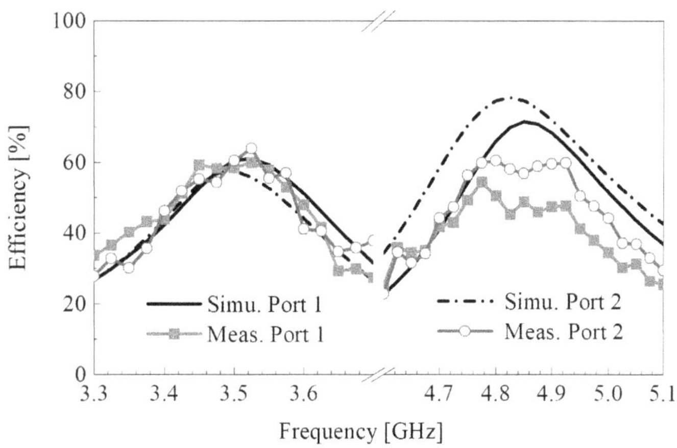

FIGURE 5.54 Radiation efficiency in low and high-frequency bands.

TABLE 5.4 Performance Comparison of the Present Design with Other Dual-Band MIMO Antenna Designs from References

References	Dual-Band Radiator Type	Antenna Pair Size (mm × mm)	Frequency Bandwidth (Frequency Range, GHz)	Port Isolation (dB)	Radiation Efficiency (%)
Reference [30]	Double-branched monopole	7 × 40	3.4–3.6, 4.8–5.1	>11.5, >15	>41, >40
Reference [37]	Double-branched IFA	6 × 42	3.3–3.6, 4.8–5.0	>12, >12	>48, >45
Reference [31]	Three-branched monopole	7 × 52	3.4–3.8, 4.8–5.0	>15.5, >19	>42, >40
Reference [9]	Double-branched monopole	7 × 22	3.4–3.6, 4.8–5.0	>17.5, >20	>50, >50
Reference [10][a]	Monopole + Slot	6 × 15	3.4–3.6, 4.8–4.9	>15, >10	>37, >40
This design	Single-branched monopole	4 × 19	3.4–3.6, 4.8–5.0	>12, >15	>41, >38

[a] Each feed port requires additional impedance matching circuitry.

Table 5.4 shows the performance comparison between this design and other reference dual-band MIMO antenna designs. From the comparison data, it can be seen that the antenna in this design has a simple radiating structure. In addition, the antenna pair in this design has the smallest size among all the designs of the same type due to the compact cell layout and decoupling circuit.

Details of several antenna designs in this chapter can be found in reference [38–41].

NOTE

1 Based on the common-mode/differential-mode decoupling theory, a single radiator is shared by two feed ports. A 180° power divider supplies the differential-mode phase, resulting in a port isolation of over 16.5 dB [7].

REFERENCES

1. Ministry of Industry and Information Technology. Circular of the Ministry of Industry and Information Technology on Matters Related to the Use of 3300–3600 MHz and 4800–5000 MHz Frequency Bands for Fifth Generation Mobile Communication Systems. (2017-11-15) [2024-05-15].
2. You X, Pan Z, Gao X, et al. 5G mobile communication development trend and some key technologies. *Science in China: Information Science*, 2014, 5: 551–563.
3. Wong KL, Chen LY. Dual-inverted-F antenna with a decoupling chip inductor for the 3.6-GHz LTE operation in the tablet computer. *Microwave and Optical Technology Letters*, 2015, 57(9): 2189–2194.
4. Ren A, Liu Y, Sim CYD. A compact building block with two shared-aperture antennas for eight-antenna MIMO array in metal-rimmed smartphone. *IEEE Transactions on Antennas and Propagation*, 2019, 67(10): 6430–6438.
5. Wong KL, Tsai CY, Lu JY. Two asymmetrically mirrored gap-coupled loop antennas as a compact building block for eight-antenna MIMO array in the future smartphone. *IEEE Transactions on Antennas and Propagation*, 2017, 65(4): 1765–1778.
6. Sun L, Feng H, Li Y, et al. Compact 5G MIMO mobile phone antennas with tightly arranged orthogonal-mode pairs. *IEEE Transactions on Antennas and Propagation*, 2018, 66(11): 6364–6369.
7. Zeng WF, Chen FC, Chu QX. Bandwidth-enhanced 5G mobile phone antenna pair with tunable electric field null. *IEEE Transactions on Antennas and Propagation*, 2023, 71(2): 1960–1964.
8. Wong KL, Lin BW, Li WY. Dual-band dual inverted-F/loop antennas as a compact decoupled building block for forming eight 3.5/5.8-GHz MIMO antennas in the future smartphone. *Microwave and Optical Technology Letters*, 2017, 59(11): 2715–2721.
9. Ren Z, Zhao A. Dual-band MIMO antenna with compact self-decoupled antenna pairs for 5G mobile applications. *IEEE Access*, 2019, 7: 82288–82296.
10. Chang L, Zhang G, Wang H. Dual-band antenna pair with lumped filters for 5G MIMO terminals. *IEEE Transactions on Antennas and Propagation*, 2021, 69(9): 5413–5423.
11. Sim CYD, Liu HY, Huang CJ. Wideband MIMO antenna array design for future mobile devices operating in the 5G NR frequency bands n77/n78/n79 and LTE band 46. *IEEE Antennas and Wireless Propagation Letters*, 2020, 19(1): 74–78.
12. Wong KL, Chen YH, Li WY. Decoupled compact ultra-wideband MIMO antennas covering 3300–6000 MHz for the fifth-generation mobile and 5GHz-WLAN operations in the future smartphone. *Microwave and Optical Technology Letters*, 2018, 60(10): 2345–2351.
13. Hei YQ, He JG, Li WT. Wideband decoupled 8-element MIMO antenna for 5G mobile terminal applications. *IEEE Antennas and Wireless Propagation Letters*, 2021, 20(8): 1448–1452.
14. Zhang A, Wei K, Zhang Z. Multi-band and wideband self-multipath decoupled antenna pairs. *IEEE Transactions on Antennas and Propagation*, 2023, 71(7): 5605–5615.

15. Chang L, Liu H. Low-profile and miniaturized dual-band microstrip patch antenna for 5G mobile terminals. *IEEE Transactions on Antennas and Propagation*, 2021, 70(3): 2328–2333.
16. Gao Y, Wang J, Wang X, et al. A low-profile broadband multimode patch antenna for 5G mobile applications. *IEEE Antennas and Wireless Propagation Letters*, 2023, 22(5): 1020–1024.
17. Zhang A, Wei K, Hu Y, et al. High-isolated coupling-grounded patch antenna pair with shared radiator for the application of 5G mobile terminals. *IEEE Transactions on Antennas and Propagation*, 2022, 70(9): 7896–7904.
18. Cheng B, Du Z. Dual polarization MIMO antenna for 5G mobile phone applications. *IEEE Transactions on Antennas and Propagation*, 2021, 69(7): 4160–4165.
19. Hu M, Li Y. Wideband back cover microstrip antenna with multiple shorting vias for mobile 5G MIMO applications. *IEEE Transactions on Antennas and Propagation*, 2023, 71(10): 8290–8295.
20. Chen X, Wang J, Chang L. Extremely low-profile dual-band microstrip patch antenna using electric coupling for 5G mobile terminal applications. *IEEE Transactions on Antennas and Propagation*, 2022, 71(2): 1895–1900.
21. Tian X, Du Z. Wideband shared-radiator four-element MIMO antenna module for 5G mobile terminals. *IEEE Transactions on Antennas and Propagation*, 2023, 71(6): 4799–4811.
22. Wong KL, Hong SE, Li WY. Low-profile four-port MIMO antenna module based 16-port closely-spaced 2× 2 module array for 6G upper mid-band mobile devices. *IEEE Access*, 2023, 16: 110796–110808.
23. Wong KL, Jian MF, Li WY. Low-profile wideband four-corner-fed square patch antenna for 5G MIMO mobile antenna application. *IEEE Antennas and Wireless Propagation Letters*, 2021, 20(12): 2554–2558.
24. Wong KL, Kao HC, Li WY. Wideband low-profile eight-port eight-wave annular-ring patch antenna based on using eight dual-shorted dual-resonant ring sectors for 8× 8 MIMO mobile devices. *IEEE Access*, 2022, 11: 18–32.
25. Vol T. Antenna diversity in mobile communications. *IEEE Transactions on Vehicular Technology*, 1987, 36(4): 149–172.
26. Wong KL, Lu JY, Chen LY, et al. 8-antenna and 16-antenna arrays using the quad-antenna linear array as a building block for the 3.5-GHz LTE MIMO operation in the smartphone. *Microwave and Optical Technology Letters*, 2016, 58(1): 174–181.
27. Wang X, Feng Z, Luk KM. Pattern and polarization diversity antenna with high isolation for portable wireless devices. *IEEE Antennas and Wireless Propagation Letters*, 2009, 8: 209–211.
28. Xu H, Gao SS, Zhou H, et al. A highly integrated MIMO antenna unit: differential/common mode design. *IEEE Transactions on Antennas and Propagation*, 2019, 67(11): 6724–6734.
29. Ren Z, Zhao A, Wu S. MIMO antenna with compact decoupled antenna pairs for 5G mobile terminals. *IEEE Antennas and Wireless Propagation Letters*, 2019, 18(7): 1367–1371.
30. Guo J, Cui L, Li C, et al. Side-edge frame printed eight-port dual-band antenna array for 5G smartphone applications. *IEEE Transactions on Antennas and Propagation*, 2018, 66(12): 7412–7417.
31. Hu W, Qian L, Gao S, et al. Dual-band eight-element MIMO array using multi-slot decoupling technique for 5G terminals. *IEEE Access*, 2019, 7: 153910–153920.
32. Wong WY. Conjoined ultra-wideband (2300–6000 MHz) dual antennas for LTE HB/WiFi/5G multi-input multi-output operation in the fifth-generation tablet device. *Microwave and Optical Technology Letters*, 2019, 61(8): 1958–1963.
33. Cui L, Guo J, Liu Y. An 8-element dual-band MIMO antenna with decoupling stub for 5G smartphone applications. *IEEE Antennas and Wireless Propagation Letters*, 2019, 18(10): 2095–2099.
34. Wong SE. High-isolation conjoined loop multi-input multi-output antennas for the fifth-generation tablet device. *Microwave and Optical Technology Letters*, 2019, 61(1): 111–119.

35. Zhao A, Ren Z. Wideband MIMO antenna systems based on coupled-loop antenna for 5G N77/N78/N79 applications in mobile terminals. *IEEE Access*, 2019, 7: 93761–93771.
36. Sun L, Li Y, Zhang Z, et al. Wideband 5G MIMO antenna with integrated orthogonal-mode dual-antenna pairs for metal-rimmed smartphones. *IEEE Transactions on Antennas and Propagation*, 2020, 68(4): 2494–2503.
37. Hu W, Liu X, Gao S, et al. Dual-band ten-element MIMO array based on dual-mode IFAs for 5G terminal applications. *IEEE Access*, 2019, 7: 178476–178485.
38. Xu Z, Deng C. High-isolated MIMO antenna design based on pattern diversity for 5G mobile terminals. *IEEE Antennas and Wireless Propagation Letters*, 2020, 19(3): 467–471.
39. Deng C. Compact broadband multi-input multi-output antenna covering 3300 to 6000 MHz band for 5G mobile terminal applications. *Microwave and Optical Technology Letters*, 2020, 62(10): 3310–3316.
40. Deng C, Liu D, Lv X. Tightly arranged four-element MIMO antennas for 5G mobile terminals. *IEEE Transactions on Antennas and Propagation*, 2019, 67(10): 6353–6361.
41. Deng C, Cao X, Li D, et al. Compact dual-band MIMO antenna with shared decoupling structure for 5G mobile terminals. *IEEE Antennas and Wireless Propagation Letters*, 2023, 22(6): 1281–1285.

CHAPTER 6

Millimeter-Wave Phased Array Antenna Design for 5G Mobile Terminals

MILLIMETER-WAVE BANDS ARE NEWLY opened frequency bands for 5G mobile communications, aiming to radically improve spectrum congestion. Typical millimeter wave commercial planning bands include N257 (26.5–29.5 GHz), N258 (24.25–27.5 GHz), N260 (37–40 GHz) and N261 (27.5–28.35 GHz). Considering the implementation difficulty, millimeter-wave bands are still not commercialized on a large scale in China, and many topics need to be further explored and demonstrated, thus researching the key technologies of millimeter-wave bands is of great significance to promote the rapid development of 5G mobile communications. This chapter will introduce the key technologies of antennas for 5G Mobile Terminals based on millimeter-wave band and give typical beam scanning phased array antenna design examples.

6.1 MILLIMETER-WAVE BAND CHARACTERISTICS AND COMMUNICATION REQUIREMENTS

The typical application scenario of millimeter-wave band is shown in Figure 6.1. A large-scale antenna array is deployed at the base station end, which can generate the corresponding number of beams and beam pointing according to the number and location of users. The antennas of mobile terminals also have beam scanning capability and can adjust the beam pointing in real time according to the position and attitude of the terminal. Antennas, as key components of beam modulation, play the role of core devices in 5G mobile communication systems. The research of 5G millimeter-wave antennas has received great attention from both academia and industry [1]. China's Antenna System Industry Alliance has been organizing the 5G Antenna and RF Technology Summit since 2018, focusing on industrial solutions for 5G antennas. Although antenna standards for millimeter-wave bands continue to evolve, the industry has reached some consensus on 5G millimeter-wave antennas.

DOI: 10.1201/9781003770039-6

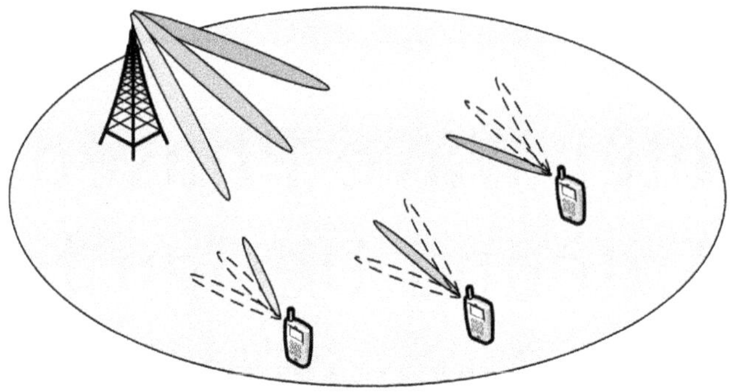

FIGURE 6.1 Typical application scenarios in millimeter-wave band.

The development of mobile terminal antennas is always in line with the development of mobile communication technology. In order to match the millimeter-wave propagation characteristics and key technologies, the mobile terminal antenna needs to take into account the line-of-sight propagation distance and spatial coverage of the signal. This makes the phased array antenna with high gain and flexible beam control capability become a commonly adopted realization scheme in the current millimeter-wave band, and also leads the R&D center of mobile terminal antenna to evolve from the traditional MIMO system in the Sub-6 GHz band to the phased array direction in the millimeter-wave band [2,3].

In addition, the mobile terminal itself is a highly integrated and complex platform and has the attributes of portability and mobility, which bring many challenges to the antenna layout and structure design. On the one hand, the internal space of the mobile terminal is extremely limited, and the continuous rise of the screen ratio of smartphones and the pursuit of users for full-screen and double-sided screen further compress the effective radiation area reserved for the antenna, which significantly increases the difficulty of antenna design. On the other hand, due to the fact that millimeter wave is more sensitive to obstacle blocking, coupled with the arbitrary orientation of mobile terminals, dual-polarized antenna design has more prominent advantages in order to reduce the polarization mismatch of the transceiver signals and to enhance the robustness of the link [4]. In addition, researchers have been devoted to exploring how to effectively reduce the user effect in handheld mobile terminals [5]. Compared with the side-fire, the end-fire can reduce the shielding effect of the user on the antenna radiation, which can help to meet the millimeter-wave antenna's threshold requirement for Equivalent Isotropically Radiated Power [6].

6.2 CURRENT RESEARCH STATUS OF PHASED ARRAY ANTENNAS FOR MILLIMETER WAVE IN MOBILE TERMINALS

According to the installation position of the phased array antenna inside the mobile terminal, the phased array antenna can be categorized into broadside phased array antenna and end-fire phased array antenna. Among them, the broadside phased array antenna is generally integrated in the back cover of the mobile terminal, and the end-fire phased array antenna is generally integrated in the bezel of the mobile terminal, and the combination of the two can realize wider beam coverage. The next section will introduce some typical broadside phased array antenna and end-fire phased array antenna realization methods.

6.2.1 Broadside Phased Array Antenna

The design method of broadside phased array antennas is more mature, and the radiating unit is generally in the form of a microstrip patch antenna. Several millimeter-wave broadside phased array antennas that can be used in 5G Mobile Terminals are shown in Figure 6.2. In Figure 6.2a [7], the antenna unit is in the form of a crossed dipole operating in the 28 GHz band, made by the low temperature co-fired ceramic process, and both orthogonally polarized beams can be beam-scanned at −45° to 45°; the unit shown in Figure 6.2b is a multilayer stacked dual-band microstrip patch antenna [8], where the high

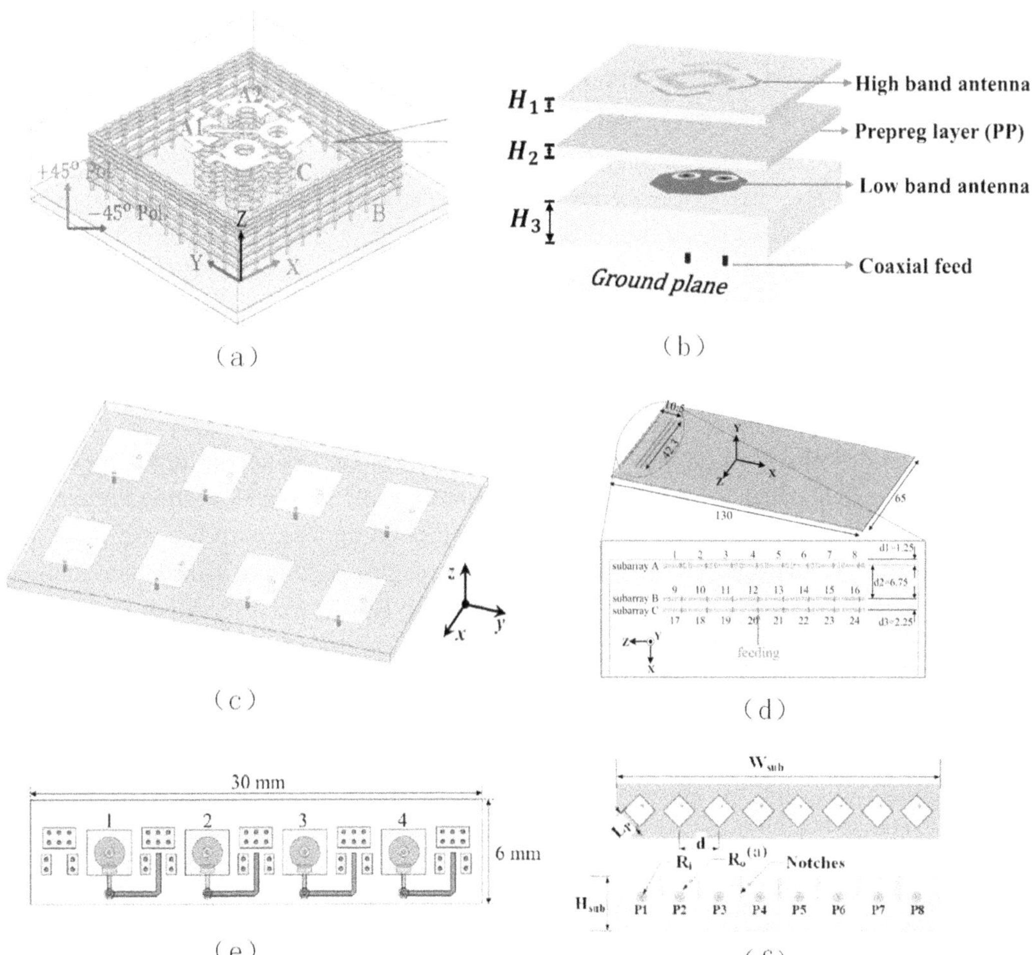

FIGURE 6.2 Several millimeter-wave broadside phased array antennas that can be used in 5G mobile terminals (a) crossed dipole (b) multilayer stacked dual-band microstrip patch antenna (c) rectangular patch (d) linear slit array (e) capacitively coupled feed square patch (f) 45° rotating patch.

band is generated by the ring microstrip in the upper layer, and the low band is generated by the rectangular tangent microstrip patch antenna in the lower layer, which can operate simultaneously in the 26 and 38 GHz bands; the cell shown in Figure 6.2c is a rectangular patch [9], which is used to expand the bandwidth by excitation of the zeroth-order modes through the shorted columns; and the cell shown in Figure 6.2d consists of three sets of linear slit arrays [9] etched on the floor. The design shown in Figure 6.2d contains three groups of linear slot arrays etched in the ground plane [10]. Since the slot elements have a bidirectional radiation characteristic, each array can achieve wide-beam scanning. The cell shown in Figure 6.2e employs capacitively coupled fed square patches [11], and each cell is internally loaded with a circular patch, the center of which is connected to the gesture recognition sensor. The patch cell shown in Figure 6.2f reduces the inter-cell inter-coupling by rotating the cell by 45°, and the main beam can be extended by short-circuit column excitation of zero-order modes. Inter-coupling and the main beam can realize beam scanning at 0°–60° [12].

6.2.2 End-Fire Phased Array Antennas

The realization forms of end-fire phased array antennas are more diverse than those of broadside phased array antennas, which attract a large number of researchers to study, and many novel designs of dual-polarized end-fire phased array antennas have been proposed. These research works can be mainly classified into two categories: one is to integrate the antenna with a metal bezel, which is used as a part of the radiator; and the other is to construct a dual-polarized end-fire phased array antenna with a multilayer printed-circuit board (PCB) stack.

Figure 6.3 shows several millimeter-wave dual-polarized end-fire phased array antennas integrated with a metal bezel. The two sets of cross-shaped slit arrays etched into the metal bezel with ±45° inclination are shown in Figure 6.3a [13].; The cell shown in Figure 6.3b is a square microstrip patch antenna with ±45° inclination [14]. The cell shown in Figure 6.3c combines horizontally-polarized and vertically-polarized dipoles, with the two polarizations designed independently [15], The cell shown in Figure 6.3d is an annular slit using a metal back cavity to form the end-fire radiation pattern [16]. The cell shown in Figure 6.3e is in the form of a chain groove [17], with the horizontal and vertical polarizations radiating in different intervals, and all the cells are connected as one long groove. The cell shown in Figure 6.3f is a bowtie-shaped slit tilted at ±45° [18], with a metal cavity loaded behind the slit for end-shooting. All these arrays produce dual-polarized end-fire beams with the advantage of low profile. However, the height of the metal bezel increases due to the need for the metal bezel to support the dual polarization.

Constructing an end-fire phased array antenna using multilayer PCB stacking can reduce the height of the metal bezel, and the related design can be found in the References [19–24].

6.2.3 Integration of Millimeter-Wave Antennas with Low-Band Antennas

The millimeter-wave band corresponds to a much smaller wavelength compared to the Sub-6 GHz band, and this difference makes it possible to integrate millimeter-wave antennas

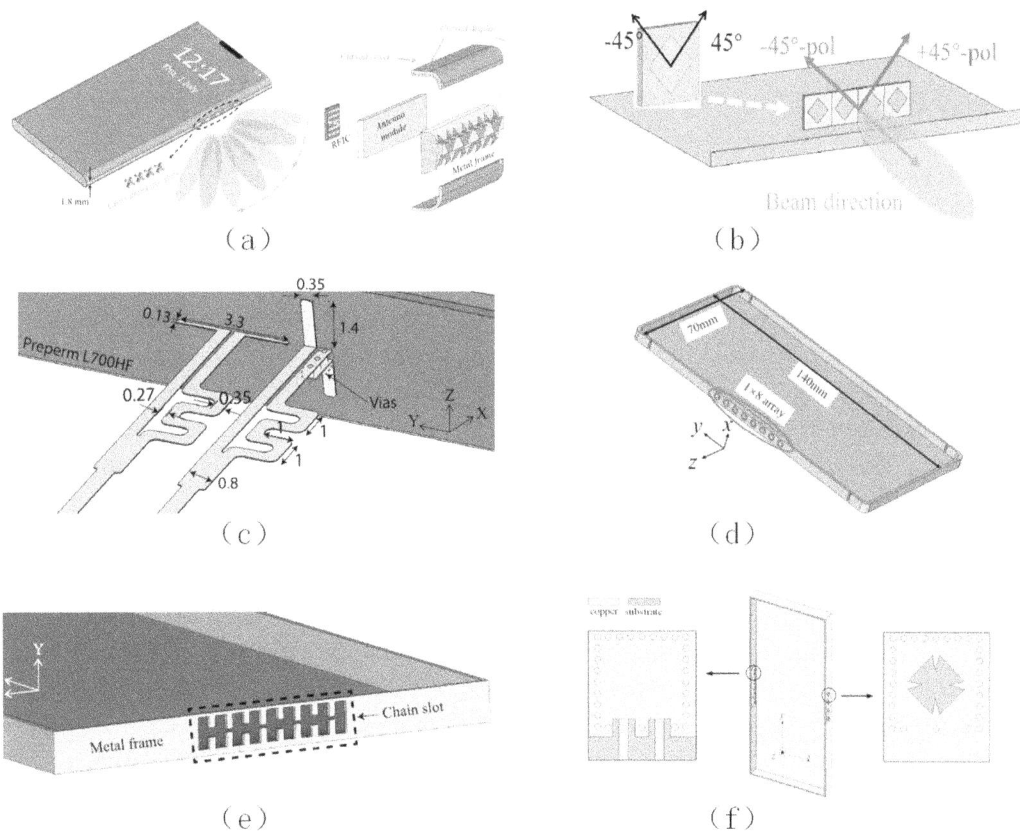

FIGURE 6.3 Several millimeter-wave dual-polarized end-fire phased-array antennas integrated with a metal bezel (a) cross-shaped slit (b) square microstrip patch antenna that is±45° tilted (c) horizontally polarized and vertically polarized dipole (d) ring slit (e) chain slot (f) bow-tie shaped slit.

into low-band antennas for more compact structural designs [25–29]. Several design examples of millimeter-wave antenna arrays integrated with Sub-6 GHz antennas are given in the reference. For example, a millimeter-wave antenna array is nested in the ground clearance of a low-band Inverted-F Antenna (IFA); the array contains four units, each of which is a binary Yagi antenna; the millimeter-wave antenna array can produce end shots in the 22–31 GHz band and the beam can be beam-scanned at −50° to 50°, and the IFA can cover the 740–960 MHz and the 1700–2200 MHz bands [25]. The antenna array integrates a long slot and two short slots [26], the long slot is used to cover the 1780–2620 MHz band, and the two short slots can be excited by a multiport to form a beam-scanning antenna in the 26–30 GHz band. A low-frequency band antenna is designed using a monopole configuration [27] (capable of operating in the 800 MHz and 2000 MHz bands), while a four-element Vivaldi antenna array is employed to form a millimeter-wave antenna array (enabling beam scanning in the 28 GHz band). The slot antenna resonates in a half-wavelength mode at low frequencies [28]. Varactor diodes are loaded in the slot to dynamically tune the resonant frequency over the 2–2.7 GHz band. This structure is reused at millimeter-wave frequencies, where beam scanning is enabled by adding multiple feed ports. The gap between the metal bezel and the ground plane is utilized to design the low-frequency antenna [29], covering the 700–980 MHz and 1700–2920 MHz bands. The millimeter-wave antenna is a folded patch array suspended above the low-frequency antenna, enabling beam scanning from 45° to 135°, while maintaining high isolation between the low-frequency and millimeter-wave antennas.

6.3 DUAL POLARIZATION END-FIRE PHASED ARRAY ANTENNA BASED ON POLARIZATION SYNTHESIS

The concept of dual-polarized end-fire phased array antenna proposed in this section originates from the combination of two circularly polarized waves [30–37]. According to the properties of electromagnetic waves, any linearly polarized wave can be decomposed into two circularly polarized waves of equal amplitude and opposite rotation. In order to obtain a specific polarization of the line polarized wave, a pair of left-spin and right-spin circularly polarized waves can be vectorially superimposed.

6.3.1 Antenna Unit Design

Based on the above analysis, the antenna is designed using a waveguide polarizer with slanted slots, and the working principle of dual polarization is shown schematically in Figure 6.4. The waveguide polarizer can be regarded as a three-port network, including two input *ports* P_{in1}, P_{in2} and one output port P_{out}. When the two input ports separately excite a line-polarized wave, the output port can generate a left-handed circularly polarized wave or a right-handed circularly polarized wave, respectively. In addition, when the two input ports do not work independently, but are excited by both in-phase and inverted signals, the output port is also capable of generating vertically and horizontally polarized dual line-polarized waves.

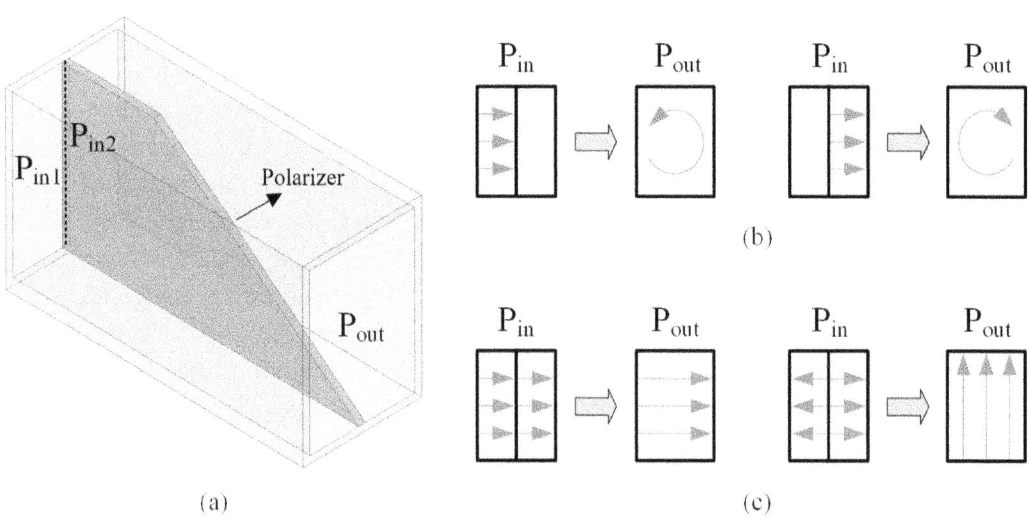

FIGURE 6.4 Schematic of the operating principle of dual polarization (a) typical structure of a waveguide polarizer (b) generation of double circularly polarized wave (c) generation of double linearly polarized wave.

Considering the more complicated processing of traditional 3D metal waveguides, a stepped slot antenna based on substrate integrated waveguide (SIW) is used as the reference antenna for the subsequent design, as shown in Antenna 1 in Figure 6.5a. The antenna consists of a two-layer stacked PCB with the stepped slot etched in the middle metal layer. Compared to the classical slant slot structure, the multilevel stepped slots provide more degrees of freedom in parameter tuning. However, the impedance matching and radiation characteristics of antenna 1 are closely related to the length of the dielectric loaded at its front end, and a longer dielectric loading is usually required to achieve a smooth transition from guided to free-space waves in the SIW, but results in a larger floor ground clearance.

The design flow of the metal ring-loaded dual-polarized end-fire phased array antenna is shown in Figure 6.5. The floor ground clearance width for each of the three cases shown in Figure 6.5 is 1.5 mm (approximately 0.14 λ_0, with λ_0 being the free-space wavelength corresponding to the 28 GHz frequency). On top of antenna 1, a closed metal ring is loaded onto the radiating port face of the antenna, as shown in antenna 2 in Figure 6.5b. This metal ring consists of metal strips printed on the top and bottom surfaces of the antenna

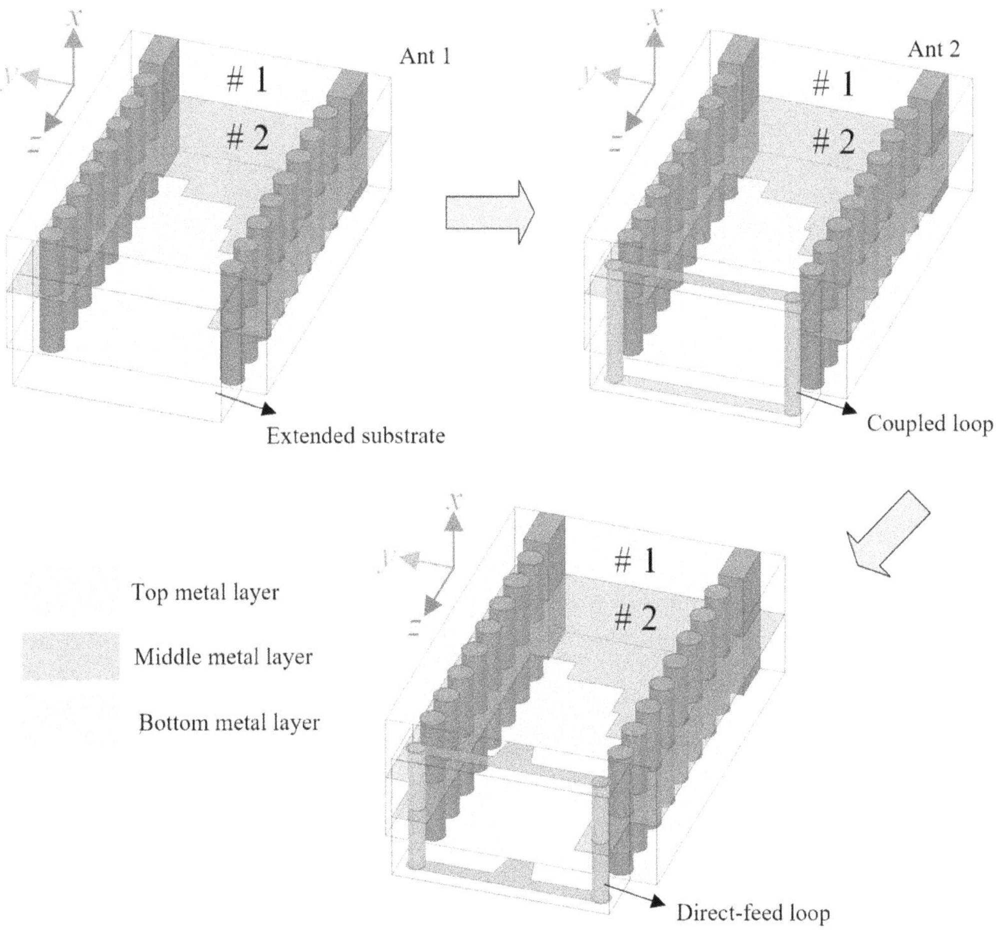

FIGURE 6.5 Design flow of the dual-polarized end-fire phased-array antenna loaded with metal ring (a) antenna 1 (b) antenna 2 (c) this design.

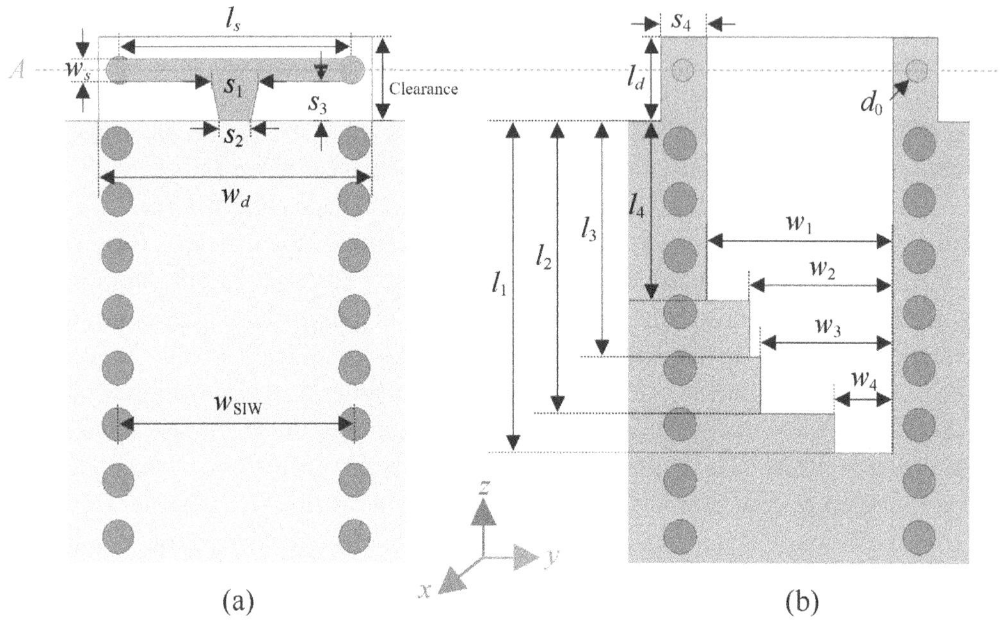

FIGURE 6.6 Structural details of the metal ring loaded dual polarized end-fire phased array antenna (a) top/bottom metal (b) middle metal.

and metal perforations on both sides. By optimizing the dimensions, the antenna can be made to resonate in the operating frequency band. Two pairs of metal connecting wires are introduced to connect the metal ring to the feed structure. Since the SIW is a balanced two-wire transmission line, the introduced connecting wires can convert the coupled excitation mode of the metal ring in Antenna 2 to a directly connected differential excitation mode, as shown in Figure 6.5c.

The structural details of the metal ring-loaded dual-polarized end-fire phased array antenna are shown in Figure 6.6, and the dielectric substrate used is a Rogers RO4003C with a thickness of 1.524 mm ($\varepsilon_r = 3.55$, $\tan\delta = 0.0027$). The width and height of the SIW feed network were adjusted to ensure that the two orthogonal modes, TE_{10} and TE_{01}, were excited in the same frequency band. The metal layer located in the middle of the two dielectric substrates effectively suppresses the energy leakage of the TE_{01} mode in the conventional SIW structure, resulting in low insertion loss for both modes. The diameter of the metal vias shown in Figure 6.6 is 0.6 mm, and the spacing between adjacent vias is 1 mm.

6.3.2 Working Mechanism and Performance

For the case of phased array or multiport excitation, the active S-parameter is an important parameter to characterize the reflection properties of the active port or cell when complex coupling effects are involved. The simulation results of the active S-parameters and gain for different antenna structures are shown in Figure 6.7 and exhibit the performance when the two input ports are fed with in-phase and inverted signals. It can be seen that antenna 1 exhibits severe impedance mismatch and low gain. This is mainly due to the more severe discontinuity effect at the interface between the medium and air at the small ground clearance

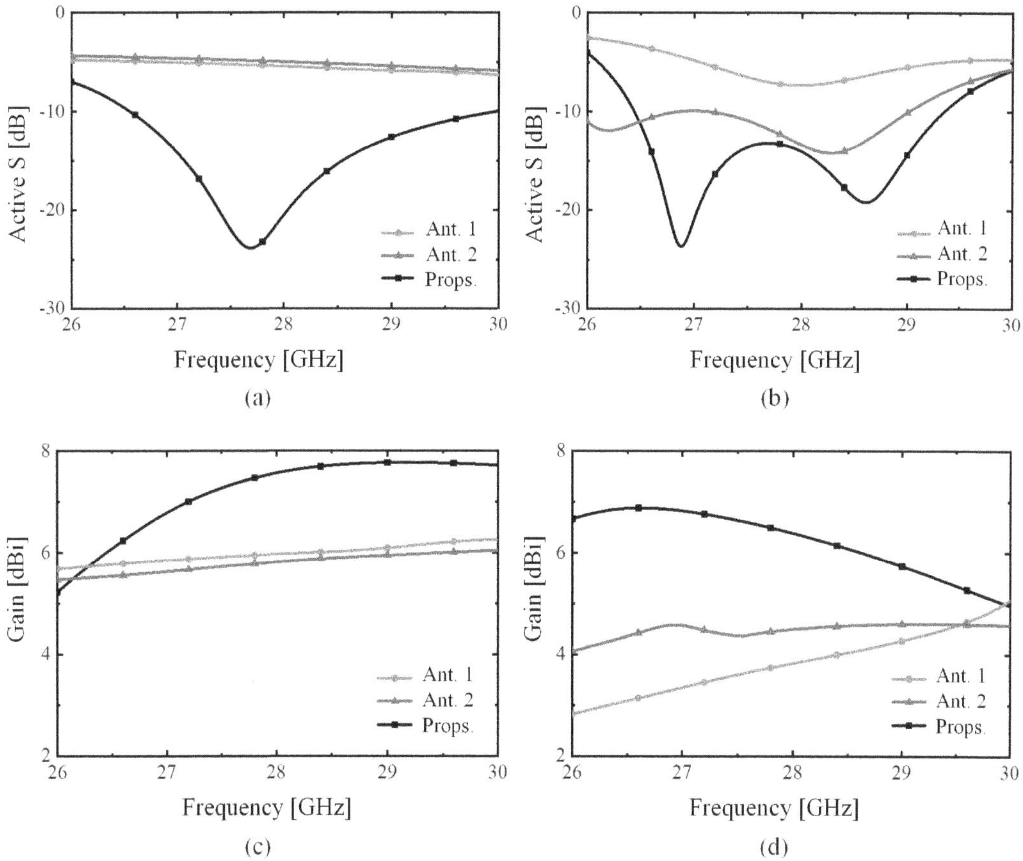

FIGURE 6.7 Simulation results of active S-parameters and gain for different antenna configurations (a) S-parameters for in-phase excitation (vertical polarization) (b) S-parameters for inverse excitation (horizontal polarization) (c) radiation gain for in-phase excitation (d) radiation gain for inverted-phase excitation.

width of 1.5 mm. With the introduction of a metal ring in antenna 2, the impedance matching for horizontal polarization becomes better, and the gain is improved. However, the structure has no significant effect on vertical polarization. In order to effectively excite the metal ring, four (two pairs) of metal connecting wires are added in this design, which significantly optimizes the reflection coefficient and gain for both polarizations. By optimizing the size of the connecting wires, the resonant frequencies of the different polarizations can be adjusted. The simulated −10 dB relative bandwidth reaches 10% (26.55–29.35 GHz). The average in-band radiation gains for vertical and horizontal polarization are 7.34 and 6.34 dBi, achieving a gain improvement of 1.38 and 2.54 dB, respectively, compared to antenna 1.

In order to clarify the working mechanism of the dual-polarized end-fire phased array antenna, the electric field distribution in the cross-section of the metal ring (cross-section identified as A and A', as shown in Figure 6.8) is given as shown in Figure 6.8. When both in-phase and anti-phase excitations are applied, orthogonal electric fields in the vertical and horizontal directions can be observed, and both polarizations share the same radiation

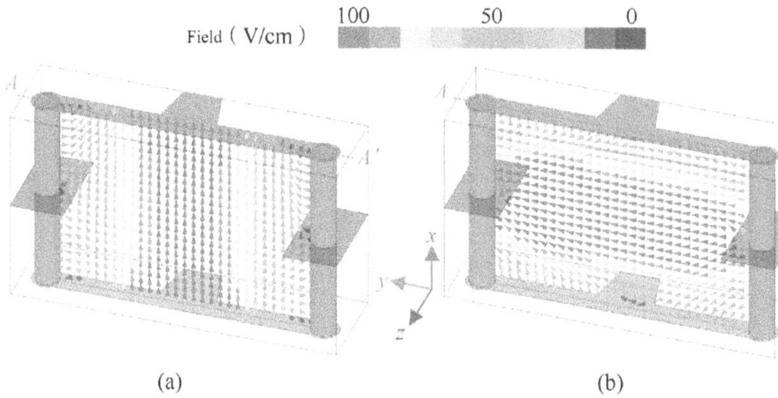

FIGURE 6.8 Electric field distribution in the cross section of a metal ring (a) vertical polarization (b) horizontal polarization.

aperture. Both the metal strip and the metal aperture operate in the one-half wavelength mode. For the vertical polarization, the resonance mainly depends on the loaded metal perforations, while the metal strips and connecting wires on the top and bottom surfaces of the antenna act as a set of balanced feed lines to direct the current conduction. For horizontal polarization, the metal perforations and the connecting wires in the middle layer form a balanced feed line whose resonance mainly depends on the metal strips.

The two pairs of connecting wires introduced are the key to achieve the dual polarization performance. Here, the effects of the widths s_1 and s_4 of the connecting lines on the active S-parameters of the two polarizations are investigated. As shown in Figure 6.9, the two pairs of connecting lines act as impedance converters, and s_1 affects the reflection coefficients only for the vertical polarization, while s_4 affects the reflection coefficients only for the horizontal polarization. The parameter scan results show that the two pairs of connecting wires are able to control the impedance of both polarizations independently.

Figure 6.10 shows the normalized radiation pattern and 3D radiation pattern of the antenna at 28 GHz. Both polarizations present symmetric radiation direction plots, and the backward radiation of vertical polarization is slightly stronger than that of horizontal polarization. The simulated gains of the two polarizations are 7.57 and 6.39 dBi, respectively, and the reasons for the gain difference can be attributed to two aspects: First, due to the inconsistency of the antenna's mouth width and height, the gains of antenna 1 in the two polarizations are already different, which indicates that the radiation characteristics of the TE_{10} and TE_{01} modes are different; second, the different radiating regions on which the two polarizations depend lead to a slight gain difference. In addition, due to the balanced excitation scheme, the antenna can achieve a cross-polarization discrimination of up to 32 dB.

6.3.3 Dual-Polarized Multibeam Endfire Phased Array Antenna Array Columns

A 1×4 end-fire phased array antenna array is constructed based on the antenna units designed in Section 6.3.2. The array layout and feed port settings are shown in Figure 6.11. Among them, the spacing between neighboring antenna units is 6.5 mm, and this value comprehensively considers the beam scanning range and unit coupling.

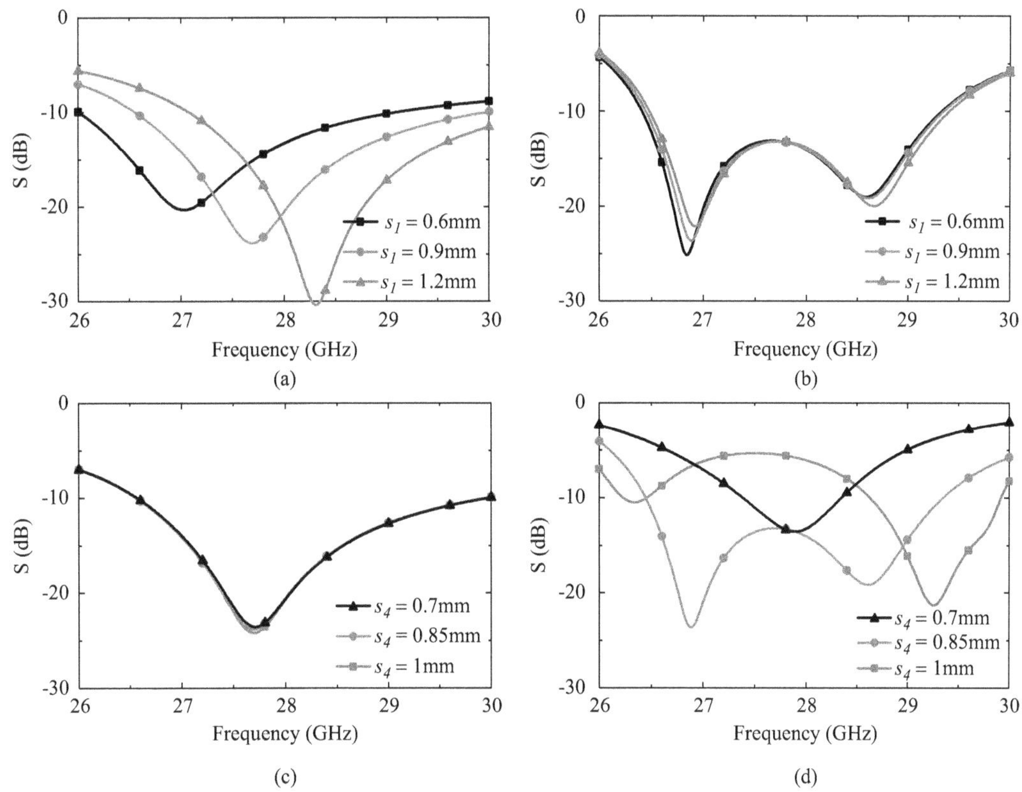

FIGURE 6.9 Parametric analysis of the antenna (a) s_1 (vertical polarization) (b) s_1 (horizontal polarization) (c) s_4 (vertical polarization) (d) s_4 (horizontal polarization).

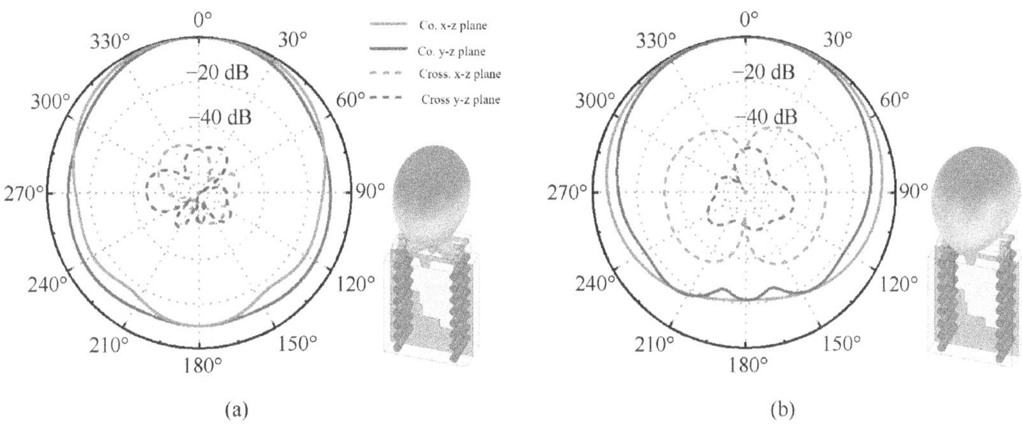

FIGURE 6.10 Normalized radiation pattern and 3D radiation pattern of the antenna at 28 GHz (a) vertical polarization (b) horizontal polarization.

In order to further analyze the working performance of the array in the phased array regime, the active S-parameter simulation results of the array at each port under two phase gradients, 45° and 135°, are given (for symmetry considerations, the results of only half

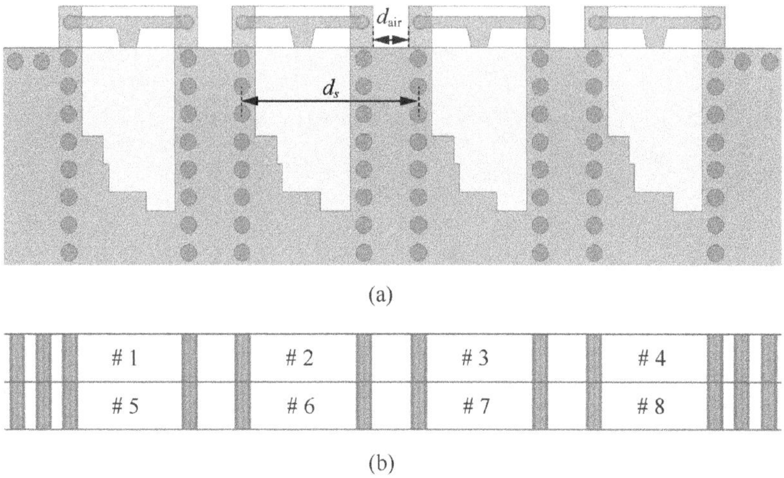

FIGURE 6.11 Array layout and feed port settings (a) top view (b) cross-section of the SIW feed network.

of the ports are given), as shown in Figure 6.12. The reflection coefficient curves of each port remain relatively stable, indicating that the array is able to maintain good impedance matching at different beam scanning angles. The beam scanning direction diagram of the dual-polarized multi-antenna end-fire phased array antenna array is shown in Figure 6.13. Under different phase gradients, the beams of the two polarizations point in the same direction, and the cross-polarization discrimination is higher than 25 dB.

5G Mobile Terminals millimeter wave antenna arrays are generally directly integrated with multi-channel transceiver chips to form Antenna-in-Package antennas. However, the process of this method is more complicated to realize, and there are two simplified methods: The first is to design a multibeam feed network to generate multiple fixed-pointing beams; the second is to use the cell radiation pattern synthesis method. The first method is used here to verify the performance of the array and a passive Butler feed network is designed.

The geometry of the antenna array and the SIW feed network is shown in Figure 6.14. The feed network can be divided into three parts, including four 180° hybrid couplers placed in parallel, a 4×4 Butler matrix, and the waveguide converter. Among them, the hybrid couplers provide the in-phase and inverted signals needed to excite the dual-polarized multi-antenna, while the Butler matrix provides the different phase gradients needed for the multibeam. The two are cascaded sequentially to form a dual-polarized multibeam feed network. Ports 1–4 and 5–8 are located at the top and bottom of this feed network to feed the vertically and horizontally polarized excitation signals, respectively.

The simulation results of the Butler matrix are shown in Figure 6.15. The average values of the reflection coefficient and mutual coupling in the 26.5–29.5 GHz band are lower than –22 dB. When the four input ports are excited separately, the adjacent output ports can obtain four stable linear phase gradients, which are 45°, –135°, 135°, and –45° in that order, and the phase fluctuation is less than 4.3°. In addition, waveguide right-angle conversion can be realized by inductive metal posts.

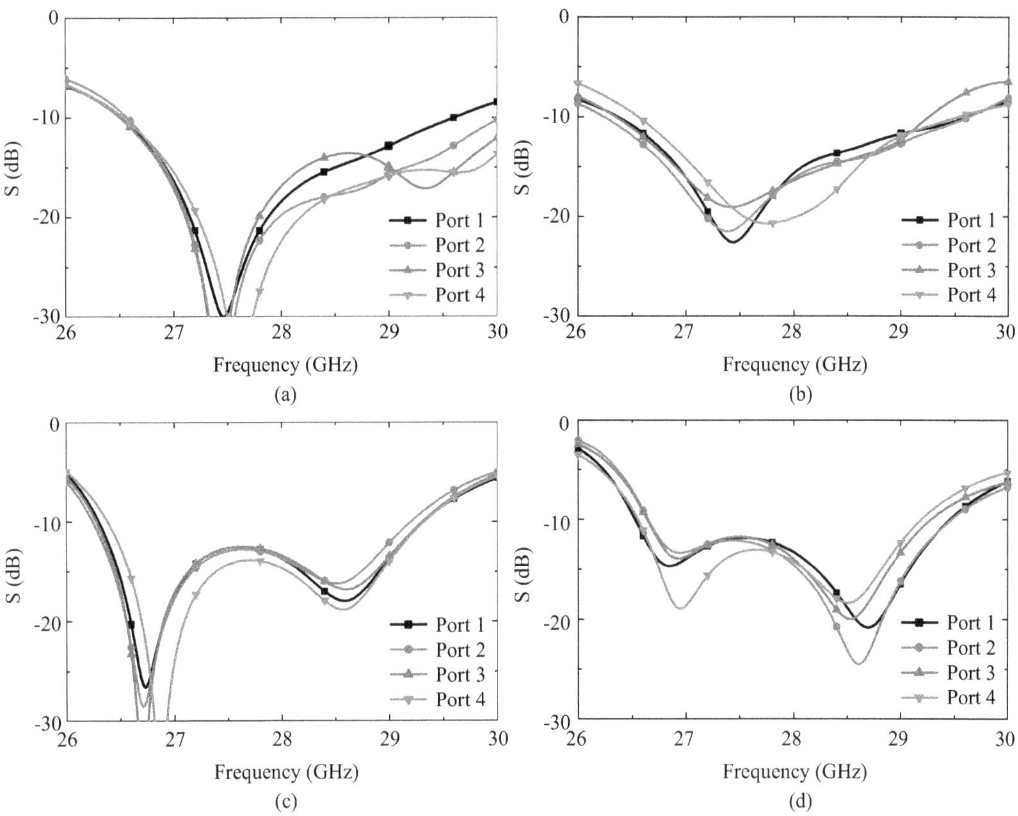

FIGURE 6.12 Active S-parameter simulation results of each port under two phase gradients of 45° and 135° (a) vertical polarization at 45° phase gradient (b) vertical polarization at 135° phase gradient (c) horizontal polarization at 45° phase gradient (d) horizontal polarization at 135° phase gradient.

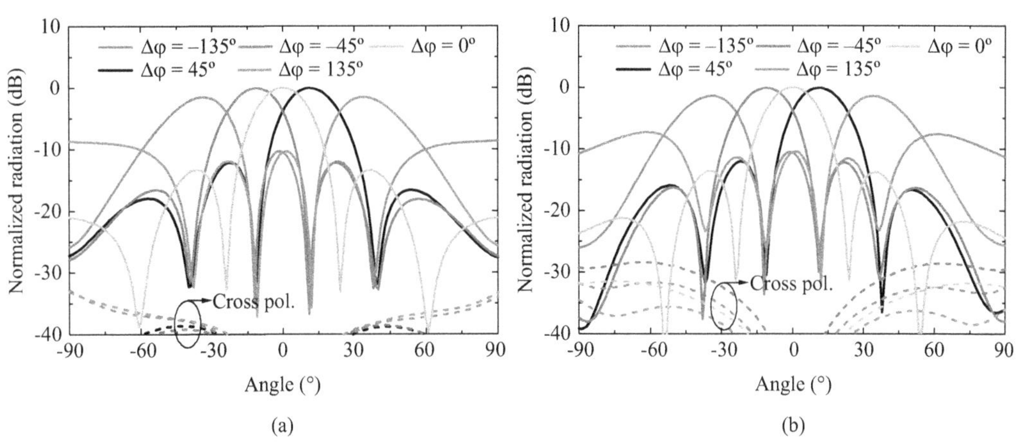

FIGURE 6.13 Beam scanning direction plots for different phase gradients (solid and dashed lines represent main and cross polarization, respectively) (a) vertical polarization (b) horizontal polarization.

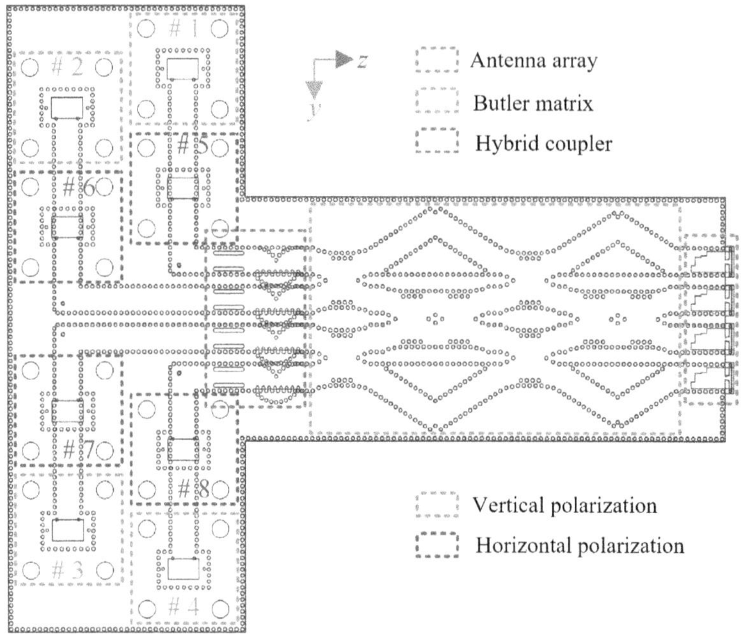

FIGURE 6.14 Geometry of the antenna array and SIW feed network.

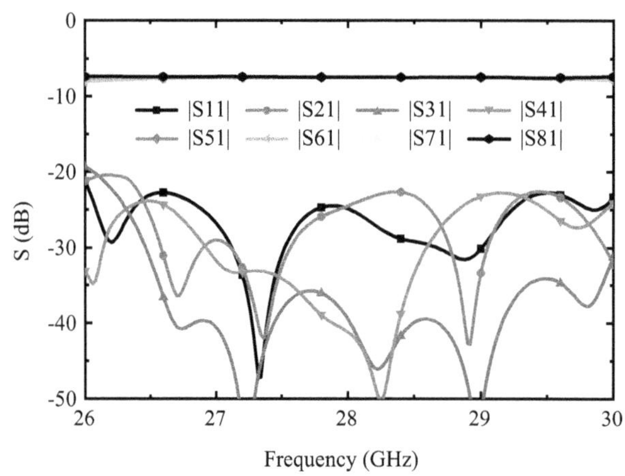

FIGURE 6.15 Simulation results of Butler matrix.

6.3.4 Simulation and Test Results

The design in Section 6.3.3 was tested by actual processing, utilizing the standard PCB process to fabricate the upper and lower dielectric substrates separately, and then tightly pressing the dielectric substrate through the fastening screws. The physical antenna after assembly is shown in Figure 6.16. Four waveguide-coaxial adapters are laid out on both sides of the antenna for easy testing. The overall size of the antenna is 124.5 mm × 105 mm × 3.048 mm.

The simulation and test results of the reflection coefficient of each port are shown in Figure 6.17. Due to the symmetry of the antenna structure, the results of the remaining four input ports are similar to Figure 6.17. The simulation and test results are highly consistent.

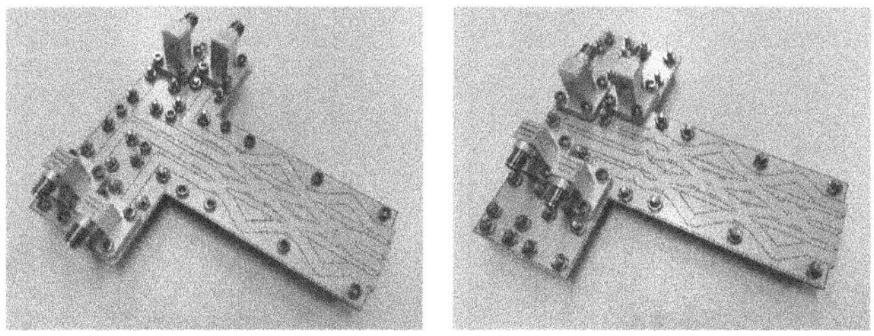

(a) (b)

FIGURE 6.16 Physical antenna after assembly (a) top view (b) elevation view.

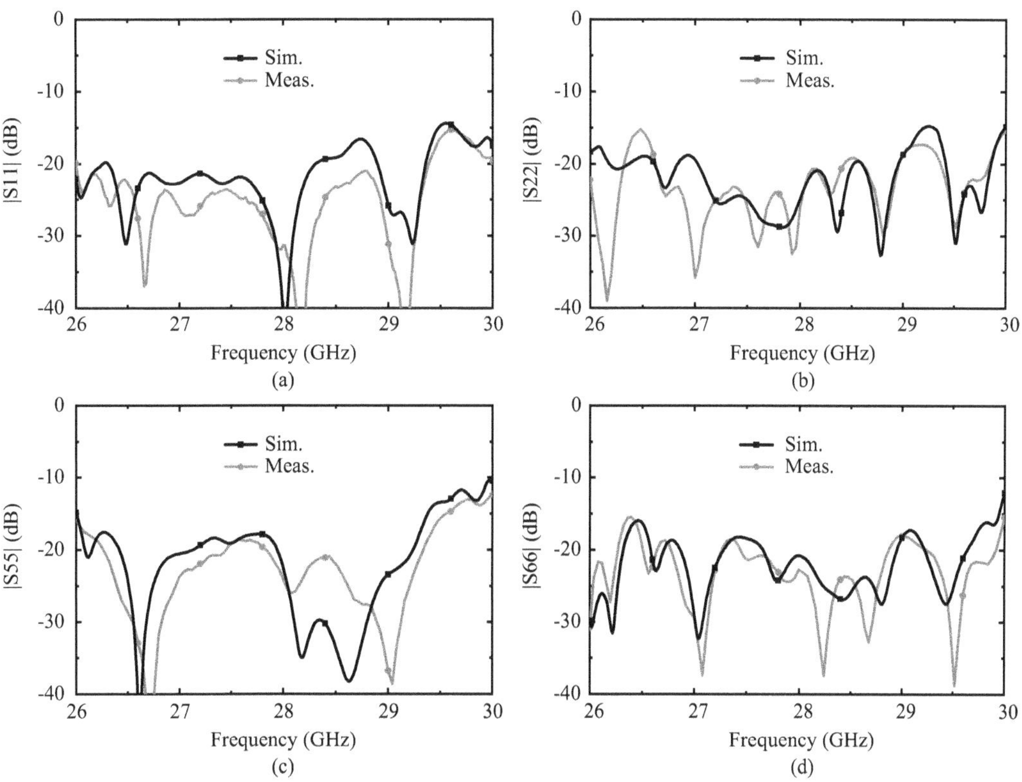

FIGURE 6.17 Simulation and test results of reflection coefficients of each port (a) S_{11} (b) S_{22} (c) S_{33} (d) S_{44}.

In the whole frequency band, the return loss of all ports is higher than 10 dB. The simulation and test results of port mutual coupling are shown in Figure 6.18. In the 26.57–29.34 GHz band, the simulated port isolation is higher than 15 dB, and in the test results, the port isolation is higher than 15 dB in the 26.5–29.5 GHz band. The deviation between the simulation and test results may come from the error in the actual machining versus manual assembly.

The simulation and test results of the beam scanning radiation pattern of the antenna array at 28 GHz are shown in Figure 6.19. The test results are able to match the simulation

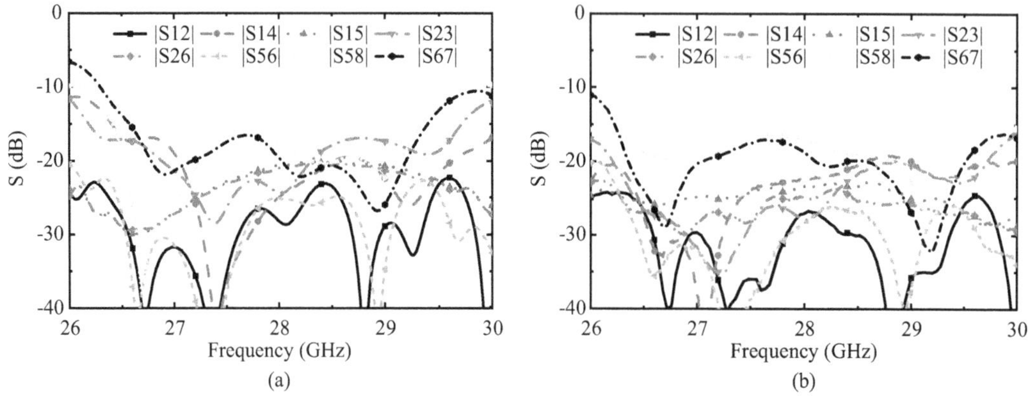

FIGURE 6.18 Simulation and test results of port mutual coupling (a) simulation results (b) test results.

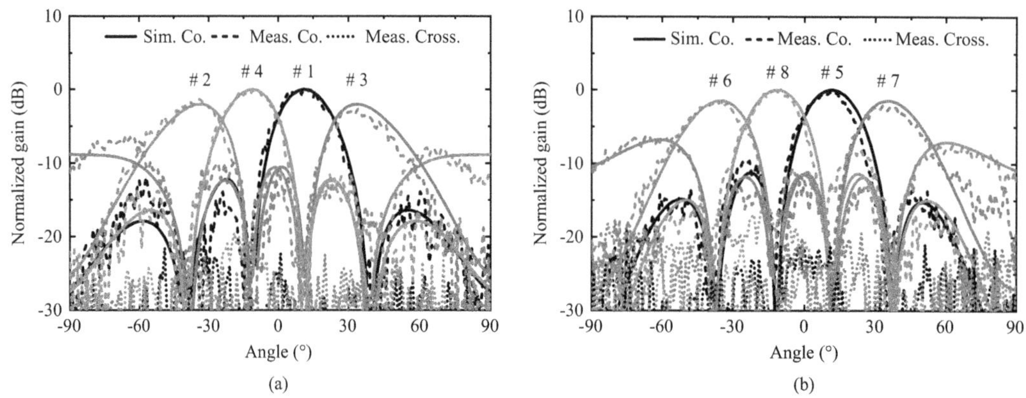

FIGURE 6.19 Simulation and test results of the beam scanning radiation pattern of the antenna array at 28 GHz (a) vertical polarization (b) horizontal polarization. Note: #1–#8 denotes beams with different pointing directions.

results well. Under the gain fluctuation of 3 dB, the range of azimuth angle that can be covered by all four directional beams of both polarizations is −48° to 48°. The cross-polarization discrimination of the tested vertical and horizontal polarizations reaches 19 and 16 dB, respectively.

Figure 6.20 shows the simulated and tested gains obtained when port 1 and port 5 are excited, respectively. The gain fluctuation in the passband can be kept relatively stable, and the peak gain of the test reaches 9.38 dBi for vertical polarization and 9.51 dBi for horizontal polarization, where the insertion loss of the waveguide-coaxial adapter has been calibrated by means of back-to-back measurements.

A comparison of the key metrics of this design with those designed in the reference is shown in Table 6.1. It can be seen that the antenna in this design has the minimum floor clearance width. Due to the effect of the floor at the end of the antenna, the antenna radiation direction diagrams in references [19,20] all exhibit asymmetric tilted beams. The cell cross-polarization discrimination is higher in this design compared to references

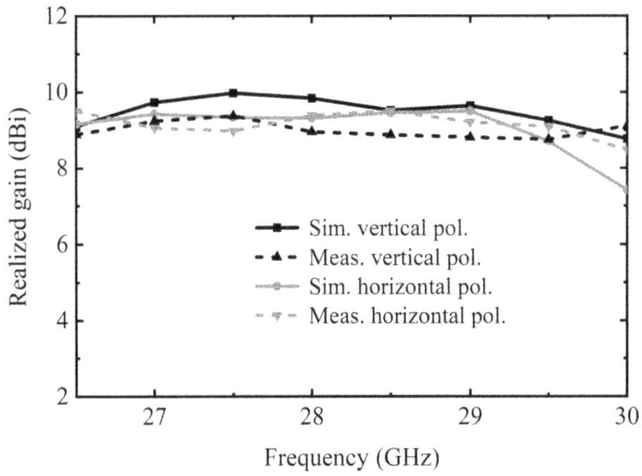

FIGURE 6.20 Simulated and tested gains when port 1 and port 5 are excited respectively.

TABLE 6.1 Comparison of key Metrics of this Design with those Designed in References

References	Array Form	Number of Layers of Dielectric Substrate	Relative Bandwidth	Beam Symmetry or Not	Unit cross-Polarization Discrimination (dB)	Array Gain (dBi)	Floor Ground Clearance Width (mm)
Reference [19]	1×4 (parallel feed)	3	25%	No	11.1	11 (average)	12.8
Reference [31]	1×8 (parallel feed)	3	18%	Yes	40	16.1 (max)	2.86
Reference [32]	1×4 (phased array)	2	7%	Yes	10	8.14 (max)	2.7
Reference [20]	1×4 (phased array)	4	19%	No	5.7	9.27 (max)	1.8
Reference [33]	1×4 (multibeam)	2	11.3%	Yes	20	9 (average)	7
This design	1×4 (multibeam)	2	10%	Yes	32	9.51 (max)	1.5

[19,20,32,33]. The dual-polarized magneto-electric dipole in the reference [31], although having a larger bandwidth, requires a larger floor clearance width and triple dielectric substrate support. It is worth pointing out that the excitation scheme in this design is not limited to passive beam fouling techniques, but other techniques can also be applied to generate the desired feed signal. The gain of the array can be further enhanced when the excitation scheme in this design is applied to a phased array regime or a low-loss beam fouling network.

6.4 DUAL-POLARIZED SMALL GROUND CLEARANCE END-FIRE PHASED ARRAY ANTENNA

In this section, a dual-polarized small ground clearance end-fire phased array antenna [38] is proposed to construct a horizontally polarized and vertically polarized radiating structure through multilayer PCB stacking to achieve a compact array design.

6.4.1 Antenna Array Structure

The array layout of the dual-polarized small ground clearance end-fire phased array antenna is shown in Figure 6.21, which consists of two layers of dielectric substrate compactly pressed vertically through the middle semi-cured sheet, both of which are modeled as Rogers RO4003C ($\varepsilon_r = 3.55$, $\tan\delta = 0.0027$), and the semi-cured sheet is modeled as Rogers RO4450F ($\varepsilon_r = 3.52$, $\tan\delta = 0.004$), and the sheet thicknesses h_s and h_p are 0.813 mm and 0.102 mm, respectively. Three metal layers (M_1–M_3) are sequentially distributed on the upper surface of dielectric substrate 1, the upper surface of dielectric substrate 2, and the lower surface of dielectric substrate 2. Four surface-mounted Mini-SMP connectors are mounted on M_1 and M_3, respectively, which are going to be used for antenna array feeding. As shown in Figure 6.21b, the vertical polarization port is defined as #1–#4, and

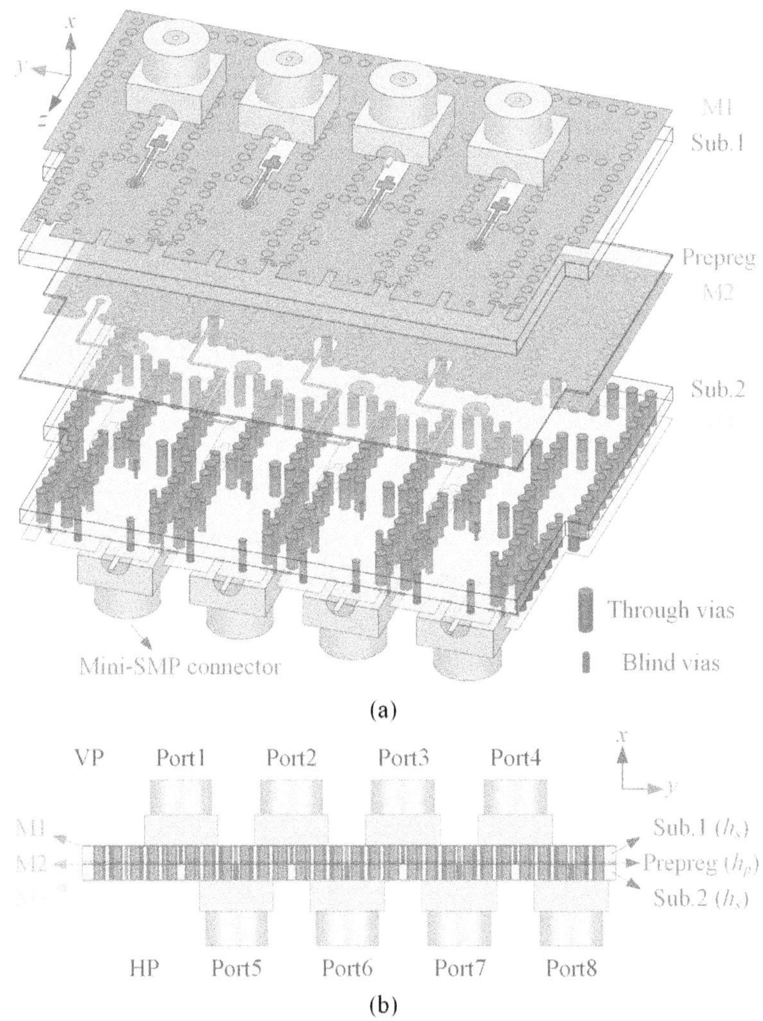

FIGURE 6.21 Array layout of dual-polarized small ground clearance end-fire phased array antenna (a) exploded view (b) side view.

the corresponding feeder blind holes connect M_1 to M_2; the horizontal polarization port underneath is defined as #5–#8, and the feeder blind holes connect M_2 to M_3; and the rest of the metal perforations are running through the whole antenna structure to realize the electrical connection of the three metal layers.

The detailed geometry of the array layers is shown in Figure 6.22. The vertical polarization is generated by the excitation of four pairs of top-hat dipoles loaded at the SIW aperture. The combination of metal apertures and rectangular patches (M_1, M_2, M_3) reduces the size of the dipole units and improves the overall compactness of the structure. Additional matching posts are provided at distance l_3 from the end of the SIW to extend the operating bandwidth of the vertical polarization. Each vertically polarized unit is excited by a capacitive metal disk loaded metal blind aperture and connected to the top Mini-SMP connector through a coplanar waveguide conversion structure.

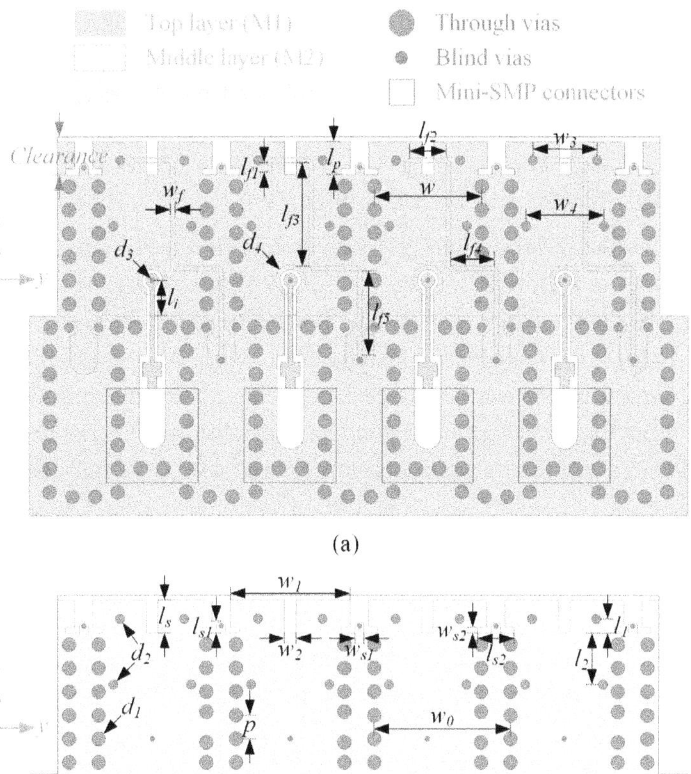

(a)

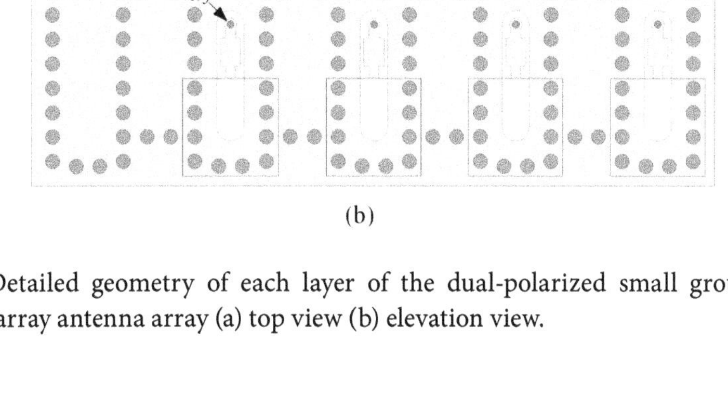

(b)

FIGURE 6.22 Detailed geometry of each layer of the dual-polarized small ground clearance end-fire phased array antenna array (a) top view (b) elevation view.

In order to produce horizontal polarization based on vertical polarization, an open slot structure is constructed using the space between the top hat dipoles. A bent strip line is introduced in M_2 and made to span both sides of the open slots to excite the horizontal current at the edges of the rectangular patch to radiate a horizontally polarized wave in the end-fire direction. Similar to the vertical polarization, the horizontal polarization is fed through the four Mini-SMP connectors located at the bottom layer and the coplanar waveguide conversion structure, and then, the energy is introduced into the bent strip line by M_3 using the gap between the metal blind aperture and the SIW.

Under this combination, the horizontally polarized structure does not destroy the original vertically polarized structure, and there is no need to additionally increase the floor ground clearance width l_d, i.e., the two polarizations can share the same radiator and coexist through the difference in the form of feed excitation, which greatly reduces the size of the antenna unit as well as the design complexity. The decoupling structure consisting of decoupling columns and decoupling slots between adjacent antenna units optimizes the port isolation and radiation characteristics of the two polarizations. The designed antenna floor clearance width is only 1.6 mm, and the overall size of the array is 28.8 mm × 16.1 mm × 1.728 mm.

6.4.2 Working Mechanism and Performance of a Single Antenna Unit

The working mechanism of the single antenna unit is analyzed below, and the design flow of the unit is shown in Figure 6.23. Based on the terminated open-circuit SIW antenna, antenna 1 is loaded with a pair of top-hat dipoles to generate vertically polarized end shots. In order to increase the antenna bandwidth, two matching posts are symmetrically inserted in antenna 2 to form a first-order inductive addition window, which improves the impedance matching without affecting the radiation performance. By introducing a multi-order addition window, a multi-stage resonant structure can also be formed, thus further extending the operating bandwidth, but the antenna length will also increase. On this basis, this design utilizes a stripline located in the middle of the two layers of SIW to form a combined feed line, which excites the loading structure to generate two orthogonally polarized waves, respectively.

The S-parameter simulation results of the above three antenna structures are shown in Figure 6.24, where the vertically polarized port is defined as port 1 and the horizontally

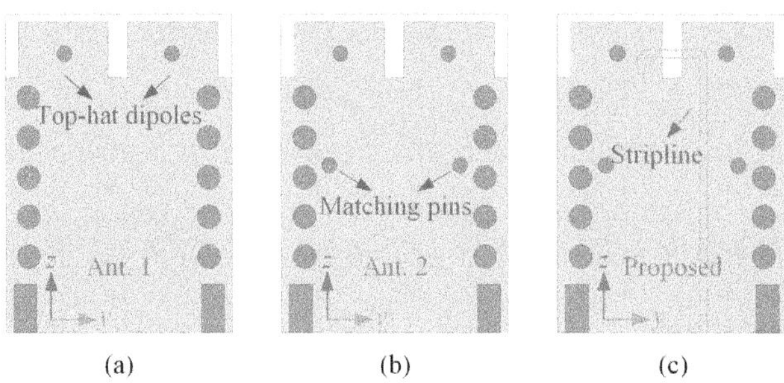

FIGURE 6.23 Design flow (a) antenna 1 (b) antenna 2 (c) this design.

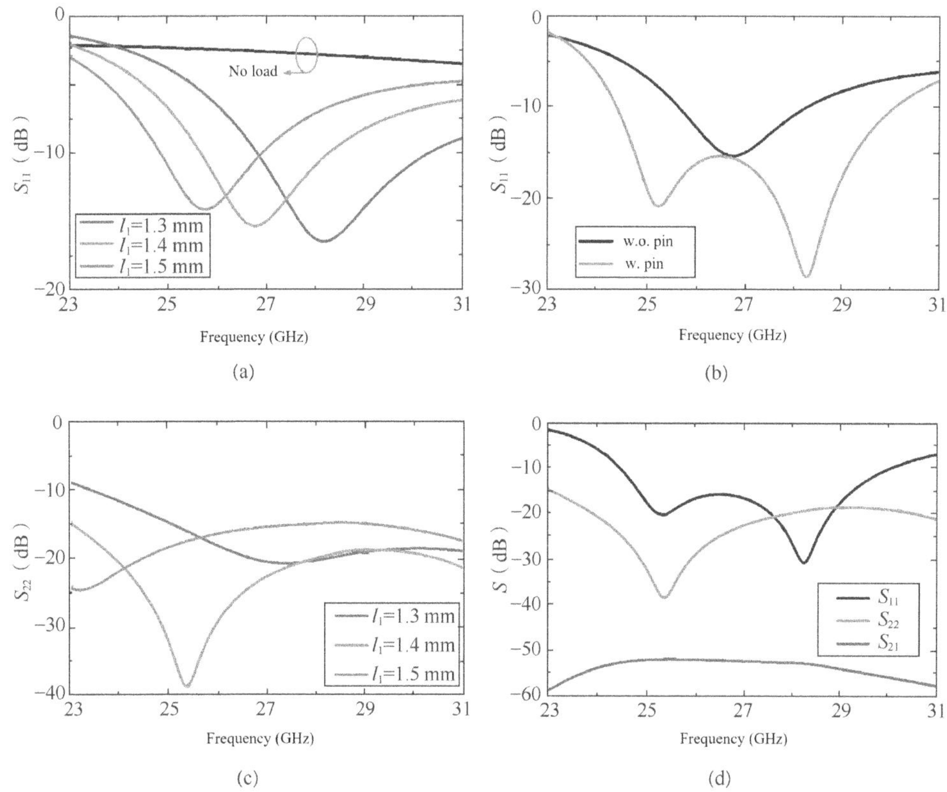

FIGURE 6.24 S-parameter simulation results for three antenna structures (a) antenna 1 (b) antenna 2 (c) this design (horizontally polarized) (d) this design (dual-polarized).

polarized port is defined as port 2. As shown in Figure 6.24a, the top-hat dipole structure is able to improve the impedance matching of the antenna efficiently compared to the SIW with unloaded structure. The patch length l_1 adjusts the resonant frequency of the dipole while keeping the dielectric substrate thickness constant. The added matching post in antenna 2 achieves wide-band vertical polarization, as shown in Figure 6.24b, the operating band of antenna 2 covers 24.41–30.07 GHz, which is wider compared to the band covered by antenna 1. At the same time, the resulting horizontal polarization can present good broadband characteristics by using the bent strip line to excite the open slot structure, as shown in Figure 6.24c, and the parameter l_1 is used to characterize the open slot length at the same time. When l_1 is increased, the operating frequency band of the horizontal polarization is shifted to lower frequencies and can cover the whole frequency band. The overlapping band for dual polarization is 24.46–30.04 GHz, as shown in Figure 6.24d. Since the transmission modes of the SIW and the stripline are TE_{10} mode and quasi-TEM mode, respectively, the two transmission lines are able to exhibit extremely low coupling, which allows the dual-polarized antenna to have an isolation of up to 50 dB.

The vector surface current distribution of the loaded structure at 27 GHz is shown in Figure 6.25. When the excitation is vertically polarized, the top-hat dipole structure

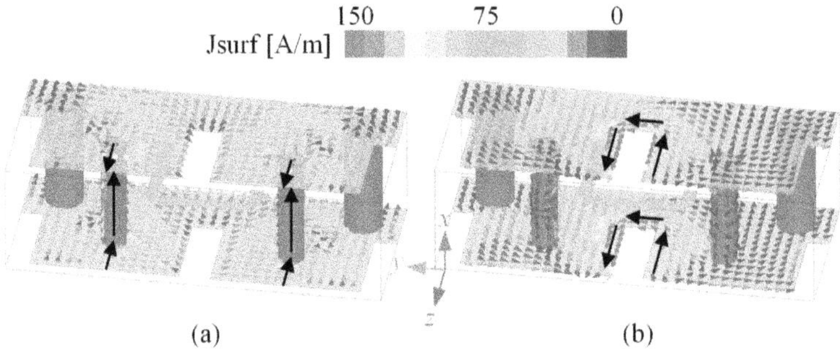

FIGURE 6.25 Vector surface current distribution at 27 GHz for the loading structure (a) vertical polarization (b) horizontal polarization.

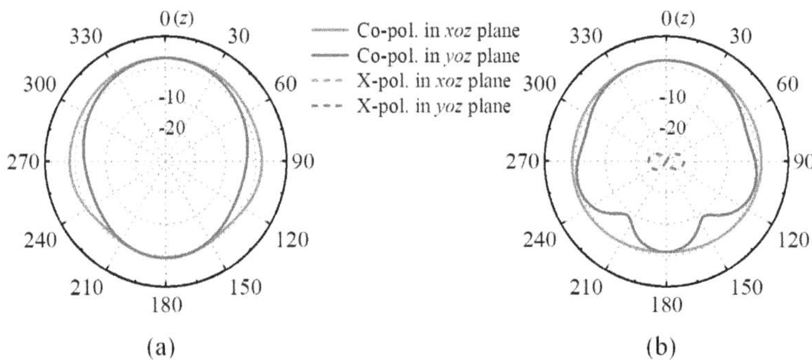

FIGURE 6.26 Radiation direction of the antenna unit at 27 GHz (a) vertical polarization (b) horizontal polarization.

operates in the one-half wavelength mode, and the current is concentrated on the two metal perforations, while the currents on the top and bottom rectangular patches are in opposite directions, indicating that the radiation at this point is dominated by the metal perforations. When the excitation is horizontally polarized, a stronger current distribution is exhibited near the open slot, and the direction of the current on the metal perforations is opposite and extremely weak, indicating that the radiation is dominated by the two-layer rectangular patch. In addition, the direction of the edge currents along the z-direction on both sides of the open slot is opposite, while the direction of the edge currents along the y-direction is the same, which gives the horizontal polarization a dipole-like radiation characteristic, and the consistency of the current distributions of the top and bottom layers ensures good symmetry of the end-shooting directional pattern.

The radiation direction diagram of the antenna unit at 27 GHz is shown in Figure 6.26. Both polarizations present symmetric radiation direction diagrams, with a gain of 3 dBi in the end-fire direction for the vertical polarization and 2.23 dBi in the end-fire direction for the horizontal polarization. It is worth pointing out that the gain of the horizontally polarized unit can be further improved with the addition of a decoupling structure on top

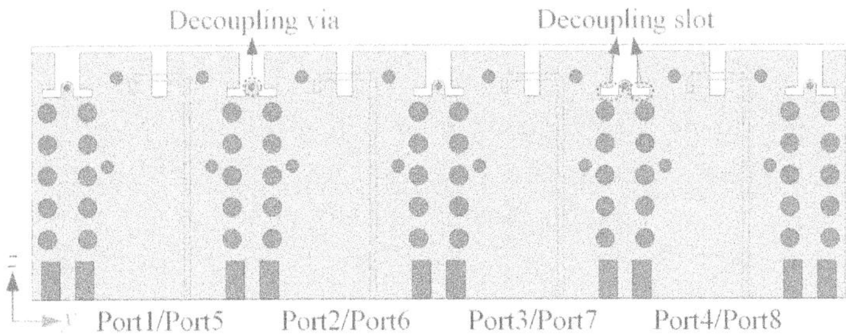

FIGURE 6.27 1×4-end-fire phased array antenna array.

of the dual-polarized array. In addition, the cross-polarization discrimination of the vertically and horizontally polarized cells are 45 and 26 dB, respectively, both of which have high polarization purity, which is due to the fact that both polarization excitation schemes are essentially balanced differential excitation.

6.4.3 Four-Cell Array and Decoupling Structure Design

The 1×4 end-fire phased array antenna array built using the antenna cells of Section 6.4.2 is shown in Figure 6.27. In order to enable the array to obtain a wider beam scanning range with smaller cell spacing, a decoupling structure consisting of decoupling columns and decoupling slots is designed between neighboring cells to reduce the inter-cell coupling. Among them, the decoupling columns are connected to the top and bottom metal plates through short metal strips to improve the port isolation for vertical polarization, and the decoupling slots are etched at the end of the feed SIW and extend to the inside of the rectangular patch in the y-direction to improve the port isolation for both polarizations at the same time; the cell spacing is set to be 6 mm (about 0.54 λ_0, with λ_0 being the wavelength of free space corresponding to the frequency of 27 GHz). In order to ensure the symmetry of the antenna structure, the lateral dimensions of the dielectric substrate are appropriately lengthened and the same decoupling structure is added to the two cells located at the edge of the array.

The effect of the decoupling structure on the port isolation for both polarizations is shown in Figure 6.28. When the decoupling structure is not added, the port isolation of the two polarizations at a cell spacing of 0.54 λ_0 is only 13 and 9 dB at maximum (obtained based on the original data). When the decoupling column is inserted, the port isolation for the vertical polarization increases by 2.5 dB. However, the decoupling column can only act as an electrical boundary on both sides of the vertically polarized cells, and cannot act on the horizontal polarization. As shown in Figure 6.28b, the port isolation for horizontal polarization is barely affected by the decoupling column. On this basis, the port isolation of both polarizations can be significantly improved by adding decoupling slot structures with choking effects.

The S-parameter simulation results of the quaternary array are shown in Figure 6.29. Due to the symmetry of the array structure, the reflection coefficient curves of four of the ports are given in Figure 6.29a, and the characteristics of the remaining ports are similar

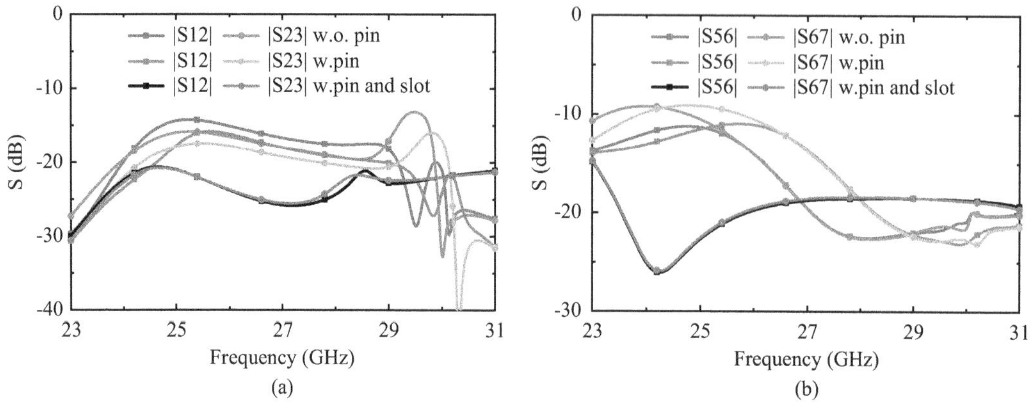

FIGURE 6.28 Effect of decoupling structure on port isolation for both polarizations (a) vertical polarization (b) horizontal polarization.

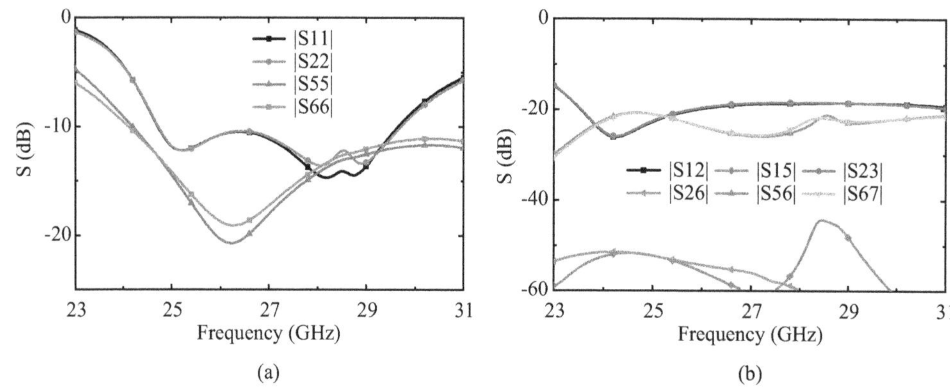

FIGURE 6.29 S-parameter simulation results of the quaternion array (a) reflection coefficient (b) port mutual coupling.

to those of these four ports. The vertical polarization has an operating band of 24.72–29.65 GHz, while the horizontal polarization has a better impedance matching with an operating band of 24.19–31 GHz. The mutual coupling curves of each port are shown in Figure 6.29b, and the isolation of each port is higher than 18 dB.

By feeding the excitation signal with linear phase gradient to each port of the quaternary array, the beam scanning direction diagram can be obtained, as shown in Figure 6.30. The peak gains of vertical and horizontal polarization are 8.59 and 10.26 dBi, respectively. Where the horizontal polarization achieves significant gain enhancement by the designed decoupling structure. Under the phase gradient of ±150°, the beam scanning ranges of the two polarizations are −46° to 46° and −44° to 44°, respectively, and the scanning gains are higher than 7 dBi, and the cross-polarization discrimination rates are all greater than 25 dB. Comparatively speaking, the gain fluctuation of the horizontal polarization is more obvious, and the peak gain of the vertical polarization is smaller, but it exhibits a better gain stability during the phased-array operation.

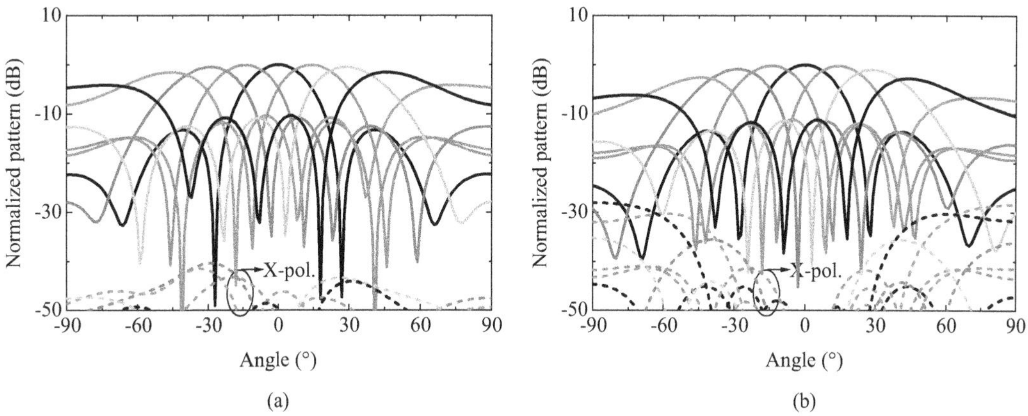

FIGURE 6.30 Beam scanning direction plot of the four-cell array at 27 GHz (solid and dashed lines represent main and cross polarization, respectively) (a) vertical polarization (b) horizontal polarization.

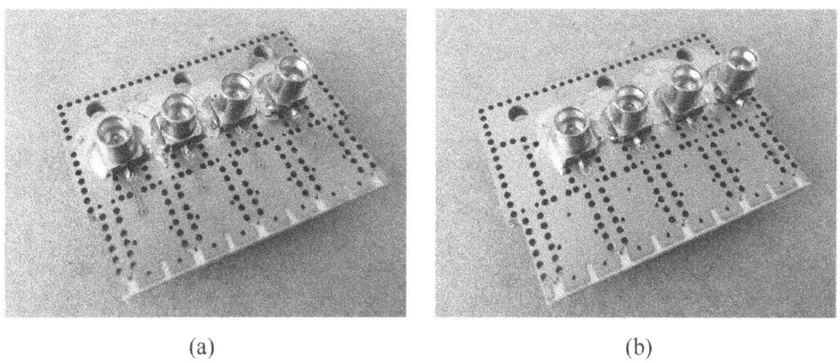

FIGURE 6.31 Antenna physical model (a) top view (b) elevation view.

6.4.4 Simulation and Test Results

In order to simulate the operating situation of the phased array, beam synthesis using test data from each unit of the array is required. On the basis of the phased array array factor, the expression form of the synthesized array orientation map can be derived, which will not be repeated here.

During antenna testing, it is easier to measure the gain radiation pattern of each unit. Accordingly, the beam scanning characteristics of the phased array can be quantitatively analyzed.

The designed antenna is physically processed by a multilayer PCB process to verify the actual performance of the dual-polarized small ground clearance end-fire phased array antenna. As shown in Figure 6.31, the antenna consists of two layers of dielectric substrate and the middle semi-cured sheet bonded by high temperature and high pressure, and eight Mini-SMP connectors are mounted on the upper and lower surfaces of the antenna to feed the dual-polarized excitation signals through the adapted test cables or adapters (SMPM 2.92 mm). The following test results have been calibrated to compensate for the insertion loss of the connecting cable.

The simulation and test results of the reflection coefficients at each port are shown in Figure 6.32. The overlapped relative bandwidth of the two polarizations is 21.4% for the simulation results and 14.6% for the test results. Comparatively speaking, the test curve of vertical polarization can match the simulation curve better, while the horizontal polarization exhibits two obvious resonance responses, and the test bandwidth is narrower than the simulation bandwidth. The difference between the simulation and test results can be attributed to the limitation of machining accuracy on the one hand, and the impedance mismatch introduced by the simulation error of Mini-SMP connectors and the positioning error of hand soldering on the other hand.

Figure 6.33 gives the simulation and test results of array port mutual coupling, and the simulation and test curves are in good agreement. The simulated and tested values of port isolation are higher than 15 and 18 dB in the operating frequency band, respectively.

The simulation and test results of the beam scanning radiation pattern of the quaternion array at 27 GHz are shown in Figure 6.34, and the solid and dashed lines in the figure correspond to the simulation and test results, respectively. The test results of the directional

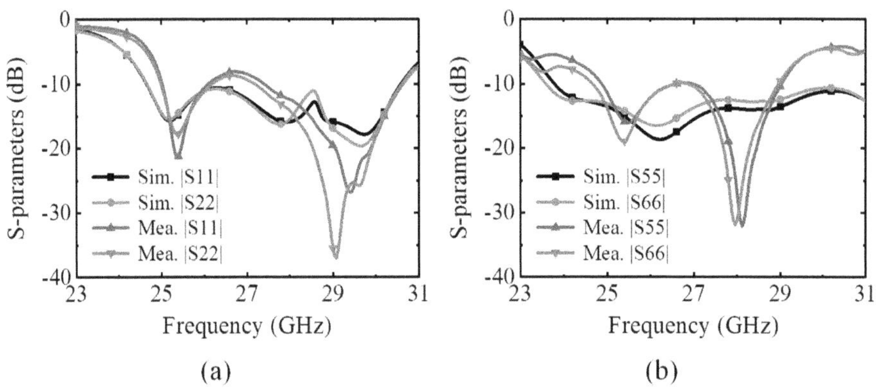

FIGURE 6.32 Simulation and test results of reflection coefficient of each port (a) vertical polarization (b) horizontal polarization.

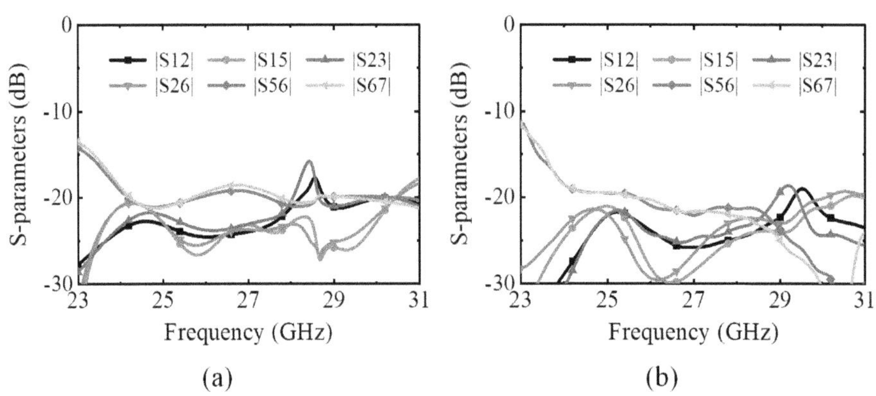

FIGURE 6.33 Simulation and test results of array port mutual coupling (a) simulated value (b) tested value.

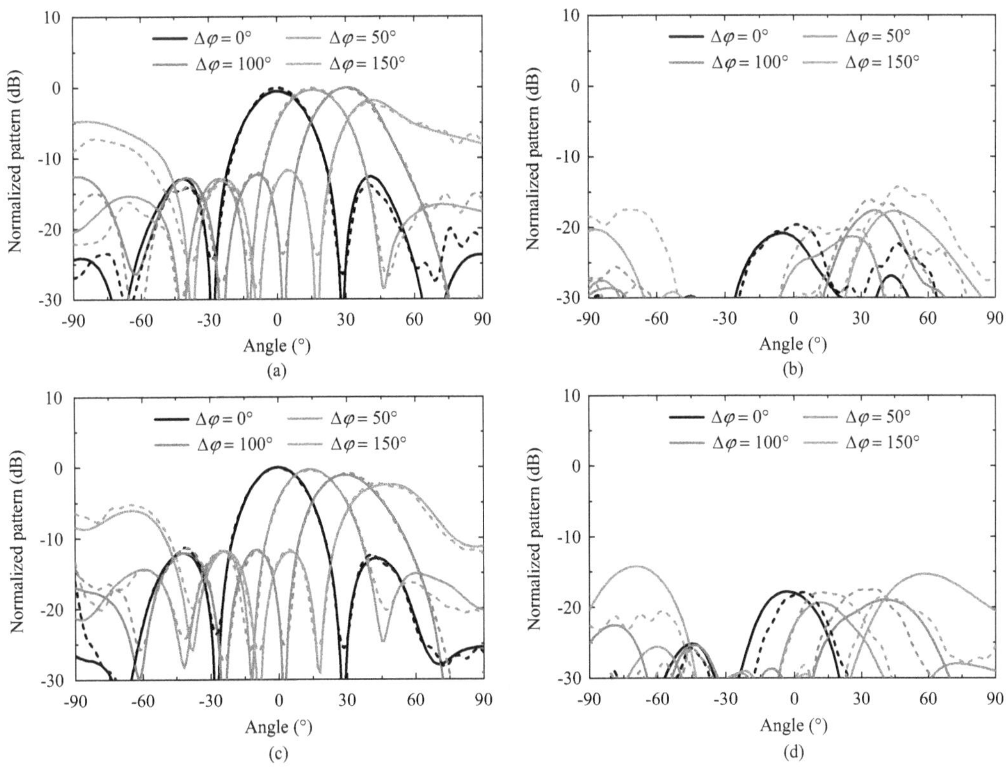

FIGURE 6.34 Simulation and test results of beam scanning direction plots of the array at 27 GHz (a) vertical co polarization (b) vertical cross polarization (c) horizontal co polarization (d) horizontal cross polarization.

map are obtained from the test values of the array directional map by beam synthesis, i.e., the far-field data of the eight ports are tested when they are excited individually, and then the scanning directional map corresponding to different excitation phases is calculated by using the array beam synthesis method. It can be seen that the simulation and test results show the same beam pointing when the phase gradient is 0°, 50°, 100°, and 150°, respectively.

The beam scanning ranges of vertical polarization and horizontal polarization are −42° to 42° and −49° to 49°, respectively, and the scanning loss is less than 2.5 dB. During the beam scanning process, the cross-polarization level of the two polarizations is less than −14 dB. Due to the influence of the feed-conversion structure, the cross-polarization discrimination rate is lower than that of the array under the ideal port excitation. However, the integration of the antenna module with the chip can further eliminate the influence of the feed structure when the mobile terminal antenna is practically applied.

The simulation and test results of the peak gain of the array are shown in Figure 6.35. The peak gains of vertical and horizontal polarization are 9.3 and 9.7 dBi, respectively. The metal shell of the Mini-SMP connector acts as a reflector to enhance the gain of vertical polarization. The gain fluctuation is less than 1.3 dB for both polarizations in the operating band.

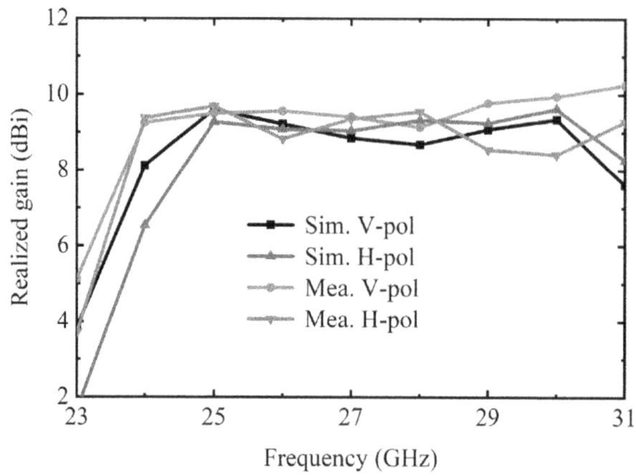

FIGURE 6.35 Simulation and test results for peak gain of the array.

TABLE 6.2 Performance Comparison between the Present Design and the Same type of Millimeter Wave Phased Array Antennas for Mobile Terminals

References	Number of Layers of Dielectric Substrate	Profile Height (mm)	Relative Bandwidth	Whether the Beam is Symmetric	Unit Cross-Polarization Discrimination (dB)	Ground Clearance Width (mm)	3 dB Beam Sweep Range
Reference [32]	2	1.1	7%	Yes	10	2.7	−40° to 40°[a]
Reference [21]	2	1.6	16.5%	Yes	13	3.4	−44° to 44°[a]
Reference [22]	2	1.1	14.3%	Yes	15	3.5	−40° to 35°[a]
Reference [20]	4	2.2	19%	No	5.7[a]	1.8	−34° to 33°
Reference [23]	2	2.2	18.5%	No	8.8[a]	4.2	−50° to 50°
This design	2	1.73	21.4%[a]	Yes	26[a]	1.6	−42° to 42°

[a] Denotes simulation results.

Table 6.2 shows the performance comparison of this design with the same type of millimeter wave phased array antennas for mobile terminals. It can be seen that this design has the smallest floor clearance width. In terms of operating frequency band, the relative bandwidth of this design is larger than that of the same type of mobile terminal millimeter wave phased array antenna. In addition, this design achieves good wide angle beam coverage.

6.5 DUAL-POLARIZED SINGLE-LAYER END-FIRE PHASED ARRAY ANTENNAS

Most of the current millimeter-wave dual-polarized end-fire phased array antennas use multilayer dielectric substrates, and the simple design method based on single-layer structure still has not made much breakthrough so far, especially in the very limited degree of freedom, the contradiction between the limitation of the single-layer design and the antenna multifunctional integration will be more prominent. The aim of this section is to overcome the technical difficulties of constructing millimeter-wave dual-polarized end-fire phased array antennas with a single-layer dielectric substrate [39] and to integrate a high-performance phased array antenna on a layer of dielectric substrate with a thickness of 1.524 mm.

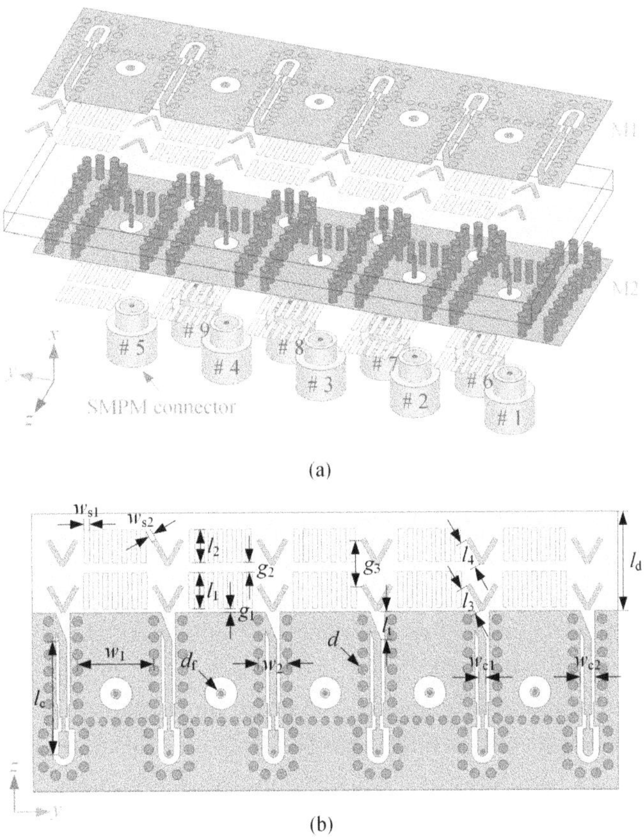

FIGURE 6.36 Array layout of a dual-polarized single-layer end-fire phased array antenna (a) exploded view (b) top view.

6.5.1 Antenna Array Structure

The array layout of the dual-polarized single-layer end-fire phased array antenna is shown in Figure 6.36, where the antenna is printed on a layer of Rogers RO4003C ($\varepsilon_r = 3.55$, $\tan\delta = 0.0027$) dielectric substrate with a thickness of 1.524 mm, and the metal layers on the top and bottom of the dielectric substrate are identified as M_1 and M_2, respectively. Five vertically-polarized and four horizontally-polarized cells are periodically staggered along the y-axis direction. The vertical polarization cell is a terminally open SIW antenna loaded with two levels of grid-type metal strips, and the horizontal polarization cell consists of a grounded coplanar waveguide (GCPW) gradient structure and two levels of V-shaped parasitic oscillators. Two additional horizontally polarized dummy elements are introduced on both sides of the array to provide symmetric boundary conditions for the edge cells. The spacing between neighboring cells is 6.5 mm, and the overall dimensions of the array are 36.3 mm × 16.8 mm × 1.524 mm. The diameter of the metal apertures used is 0.6 mm and the spacing between neighboring apertures is 1 mm.

Both polarizations are fed with coaxial probes at the bottom of M_2. A 9-optical-hole plug-in SubMiniature Push-on Micro (SMPM) coaxial connector was chosen for testing purposes.

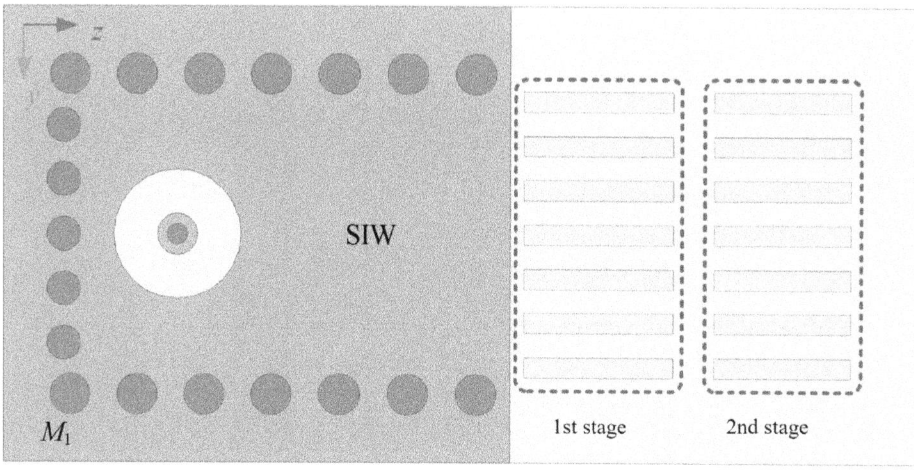

FIGURE 6.37 Geometry of the vertical polarization unit.

The feed ports for both polarizations are co-aligned along the y-axis direction, with the vertical polarization ports defined sequentially as #1–#5 and the horizontal polarization ports defined sequentially as #6–#9.

The vertical polarization unit is constructed based on the terminated open-circuit SIW antenna, and its geometry is shown in Figure 6.37. Due to the restricted profile height of the dielectric substrate, the dielectric-filled open-ended waveguide structure tends to have a more severe impedance mismatch with significant backward radiation. Therefore, loading the transition structure at the antenna radiation aperture is an effective method to enhance the performance at low profile.

The designed vertical polarization unit adopts a two-stage grid-type metal strip loading structure. The lengths of the two-stage metal strips are l_1 and l_2, respectively, to form a set of cascaded coupled resonators, which generate two neighboring resonance points in the band, respectively, to extend the operating bandwidth. By adjusting the spacing of the metal strips and the distance between the metal strips and the end of the SIW, the two-stage loading structure can present a similar field strength distribution and significantly suppress the backward radiation of the end-fire phased array antenna. The antenna is fed by a coaxial probe, and capacitive loading is introduced by designing the ring-slot structure at M_1 to achieve good impedance matching and mode conversion.

The effect of the two-stage metal strip on the performance of the vertically polarized unit is shown in Figure 6.38. When the SIW front end is not loaded with the metal strip structure, the Dielectric Resonator Antenna alone cannot achieve good impedance matching with S_{11} higher than −4 dB and the front-to-back ratio is only 1.3–5.2 dB. With the introduction of the 1st stage metal strip, the Dielectric Resonator Antenna can be resonated around 25.3 GHz and the −10 dB bandwidth is already able to cover the 24.04–27.59 GHz band with an improved gain. However, the effect of this structure on the front-to-back ratio is still weak. Compared to the unloaded structure, the front-to-back ratio is improved by 2.56 dB on average.

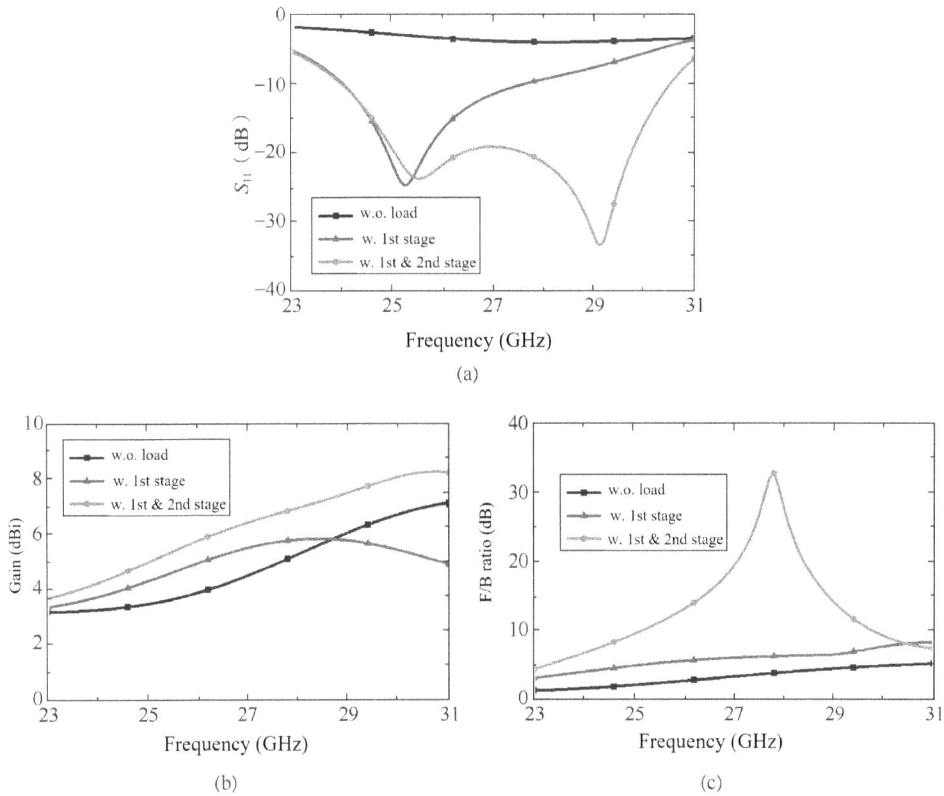

FIGURE 6.38 Effect of two-stage metal strip on the performance of vertically polarized unit (a) $S_{(11)}$ (b) gain (c) front-to-back ratio.

On this basis, by loading the 2nd stage metal strip, the antenna is able to generate a high-frequency resonance near 29.1 GHz, and this high-frequency resonance point together with the low-frequency resonance point at 25.3 GHz achieves a good impedance matching in a wide frequency band. At this point, the −10 dB relative bandwidth of the antenna can reach 23.8%, covering the 24.01–30.51 GHz band. Moreover, the antenna gain and front-to-back ratio have been significantly improved. In the operating band, the average gain is 6.43 dBi and the peak of the front-to-back ratio is as high as 33 dB.

The radiation direction diagram of the vertical polarization cell at 27 GHz is shown in Figure 6.39. The radiation direction plot shows good symmetry, with a gain of 6.41 dBi in the end-fire direction, a front-to-back ratio of 20 dB, and a cross-polarization discrimination of 35 dB.

In order to enable the single-layer dielectric substrate to support beam scanning for both polarizations simultaneously, miniaturized design and compact integration of the antenna units need to be achieved. The horizontal polarization units in this section are compactly arranged between adjacent vertical polarization units based on the GCPW transmission line, and the geometrical structure is shown in Figure 6.40. Due to the opposite direction of the electric field on both sides of the GCPW center guideband, in order to avoid the split of the end-fire beam, it is proposed to construct a gradient structure

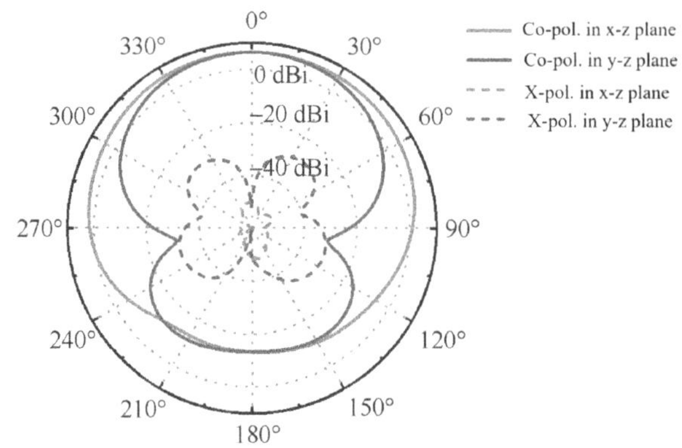

FIGURE 6.39 Radiation direction plot of vertically polarized unit at 27 GHz.

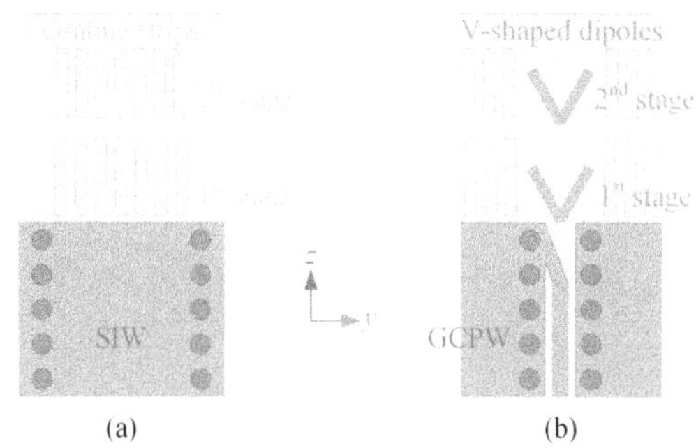

FIGURE 6.40 Horizontal polarization unit geometry.

with a bent GCPW. In this case, the guideband is bent along the y-direction and short-circuited to the ground plane on one side of the GCPW at the end, so as to suppress the inverse-phase electric field of the GCPW, and to form a horizontally-polarized field distribution by the asymptotic radiation. At the same time, a two-stage V-shaped parasitic oscillator is introduced into M_1 to improve the end-fire directivity. The two arms of the parasitic oscillator are clamped at an angle of 60°, and the V-bending allows it to operate in the one-half wavelength mode. By optimizing the spacing between the two V-shaped parasitic oscillators, the gain stability can be significantly improved. The antenna is connected to the coaxial probe via an impedance transformation section and a GCPW tapering structure, with the feed point located to the left of the vertical polarization unit to accommodate the SMPM coaxial connector.

The effect of the V-shaped parasitic oscillator on the performance of the horizontal polarization unit is shown in Figure 6.41. Since the horizontal polarization unit is arranged more compactly and is easily affected by adjacent units, the vertical polarization units on both sides and the loaded two-stage metal bars are retained in the analysis process to make the analysis results closer to the actual situation. It can be seen that the designed V-shaped parasitic vibrator as a parasitic structure does not significantly affect the impedance matching of the antenna, and the antenna without vibrator structure using the GCPW gradient structure can effectively support the horizontal

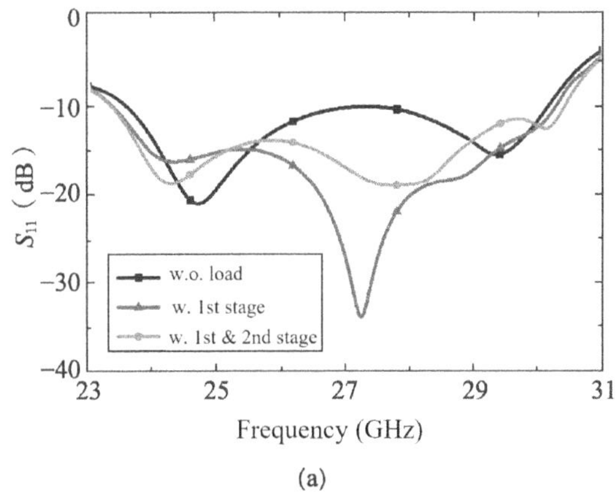

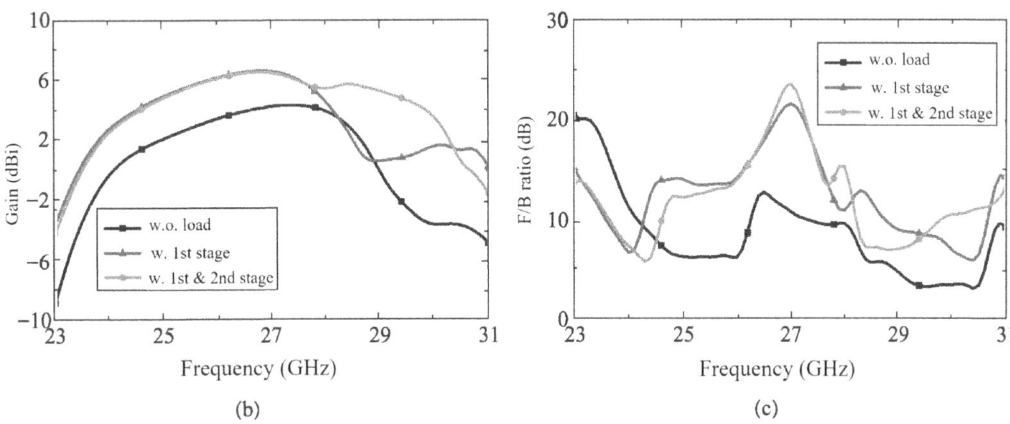

FIGURE 6.41 Effect of V-shaped parasitic oscillator on the performance of horizontally polarized cell (a) $S_{(11)}$ (b) gain (c) front-to-back ratio.

polarization and cover a wider frequency band interval. However, the gradient structure itself has limited radiation capability, with a maximum end-fire direction gain of 4.39 dBi and a front-to-back ratio of only about 10 dB in the narrower band around 27 GHz. When the 1st stage V-shaped parasitic oscillator is introduced, the antenna gain and the front-to-back ratio are significantly improved: the maximum gain is 6.67 dBi and the peak front-to-back ratio reaches 22 dB. However, the gain fluctuation of this structure is more drastic at high frequencies, and the bandwidth of the 3 dBi gain can only cover the frequency band of 24.31–28.2 GHz, which is obviously narrower than the antenna's working frequency band.

In contrast, the two-stage V-shaped parasitic oscillator loading structure achieves superior radiation performance. The antenna has a −10 dB relative bandwidth of 26.4% and a 3 dBi gain bandwidth covering 24.35–29.99 GHz, with a flatter and more stable gain than that of a V-shaped parasitic oscillator loaded with only the 1st stage. The antenna achieves a peak gain of 6.58 dBi with a large front-to-back ratio.

A plot of the radiation direction of the horizontally polarized unit at 27 GHz, exhibiting some beam tilt, is shown in Figure 6.42. Within the beam scanning plane (*yoz* plane), the

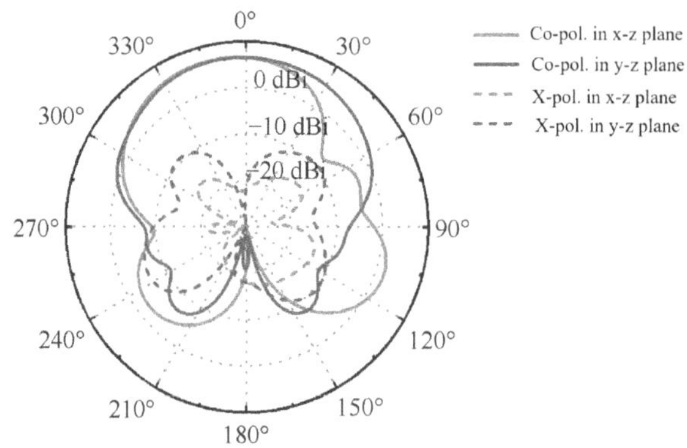

FIGURE 6.42 Radiation direction of horizontally polarized unit at 27 GHz.

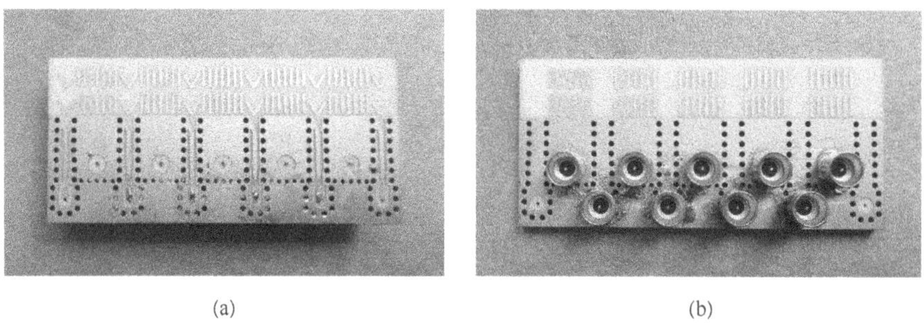

FIGURE 6.43 Antenna physical prototype (a) top view (b) elevation view.

antenna still has good directional radiation characteristics, with a gain of 6.51 dBi in the end-fire direction and a cross-polarization discrimination higher than 16 dB in the upper half plane.

6.5.2 Simulation and Test Results

The proposed dual-polarized single-layer end-fire phased array antenna is processed and experimentally verified with a physical prototype as shown in Figure 6.43, using only a single-layer dielectric substrate. Two rows of SMPM coaxial connectors are soldered to one side of the dielectric substrate and connected to the test equipment through SMPM 2.92 mm adapters and cable assemblies to achieve the feed excitation for the two polarizations respectively.

The reflection coefficient curves for each port of the antenna array are shown in Figure 6.44. Due to the symmetrical distribution of the array structure, only the S-parameters of the vertically polarized ports (#1–#3) and the horizontally polarized ports (#6, #7) are given here, and the characteristics of the remaining ports are similar to theirs. As can be seen in Figure 6.44, good impedance matching is achieved for both polarized ports. The operating band of the vertical polarization is slightly shifted toward the high frequencies compared to the results when the antenna units are operated individually. The −10 dB relative bandwidth of the two polarized ports overlapping can be up to 21.3%, which is able to cover the N257, N258, and N261 bands.

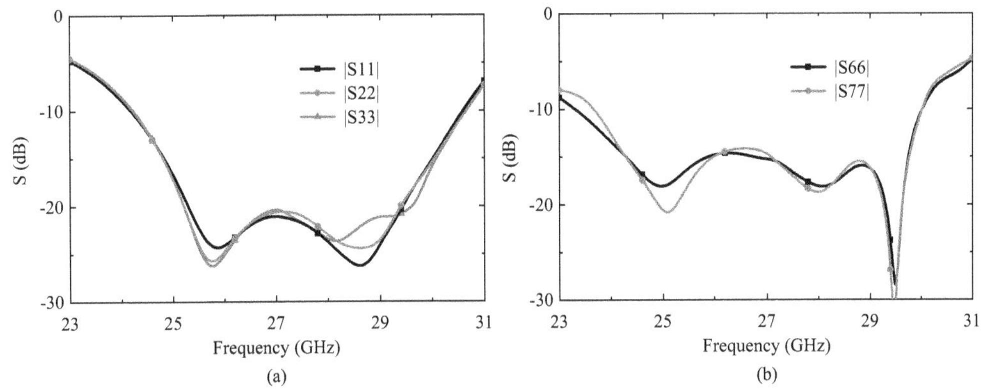

FIGURE 6.44 Reflection coefficient curves for each port of the antenna array (a) vertical polarization (b) horizontal polarization.

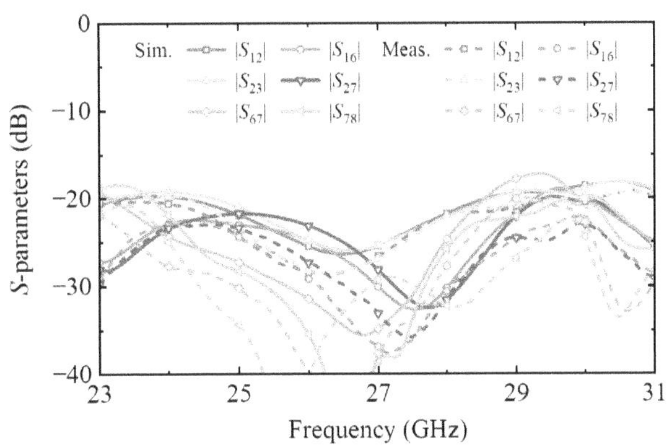

FIGURE 6.45 Port mutual coupling curve.

The port mutual coupling curves are shown in Figure 6.45. In the operating frequency band, there is a high isolation between the two polarized ports and between the adjacent polarized ports, and the mutual coupling level of any port is lower than −18 dB. This is to some extent attributed to the high gain directional radiation of the two polarized ports, and it also creates the conditions for the beam scanning of the phased array antenna.

The beam sweep direction of the antenna array at 27 GHz is shown in Figure 6.46. The gains in the maximum end-fire direction are 12.53 dBi and 12.39 dBi for vertical and horizontal polarization, respectively. Within the 3 dB gain fluctuation, the beam scan range is −43° to 43° for vertical polarization, and the beam scan range is slightly smaller than that of the vertical polarization at −36° to 38°. This situation occurs due to the fact that the metal strips loaded on the front end of the SIW affect the horizontal polarization, whereas the V-shaped parasitic oscillator structure does not significantly act on the vertical polarization, thus limiting the beam scanning range of the horizontal polarization to some extent.

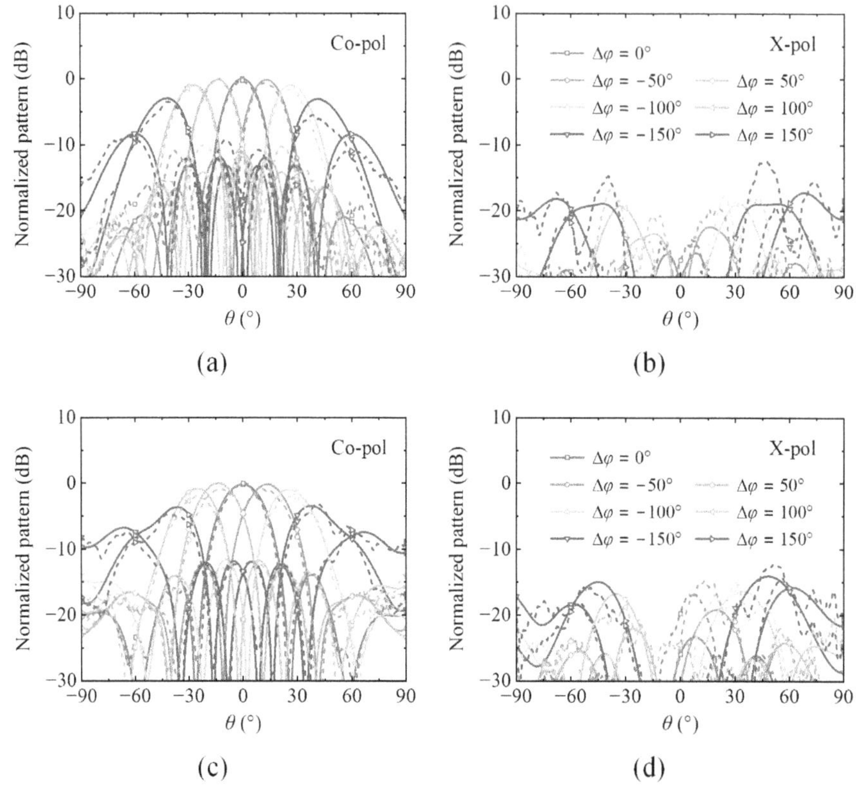

FIGURE 6.46 Beam scan direction diagram of the antenna array at 27 GHz (solid and dashed lines represent main and cross polarization, respectively) (a) vertical co polarization (b) vertical cross polarization (c) horizontal co polarization (d) horizontal cross polarization.

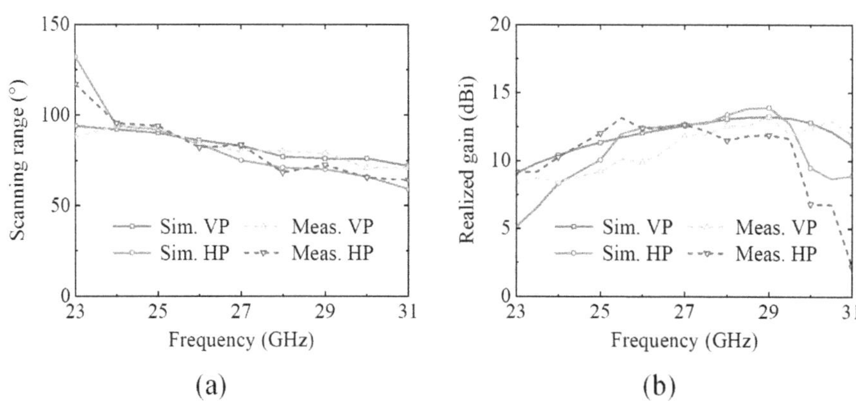

FIGURE 6.47 Peak gain curve of antenna array.

The peak gain curve of the antenna array is shown in Figure 6.47. From Figure 6.47, it can be seen that vertical polarization presents higher gain stability. The maximum gains of vertical and horizontal polarization are 13.23 and 13.65 dBi, respectively. In the operating frequency band, the gains of both polarizations are higher than 8.6 dBi.

TABLE 6.3 Comparison of key Metrics Between the Present Design and the Designs in the Reference

References	Profile Height (mm)	Operating Frequency Band	Isolation (dB)	Cell Spacing (mm)	Whether the Beam is Scanning	Scanning Range
Reference [34]	1.27	24–32 GHz	18.85	10	No	–
Reference [24]	1.575	28/38 GHz	14	6.6	Yes	–
This design	1.524	24.24–30.02 GHz	18	6.5	Yes	−36° to 38°

Since the design of dual-polarized single-layer end-fire phased array antennas is significantly more difficult than that of multilayer structures, the relevant reference and reports are extremely limited. In order to show the performance advantages of the proposed structure, the key metrics of the present design are compared with those of the single-layer designs in the two currently published references, as shown in Table 6.3. The design in the reference [34] has only one dielectric substrate but the cell spacing is close to one wavelength. Therefore this design only generates a directional end-fire beam and has no beam-scanning capability. In addition, although the wide-band phased array design in the reference [24] is realized with a compact cell arrangement, a thicker dielectric substrate is required to suppress the intensity of the signal zero of the vertical polarization in the end-fire direction, and the beam-scanning performance of the antenna array is not given for both polarizations. In contrast, the present design achieves a wide band, low profile with high port isolation with a single layer structure and is able to achieve good wide angle beam coverage.

6.6 DUAL-POLARIZED PATCH AND DIELECTRIC ROD PHASED ARRAY ANTENNA

The introduction of the metal bezel enhances the texture of the whole machine and provides more design strategies for the layout of the terminal antenna. The open-window design of the metal bezel shown in Figure 6.48 is a possible application scenario for the dual-polarized end-fire phased array antenna inside 5G Mobile Terminals. In this section, two prototypes of metal-bezel integrated antennas with different radiating structures are discussed from the co-design of millimeter-wave dual-polarized end-fire phased array antennas with metal bezels.

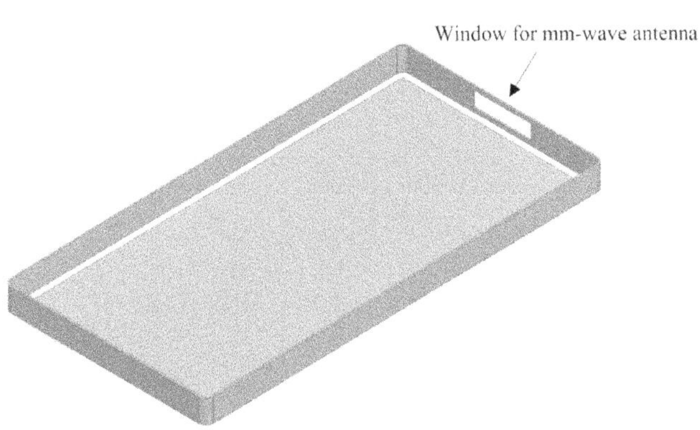

FIGURE 6.48 Open-window design with metal bezel.

First, based on the low-profile advantage of microstrip patch antennas, a ±45° millimeter-wave dual-polarized patch phased-array antenna array [40] is designed to compare the differences in impedance matching and coupling performance between the two different dual-polarization excitation schemes and to propose an easy-to-integrate metal encapsulation scheme. Then, to address the shortcomings of the traditional metal waveguide-excited dielectric rod antenna that is difficult to be compatible with the existing planar integration technology, a planar slit-excited millimeter-wave dual-polarized dielectric rod phased array antenna [41] is designed to give a compact single-layer microstrip-slit-coupled feeding scheme, and the performance of the antenna array is verified in the end-device integration environment.

6.6.1 Millimeter-Wave Dual-Polarized Patch Phased-Array Antenna Array Columns

From the point of view of cost and performance, the quaternary array is a more suitable choice. To realize the design of quadratic dual-polarized patch phased-array antenna array, a feed point is introduced in two orthogonal directions of the square patch using dual-probe based double-feed technique. The excitation schemes for two types of dual-polarized patch phased-array antenna arrays are shown in Figure 6.49, respectively; Figure 6.49a shows the feed points placed in the planes of φ (azimuth) = 0° and φ = 90° to excite the vertical and horizontal polarization, and Figure 6.49b shows the feed points placed in the planes of φ = ±45° to excite the ±45° dual polarization. The constructed metal patches were printed on a Rogers RO4003C (ε_r=3.55, tanδ = 0.0027) dielectric substrate with a thickness of 0.813 mm. In particular, the size of the patches is 2.5 mm×2.5 mm, and the spacing between the centers of the patch cells is 5 mm (approximately 0.47 λ_0, with λ_0 being the free-space wavelength corresponding to the 28 GHz frequency). After optimized tuning, each patch cell operates in the 28 GHz band with impedance matched to 50 Ω. The selected metal floor has dimensions of 20 mm×5 mm.

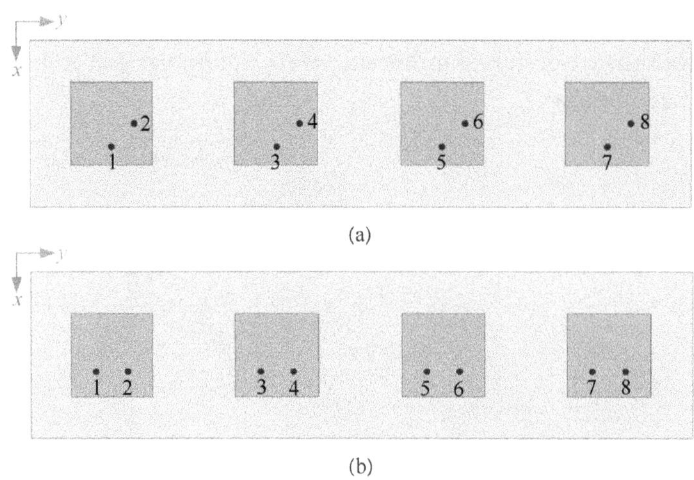

FIGURE 6.49 Excitation schemes for two types of dual-polarized patch phased-array antenna arrays (a) vertical/horizontal polarization (b) ±45° dual polarization.

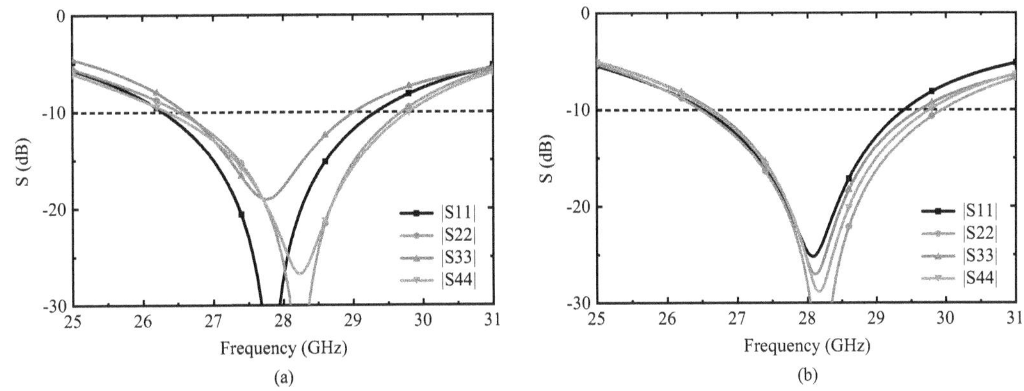

FIGURE 6.50 Reflection coefficient curves for two excitation schemes (a) vertical/horizontal polarization (b) ±45° dual polarization.

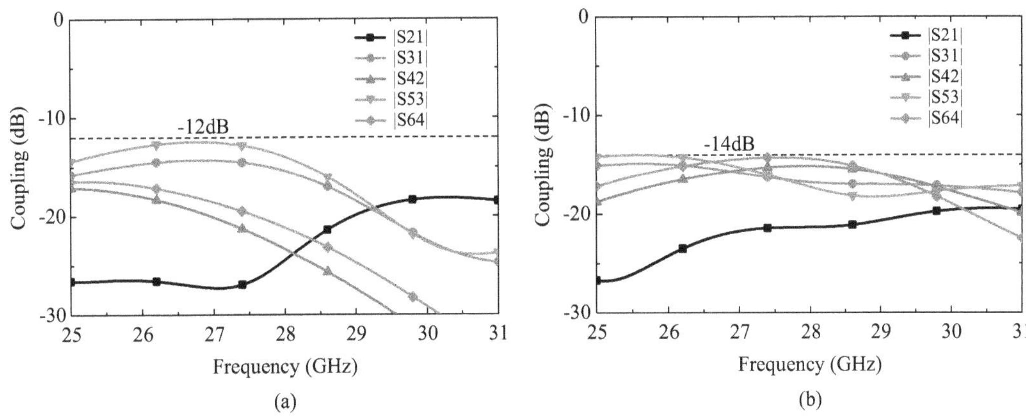

FIGURE 6.51 Port mutual coupling curves for two excitation schemes (a) vertical/horizontal polarization (b) ±45° dual polarization.

Figures 6.50 and 6.51 compare the reflection coefficient and port mutual coupling curves for the two excitation schemes, respectively. From Figure 6.50a, it can be seen that the resonant frequencies of the four patch cells have deviated from the set 28 GHz center frequency, while the reflection coefficient curves shown in Figure 6.50b can be kept stable. In addition, it can be seen from Figure 6.51a that the vertical/horizontal polarization excitation scheme corresponds to a maximum port isolation of only 12 dB, while it can be seen from Figure 6.51b that the ±45° dual polarization excitation scheme corresponds to a port isolation of up to 14 dB.

From the above comparison results, the two dual-polarization excitation schemes differ in array performance, i.e., ±45° dual-polarization contributes to the stability of the resonant frequency and presents high isolation. Therefore, subsequent designs will use this excitation scheme to generate the desired dual polarized waves.

Considering the integration of the proposed antenna array with a metal bezel, a packaging structure consisting of a metal fence is built on top of the ±45° dual-polarized patch

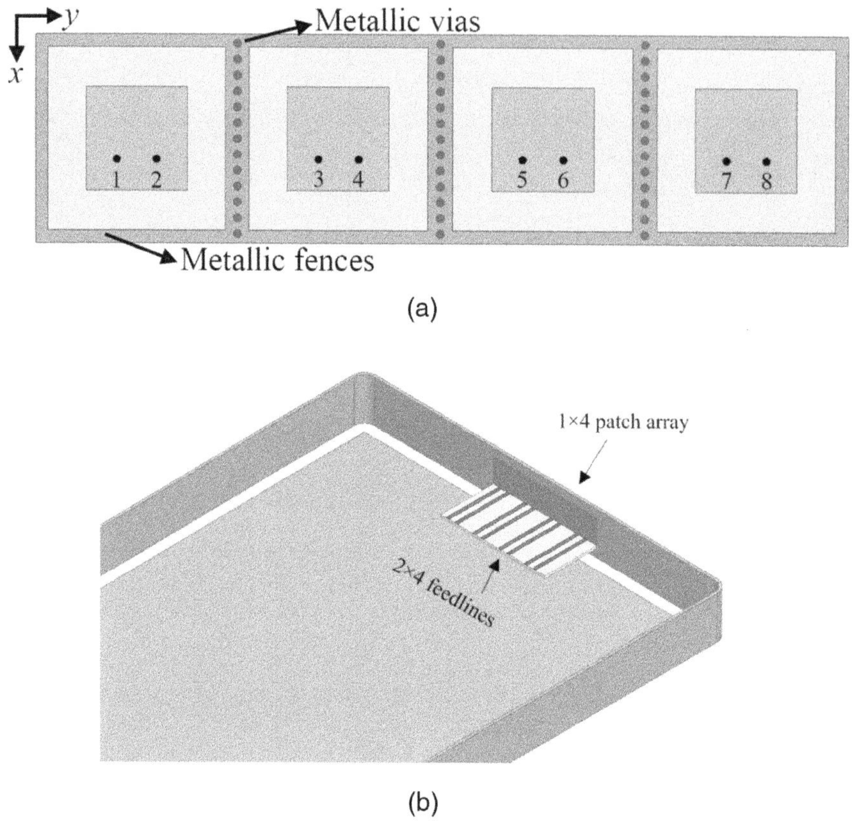

FIGURE 6.52 Dual-polarized patch phased-array antenna array (a) antenna-in-package antenna array in metal fence (b) antenna module integration layout inside mobile terminal.

phased-array antenna array, as shown in Figure 6.52a, with a width of 0.3 mm. for ease of processing and fabrication, three rows of metal overhangs are introduced as virtual electric walls between adjacent patch units in place of the fully metalized fence. Figure 6.52b shows the integrated layout of the antenna module inside the mobile terminal. The antennas are placed vertically behind the metal bezel openings to capture the end-firing beams directed toward the outside of the device. A set of 2×4 microstrip lines are printed on another single-layer dielectric substrate, which are vertically connected to each of the eight feed ports. Since the eight feed ports of the antenna array are in the form of a common line arrangement, the electrical connection between the ports and the antenna can be realized by using only the same layer of microstrip lines on a single feed plate. However, for the vertical/horizontal polarization excitation scheme, two layers of microstrip lines are required because the feed ports for vertical and horizontal polarization are located on two parallel lines. Therefore, applying dual-polarized excitation along the $\varphi=\pm 45°$ plane not only improves the port isolation but also greatly simplifies the complexity of the feed circuit.

Figures 6.53 and 6.54 show the reflection coefficient and port mutual coupling curves of the encapsulated structure, respectively. It can be seen that both the metal fence and the metal aperture help to stabilize the resonance frequency so that each port of the array

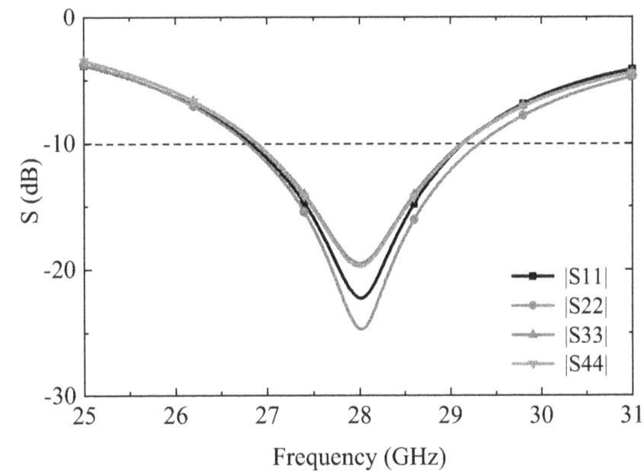

FIGURE 6.53 Reflection coefficient curves for packaged structures.

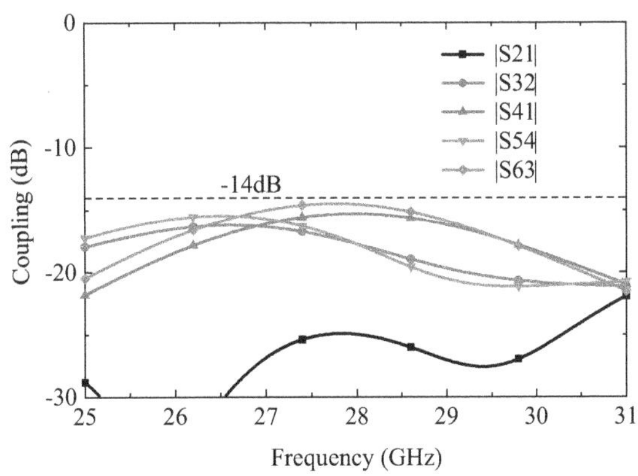

FIGURE 6.54 Port coupling curves for the package structure.

resonates at the 28 GHz band. The −10 dB relative bandwidth of each port coincidence is 7.9%, which provides full coverage of the N261 band. The antenna has up to 14 dB of port isolation.

The radiation direction diagram of the patch cell in the $\varphi=\pm45°$ plane is shown in Figure 6.55. Due to the good symmetry of the ±45° dual-polarized structure, only the simulation results of the +45° polarization in the $\varphi=\pm45°$ plane are given here. The maximum gain of the antenna array is 6.6 dBi, and the cross-polarization level is below −17 dB. In addition, loading the metal fence improves the cross-polarization characteristics, which is mainly due to the fact that the metal fence can effectively suppress the radiation from the probe.

The beam scanning performance of the antenna array can be further analyzed by feeding each input port with an excitation signal of the same amplitude and different linear phase gradients. The beam scanning direction diagram corresponding to −45° polarization is shown in Figure 6.56. The maximum gain of the main beam is 11.2 dBi. Within the gain fluctuation of 3 dBi, the beam scanning range of the array is −54° to 54°, and the side

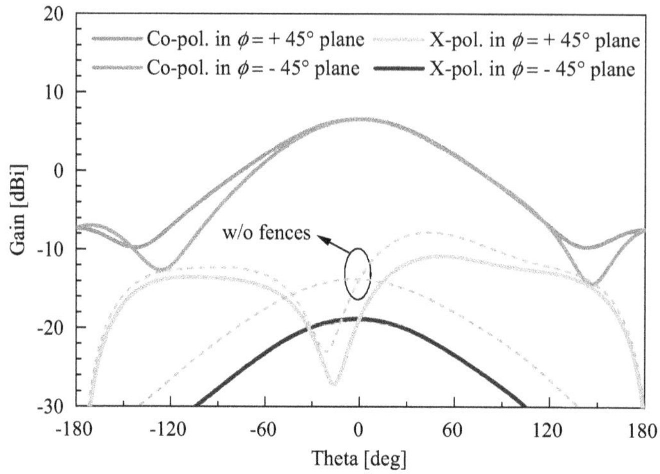

FIGURE 6.55 Radiation direction of the chip unit in the $\varphi=\pm 45°$ plane.

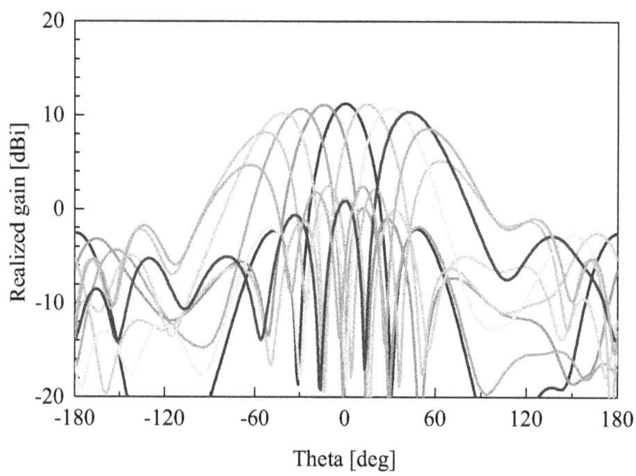

FIGURE 6.56 −45° polarization corresponding to the beam scanning direction.

lobe level is within a reasonable range. The beam sweep performance for +45° polarization is similar to that for −45° polarization.

6.6.2 Millimeter-Wave Dual-Polarized Dielectric Rod Phased Array Antenna Array Columns

The structure of the dual-polarized dielectric rod phased-array antenna array is shown in Figure 6.57, including a two-layer stacked dielectric substrate, four dome-shaped dielectric rod units, a metal floor with engraved coupling slits, and a set of 2×4 50 Ω microstrip lines for dual-polarization excitation. The thickness of the bottom dielectric substrate is 0.254 mm, the plate is chosen from Rogers RT/duroid 5880 (ε_r=2.2, tanδ = 0.0009), and the dielectric rod material is high-density polyethylene plastic (ε_r=2.3, tanδ = 0.0005). In order to improve the alignment accuracy of the dielectric rod and the feeder dielectric substrate during the actual assembly process, an additional high-density polyethylene substrate with

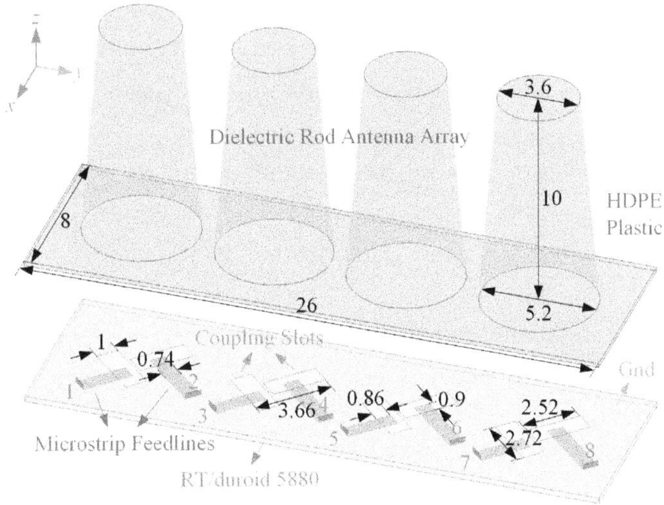

FIGURE 6.57 Structure of the dual-polarized dielectric rod phased array antenna array.

a thickness of 0.2 mm was added underneath the radiator, and the two can be molded with the help of 3D printing technology to achieve one-piece processing. In view of the advantages of ±45° dual polarization over vertical/horizontal polarization in terms of port isolation and radiation characteristics as described in Section 6.6.1, the microstrip lines and coupling slits are also arranged along the $\varphi = \pm 45°$ plane here.

The HE_{11} mode of the dielectric rod can be efficiently excited using the proposed planar coupling-feeding scheme, thus replacing the metal waveguide and the corresponding feed-conversion structure required for the classical dielectric rod antenna excitation, and dramatically improving the planar integration of the structural design. The spacing between neighboring dielectric rod units in the built antenna array is 6 mm (about 0.56 λ_0, with λ_0 being the free-space wavelength at the frequency of 28 GHz), and the dimensions of the metal floor used are 26 mm × 8 mm.

The simulated S-parameters of the antenna array are shown in Figure 6.58. The antenna presents good impedance matching and a −10 dB relative bandwidth of 15.1%, which is sufficient to cover the N257 and N261 bands. The port isolation is higher than 19 dB in all the operating bands.

The radiation direction diagram of the antenna array at 28 GHz is shown in Figure 6.59. Due to the offset of the coupling slit position with respect to the center of the dielectric rod cell, a slight tilt of the beam can be observed. In the zenith direction, the gain is 8.1 dBi for +45° polarization and 7.78 dBi for −45° polarization. The cross-polarization discrimination of the antenna array is higher than 18 dB.

The beam scanning direction diagram of the antenna array at 28 GHz is shown in Figure 6.60. The ±45° polarization excitation scheme makes the radiation characteristics more similar for both polarizations. The maximum gain is 12.17 dBi for +45° polarization and 12.02 dBi for −45° polarization. The beam sweep range is −40° to 40° for +45° polarization and −38° to 43° for −45° polarization.

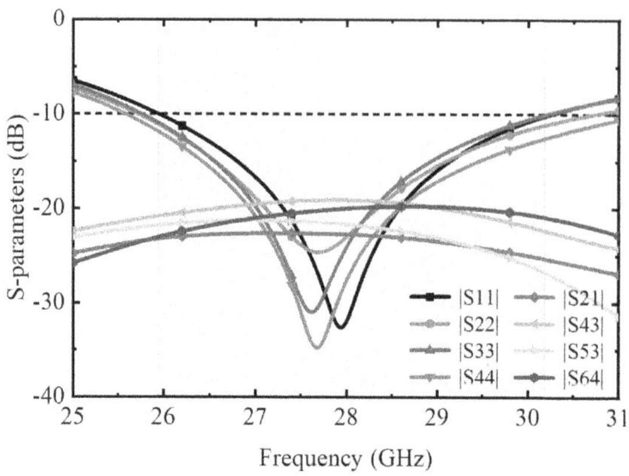

FIGURE 6.58 Simulated S-parameters of the antenna array.

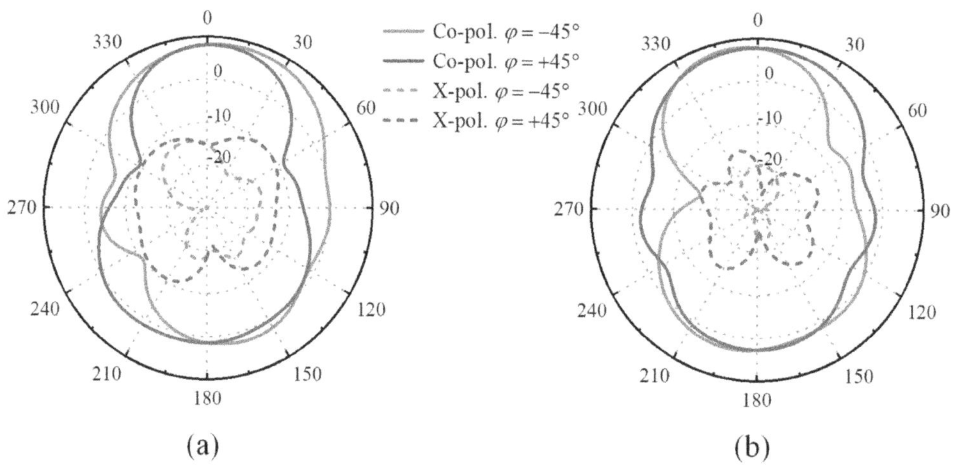

FIGURE 6.59 Radiation direction plot of the antenna array at 28 GHz (a) +45° polarization (b) −45° polarization.

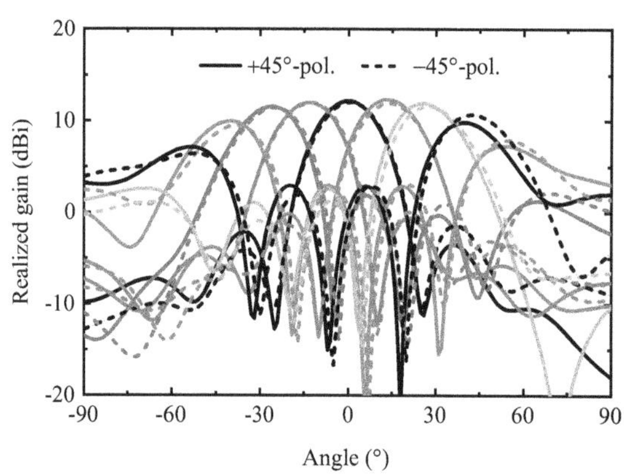

FIGURE 6.60 Beam scan direction diagram of the antenna array at 28 GHz (solid and dashed lines represent +45° polarization and −45° polarization, respectively).

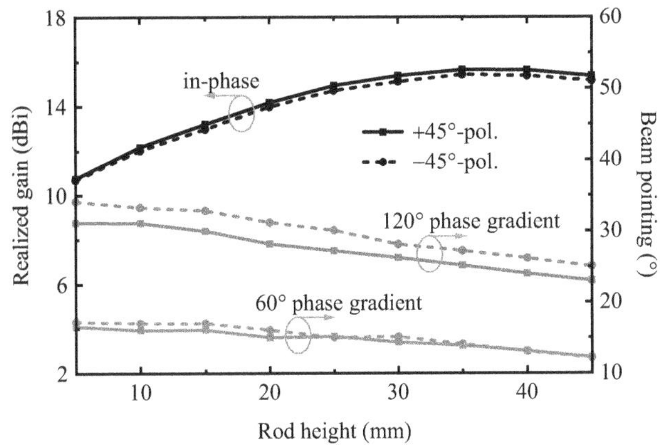

FIGURE 6.61 Gain and beam pointing corresponding to different rod heights (solid and dashed lines represent +45° polarization and −45° polarization, respectively).

The rod height of dielectric rod is a key parameter for antenna design. For a conventional single dielectric rod, the antenna can maintain good traveling wave characteristics within a certain height range, and the gain is approximately positively correlated with the rod height. Figure 6.61 shows the gain and beam pointing corresponding to different rod heights, and analyzes the variation of array performance with rod height. When the bar height is varied over several wavelengths, the array gain increases with the bar height. However, a continued increase in height leads to a larger spatial coupling strength between the dielectric rod units, and the coupling effect is more pronounced when the scanning angle increases, thus limiting the overall spatial coverage of the array. Considering the gain, beam pointing and the limited antenna size inside the mobile terminal, the selected rod height is 10 mm.

In order to further evaluate the advantages of antenna arrays for real-world mobile handheld devices, the focus here is on analyzing the metal-bezel integrated design scheme shown in Figure 6.62. The modeled outline of the mobile terminal is simplified and a 30 mm×7 mm rectangular radiating window is etched into the metal bezel to create the end shot. For aesthetic and practical reasons, the radiating window is filled with polytetrafluoroethylene material with a thickness of 0.5 mm (ε_r=2.1, tanδ = 0.001), which is also used as the dielectric cover layer of the millimeter-wave antenna.

The S-parameters of the integrated array are shown in Figure 6.63. It can be seen that the integrated environment has less impact on the designed array and the antenna can still maintain the original good matching performance. The operating band of −10 dB for each port overlap is 25.91 to 30.46 GHz, the relative bandwidth is about 16.1%, and the port isolation is higher than 20 dB.

The beam sweep direction of the integrated array at 28 GHz is shown in Figure 6.64. The beam sweep range of the integrated array is optimized with respect to the parameter of the side-lobe level as compared to the array working alone, which is mainly due to the expanded radiating mouth surface by the dielectric filling at the metal bezel openings. The maximum gain for the +45° and −45° polarizations is 12.11 and 12.04 dBi, respectively.

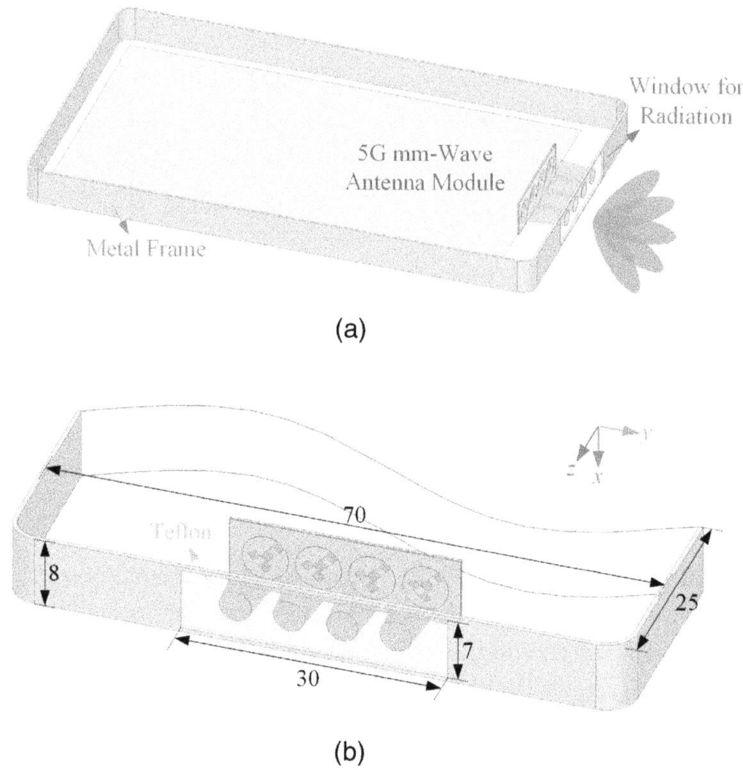

FIGURE 6.62 Integrated design scheme for metal bezel.

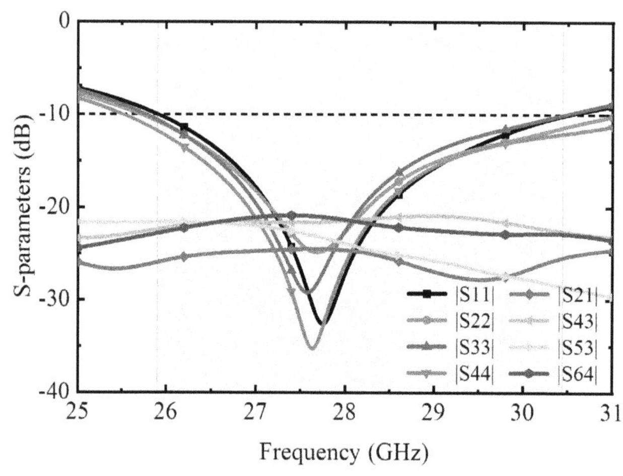

FIGURE 6.63 S-parameters of the integrated array.

The beam sweep range for both polarizations is −39° to 41° for a 3 dBi gain swing. The peak gain curves for the integrated array are shown in Figure 6.65. In the operating band of interest (25.91–30.46 GHz), gains greater than 9.7 dBi are achieved for both polarizations.

Table 6.4 shows the performance comparison between the two designs in Section 6.6 and the internationally reported metal-bezel integrated antennas. In comparison,

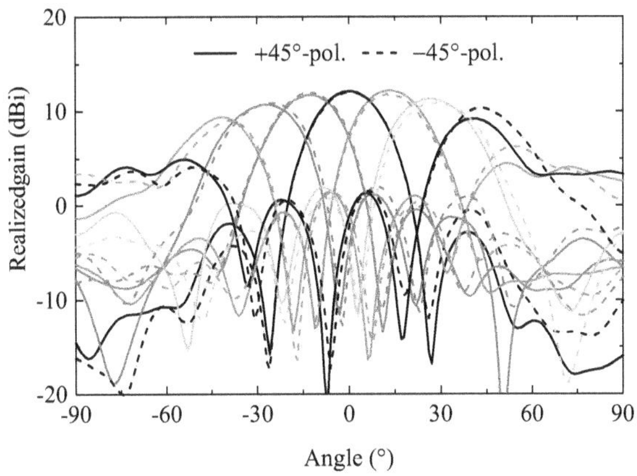

FIGURE 6.64 Beam sweep direction plot of the integrated array at 28 GHz (solid and dashed lines represent +45° polarization and −45° polarization, respectively).

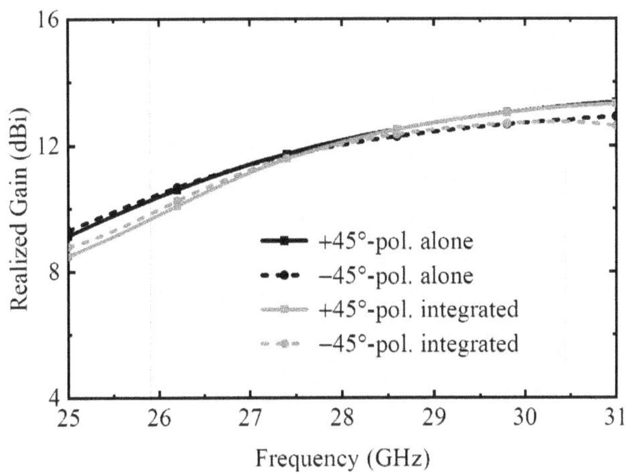

FIGURE 6.65 Peak gain curve for integrated array.

the dual-polarized patch phased-array antenna array in this section is structurally simpler, while the dual-polarized dielectric rod phased-array antenna array can also be realized by low-cost processing. In terms of typical technical specifications, the first design has a wider operating band than the two back-cavity structures in the references [16,18] and is able to achieve port isolation comparable to the designs in the references [15,17,18]. The second design further improves the operating performance with better relative bandwidth or port isolation than the results in the above reference. Although the Dielectric Resonator Antenna proposed in the references [35] achieves 16% relative bandwidth with 20 dB port isolation, the array gain is lower. In addition, there are different ways of defining the beam scan in the above reference, such as requiring a gain greater than 7 dBi. While Section 6.6 defines the beam scanning range with a more stringent 3 dB scanning loss, the arrays designed accordingly can not only achieve good beam scanning performance but also provide a more stable wireless link.

TABLE 6.4 Antenna Performance Comparison

References	Antenna Type	Operating Frequency Band (GHz)	Port Isolation (dB)	Array Gain (dBi)	Beam Sweep Range
Reference [16]	Back cavity ring slot antenna (1×8)	27.5–28.35	20	4.86	−50° to 50° (>12.1 dBi)
Reference [18]	Back cavity slot antenna (1×4)	26.6–27.8	13	7	−69° to 69° (>7 dBi)
Reference [35]	Dielectric Resonator Antenna (1×4)	25.3–29.8	20	6.38	−55° to 55° (>7 dBi)
Reference [15]	Dipole antenna (1×3)	28–33	15	-	−40° to 40° (>6 dBi)
Reference [17]	Chain slot antenna (1×4)	24.5–29.5	15		−40° to 40° (3 dB scanning loss)
This design	Patch phased-array antenna (1×4)	26.89–29.12	14	6.6	−59° to 59° (3 dB scanning loss)
	Dielectric stick phased array antenna (1×4)	25.91–30.46	20	8.1/7.78	−40° to 40° (3 dB scanning loss)

6.7 PHASED ARRAY ANTENNA WITH INTEGRATED PIN DIODE

In this section, a phased-array antenna design [42] is proposed without the use of a conventional phase shifter. By introducing a PIN diode to dynamically reconfigure the feed branches leading out of the traveling-wave transmission line, a 2-bit phase shifter and a serial power-division network are constructed, which provides a new way of thinking about designing a mobile terminal antenna in a low-cost manner.

6.7.1 Serial-Fed Beam Scanning Antenna Design

The geometry of the series-fed beam-scanning antenna is shown in Figure 6.66 and consists of two closely stacked dielectric substrates. The top substrate is made of Rogers RT/duroid 5880 (ε_r=2.2, tanδ=0.0009) and has a thickness of 0.787 mm. The bottom substrate is made of Rogers RO4003C (ε_r=3.55, tanδ=0.0027) and has a thickness of 0.203 mm. Three metal layers are printed on both substrates. The patch array is located in the top metal layer, which has 10 uniformly distributed antenna units. The floor is located in the middle metal layer, which has two rows of symmetrically etched slits along the y-axis. The feed network is located in the bottom metal layer and consists of a microstrip line and 10 periodically distributed structures. Each structure has four drive branches extending from both sides of the microstrip line and two parasitic branches suspended below the slits. Binary RF switches are used to control the connection and disconnection of the driver and parasitic branches. The periodic distribution structure is equivalent to a series of 2-bit phase shifters. The design has the advantage of a flexible and adjustable structure with the radiating array and the feeder network separated by the floor and the energy coupled through the slit.

The radiating structure of each antenna unit is a patch coupled through a slit. The radiation pattern of the patch is mainly determined by the microstrip line below the slit. If a switch on the microstrip line is closed, that microstrip line branch is connected to the

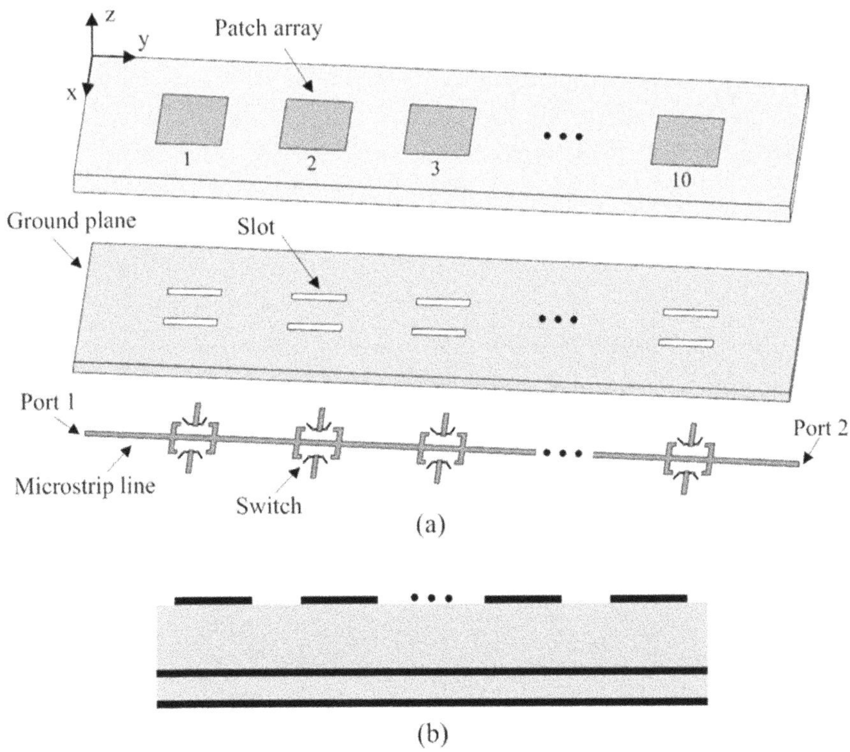

FIGURE 6.66 Geometry of a series-fed beam-scanning antenna (a) 3D decomposition view (b) side view.

parasitic branch. The signal will flow along the conducting current path and the energy will be coupled to the patch through the slit above the path. In order to obtain a map of the direction of broadside radiation, only one switch is closed during the reconfigurable process. The energy coupled from the microstrip line is determined by the branch line width. As the line width increases, the impedance of the branch decreases and the branch will receive more energy from the microstrip line, i.e., the leakage rate increases. However, the increase in leakage rate will deteriorate the impedance matching of the microstrip line. The simulated S-parameters of one antenna cell in the array when one switch is closed are shown in Figure 6.67. The results show that the leakage rate of this antenna cell is within −1 dB and the input impedance matching is good. To simplify the analysis, the switch's closed and disconnected states are assumed to be ideal in the simulation.

When any one of the switches below the antenna unit is closed and the remaining three switches are disconnected, the branch where the closed switch is located will be connected to the main circuit. The four switches correspond to four states of connectivity, and although the antenna unit produces the same directional map of the side-emitting radiation in all four states, the phases of the radiated fields are different. Figure 6.68 reveals the near-field characteristics of the antenna unit, observing the near-field above the patch to depict the phase change. As shown in Figure 6.68a, the observation point is located on the z-axis, approximately 0.2 λ_0 from the patch (λ_0 is the free-space wavelength corresponding

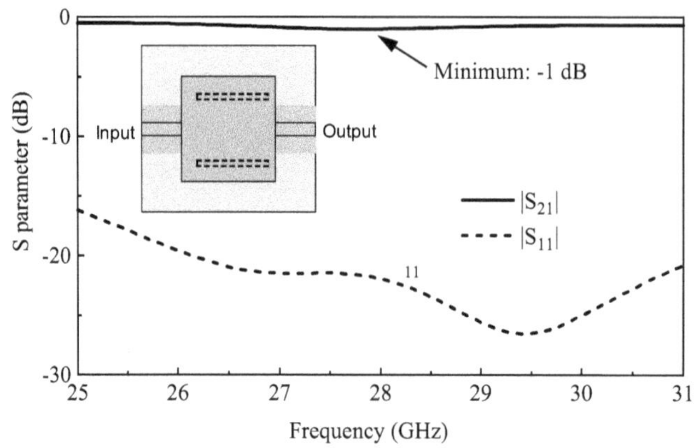

FIGURE 6.67 Simulated S-parameters of one antenna cell in the array when one switch is closed.

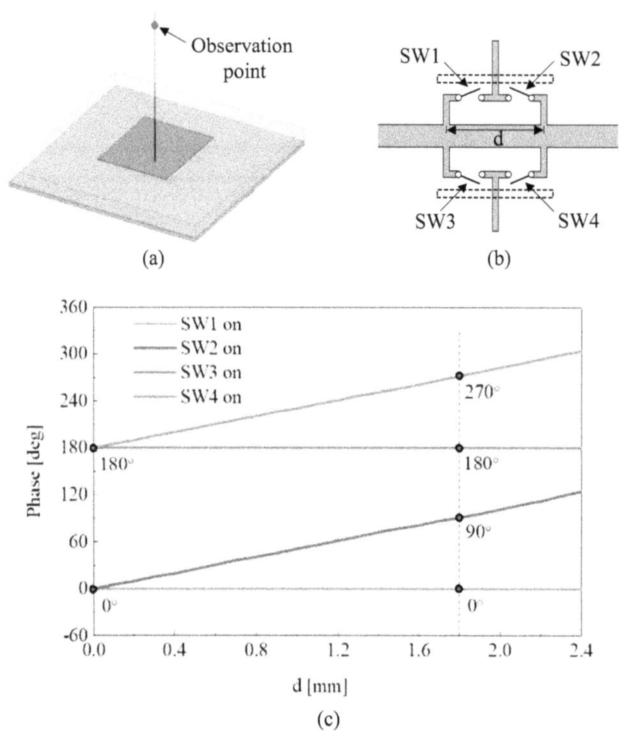

FIGURE 6.68 Near-field characteristics of the antenna unit (a) observation point (b) 2-bit phase shifter (c) near-field phases with different offset distances.

to the 28 GHz frequency). The four switches in Figure 6.68b can be divided into two groups. sw_1 and sw_2 are on one side of the transmission line, and sw_3 and sw_4 are on the other side of the transmission line. The offset distance between two switches in each group has a significant effect on the phase. Figure 6.68c shows the near-field phase with different offset distances. The reconfigurable network has four states, in each state one switch is disconnected and the other three are closed. When the offset distance is 0, only two switches are needed, and one switch is closed while the other is disconnected, for a total of two states. The results show that there is a 180° phase difference between the two states, which means that the two switches form a 1-bit phase shifter. As the offset distance increases, the phase difference between the same set of switches becomes larger, while the phase difference between the two sets of switches remains 180°. When the offset distance is 1.8 mm, the four states correspond to the phases of 0°, 90°, 180°, and 270°, respectively, constructing an equivalent 2-bit phase shifter.

The principle of operation of a 2-bit phase shifter is shown in Figure 6.69, where the electric fields on both sides of the microstrip line are in opposite directions. At the same time, the phase of the transmission line shows a periodic distribution. When the propagation distance between two points on the transmission line is $\lambda_g/4$ (λ_g is the waveguide wavelength), there is a 90° phase delay between the two points. Therefore, four branches can be introduced at four positions in the figure to generate four phase distributions of 0°, 90°, 180° and 270° without the use of a conventional phase shifter.

Based on the proposed reconfigurable cell, the radiation pattern of the array is calculated. The electromagnetic wave propagation model of a one-dimensional phased array is shown schematically in Figure 6.70. According to the phased array theory, the electromagnetic waves on the dashed line should have the same phase delay to ensure that the main beam points to the specified angle θ. Considering that mobile terminals require less accuracy in the beam direction, a 2-bit phase shifter is a more appropriate choice in order to reduce the cost.

The antenna cell spacing in this design is about 0.44 λ_0, and the corresponding electrical length in the medium is 3 $\lambda_g/4$. Figure 6.71 shows the beam scanning performance of a one-dimensional phased array (10-cell array). The scanning step is 10°, the theoretical scanning range is set to −80° to 80°, and the simulated scanning range is −65° to 65°. When the beam is close to the zenith direction, the simulated beam direction agrees well with the theoretically calculated value. When the beam is turned to the end-shooting direction, the beam tilt becomes difficult and the simulated beam scanning range is smaller

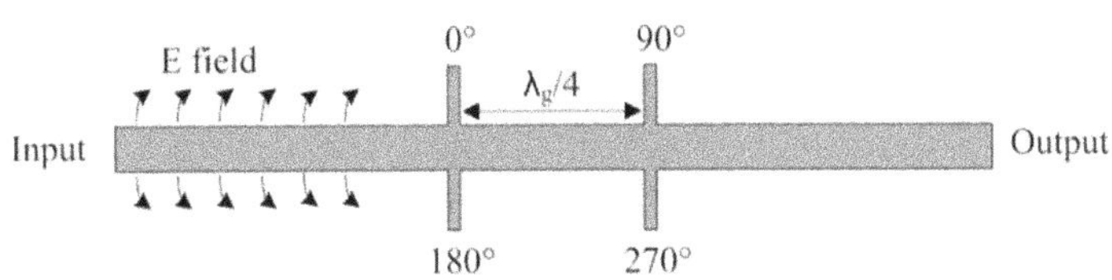

FIGURE 6.69 Working principle of 2-bit phase shifter.

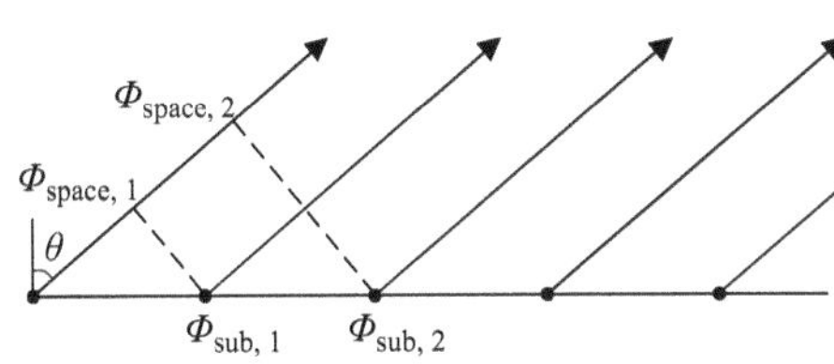

FIGURE 6.70 Schematic electromagnetic wave propagation model of 1D phased array.

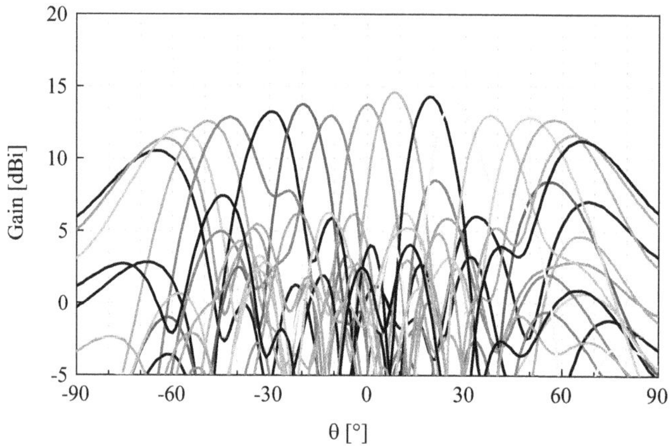

FIGURE 6.71 Beam scanning performance of a 1D phased array (10-cell array) (solid lines are simulation results, vertical dashed lines are theoretical calculations).

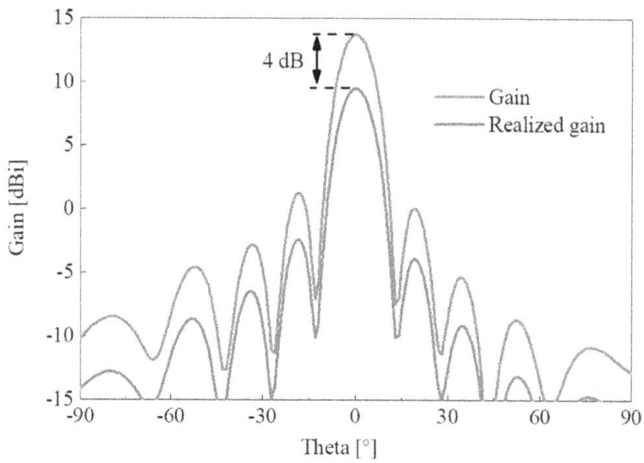

FIGURE 6.72 The gain and the realized gain of the array with broadside beam showing the effect of significant impedance mismatch that occurs when the beam is directed along broadside direction.

than the theoretically calculated beam scanning range. The reason for this situation is that as the main beam shifts from the zenith direction to the end-fire direction, the coupling between the antenna units starts to become stronger, which gradually restricts the beam from deflecting to the end-fire direction.

It is well known that leaky wave antennas suffer from a sharp gain degradation in the zenith direction due to the fact that all the antenna units are excited in-phase, and the power reflected from each antenna unit will be coherently superimposed in-phase at the input port, leading to a strong impedance mismatch phenomenon. This behavior is also observed in this design. Figure 6.72 shows the results of the ideal gain and actual gain for a fixed frequency leaky antenna generating a zenith beam. When impedance matching is considered, the actual gain drops by about 4 dBi compared to the ideal gain.

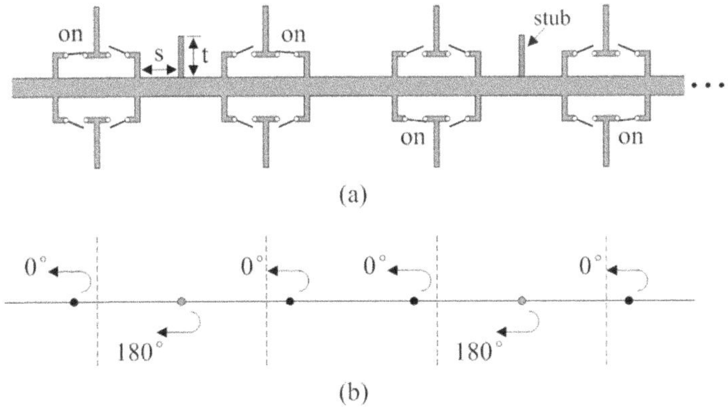

FIGURE 6.73 Introduction of a compensation structure to compensate for the mismatch of in-phase excitation (a) improved feed network (b) schematic of the mechanism of reflection cancelling.

To overcome the gain degradation due to the zenith beam, a compensation structure was introduced to compensate for the mismatch of the in-phase excitation. Figure 6.73a shows a modified feeder network with five short cutoffs added to the 10 phase shifters. The reflections generated by these short cutoffs are equal in amplitude and opposite in phase to the reflections generated by the phase shifters. The mechanism of reflection canceling is shown schematically in Figure 6.73b. In order to achieve optimal reflection cancellation, the parameters were scanned and found to be most effective with parameters $s=1.8$ mm and $t=0.6$ mm, and the switch was configured to be in the broadside beam state. It is important to note that the positions of the short cutoff lines are not uniformly distributed due to the spacing of neighboring reflection units of $\lambda_g/2$ or λ_g. If the spacing is λ_g, a truncation line is added in the middle of the two reflection units, which introduces additional reflections with a phase of 180°. The reflections from the truncation line and the reflection unit are in opposite phases, and the cancellation of both reflections helps to reduce the impedance mismatch at the input port.

The reflection amplitude generated by the short cutoff is controlled by varying the length (t) of the short cutoff. The effect of the length of the short cutoff line on the broadside gain at 28 GHz is shown in Figure 6.74. The results show that the broadside gain increases as the length of the truncated line increases. It should be noted that the introduction of a short cutoff line will affect the phase distribution on the transmission line. If the reflection from a single short cutoff line is strong, the phase gradient and beam pointing will be affected. After optimization, 0.6 mm is finally used as the length of the short cutoff line.

To evaluate the effect of switching on performance, the antenna cell model with PIN diodes and bias network is investigated, as shown in Figure 6.75. Each antenna cell has four PIN diodes, and two PIN diodes on the same side of the transmission line share the same DC bias circuit. In this design, an open microstrip line of length $\lambda_g/2$ is used as the DC bias line. The shunt point of the DC bias line is placed at the actual short circuit point at a distance of $\lambda_g/4$ from the open end of the DC bias line.

According to the equivalent circuit, the PIN diode is equivalent to a resistor when the bias voltage is positive and a capacitor when the bias voltage is negative. The values of the

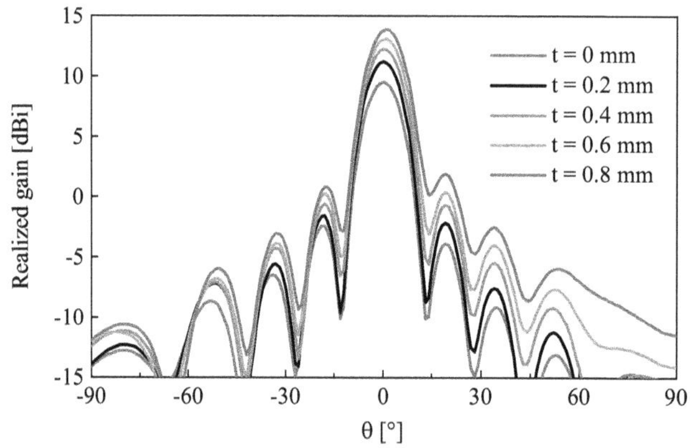

FIGURE 6.74 The effect of the length of the reflection cancellation stubs on realized broadside gain at 28 GHz.

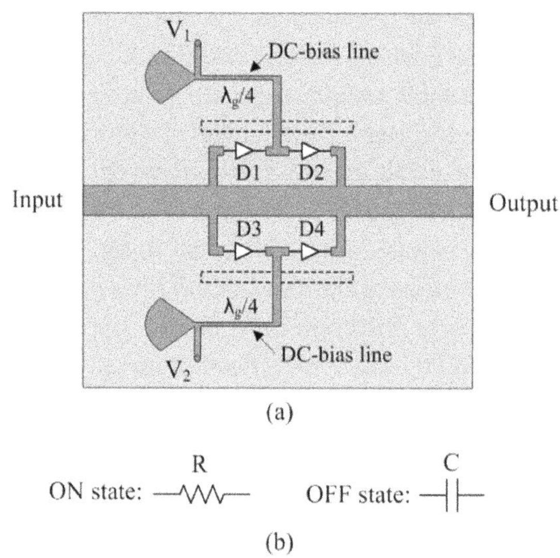

FIGURE 6.75 Element with diodes and bias network (a) practical model with four diodes (b) equivalent circuit of the diode.

two equivalent components are key factors in determining the forward insertion loss and reverse isolation level of the PIN diode. According to the datasheet for the MA4AGP907 PIN diode, the typical value of the equivalent resistance is 5 Ω, which gives an insertion loss of about 0.3 dB at 28 GHz; the typical value of the equivalent capacitance is 0.02 pF, which gives an isolation level of about 12 dB at 28 GHz.

The designed array was re-simulated using the equivalent circuit with the PIN diode switch closed and disconnected. The beam scanning performance of the one-dimensional phased array with PIN diode switch at 28 GHz is shown in Figure 6.76. The beam sweep width of this model is reduced by about 10° compared to the model with ideal switches. The total radiation efficiency of the beam is 54%–75%. This means that the overall losses

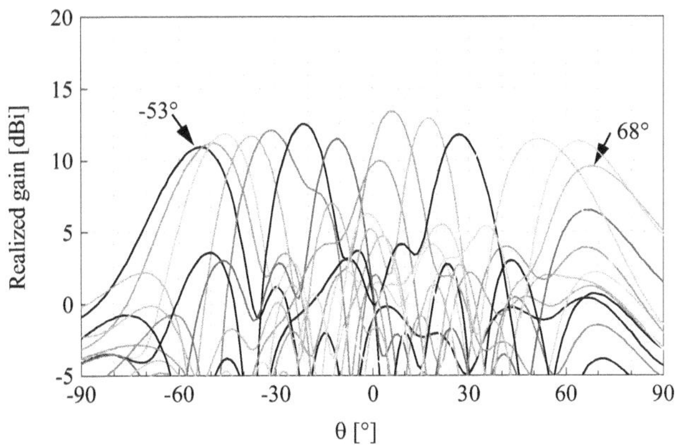

FIGURE 6.76 Beam scanning performance of 1D phased array.

of the array, including mismatch losses, residual losses, ohmic and dielectric losses, and switching losses, are less than 3 dB over the scanning range.

6.7.2 Spatial Coverage in 5G Mobile Terminals

This section analyzes the spatial coverage of the series-fed beam scanning antenna array in 5G Mobile Terminals. Figure 6.77 shows two typical layouts of the array in 5G Mobile Terminals: one is to place the array horizontally on the top of the cell phone floor and align it with the short edge; the other is to place the array vertically in the end-fire direction of the cell phone floor, which facilitates the design of integration with the cell phone bezel. In both typical layouts, the distance between the array and the cell phone floor is 2 mm, and the size of the cell phone floor is 140 mm × 70 mm.

Due to the strong directionality of the designed patch arrays, at least two arrays are required to provide full-space signal coverage. Considering the symmetry, only one array in the hemispherical coverage is discussed in this section to calculate the spatial coverage efficiency of the two typical layouts in the upper hemispherical plane. To avoid frequent switching between different beams, the scanning step is set to 10° with 19 beams in the upper hemispherical plane. However, considering the maximum available scanning range in practical applications, only 17 beams were used, corresponding to a scanning angle of −80° to 80°.

The field strength at each point in space is chosen to be the maximum of the 17 beams, resulting in a synthetic radiation pattern of the hemisphere. The synthetic radiation patterns at 28 GHz for two typical layouts are shown in Figures 6.78. In Figure 6.78, $\varphi = 90°$ and $\varphi = 270°$ represent the scanning planes. It can be observed that the signal is strong as the scanning angle approaches the side-shooting direction. As the scanning angle approaches the end-fire direction, the signal becomes weaker. The maximum gain of the array placed horizontally is greater than the maximum gain of the array placed vertically.

In order to quantitatively assess the spatial coverage efficiency of the two typical layouts, a representation of the cubic angle coverage efficiency is used:

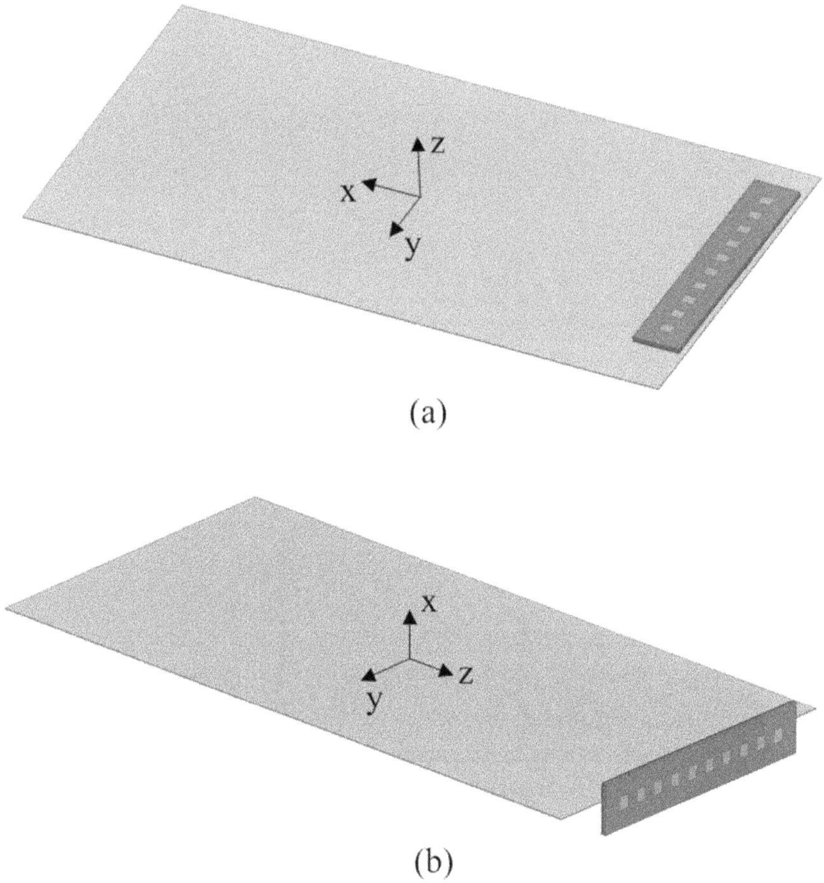

FIGURE 6.77 Two typical layouts of the array in 5G mobile terminals (a) horizontal placement (b) vertical placement.

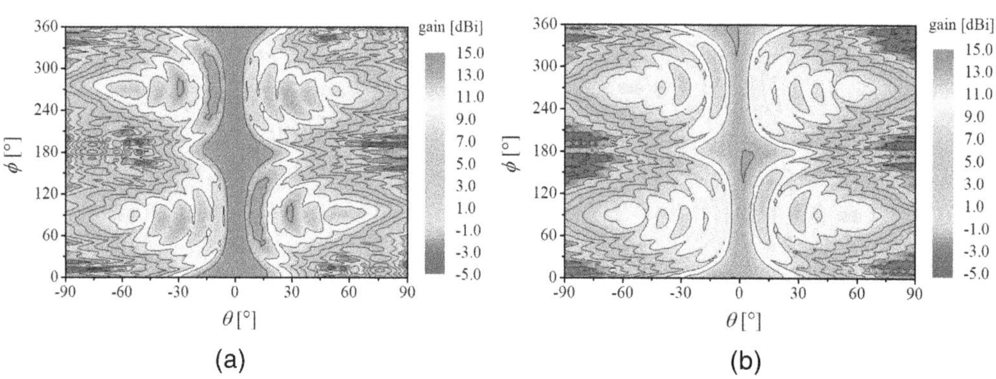

FIGURE 6.78 Synthesized radiation direction at 28 GHz for two typical layouts (a) horizontal placement (b) vertical placement.

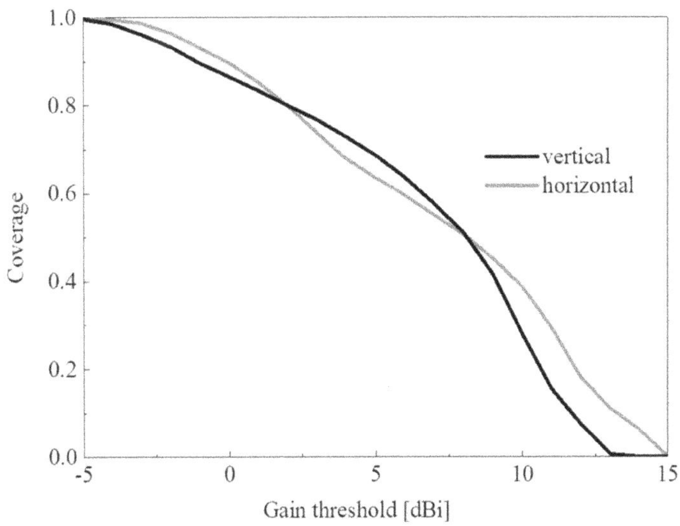

FIGURE 6.79 Spatial coverage efficiency for two typical layouts.

$$\text{Efficiency} = \frac{\text{Steerable solid angle}}{\text{Maximal solid angle}}\bigg|_{\text{gain threshold}} \tag{6.1}$$

In this design, since only half a sphere is considered, the maximum stereo angle is 2π. The spatial coverage efficiency can be calculated based on the data in Figure 6.78, and the relationship between the spatial coverage efficiency and the gain threshold is shown in Figure 6.79. It can be observed that the two curves have similar trends. Both typical layouts have excellent spatial coverage. When the desired gain threshold is high (>8 dBi), the horizontal placement of the array has better spatial coverage. Vertical placement of the array provides better spatial coverage when the desired gain threshold is medium sized (>3 dBi). Thus, this evaluation informs the choice of antenna array placement within the cell phone.

6.8 SUB-6 GHz ANTENNA INTEGRATED WITH MILLIMETER WAVE ANTENNA

This section presents the design of a Sub-6 GHz antenna integrated with a millimeter-wave antenna [43]. The geometry of the designed antenna is shown in Figure 6.80, which consists of a piece of cell phone floor and a unbroken uncut metal bezel, where the cell phone floor is flush with the bottom of the metal bezel. A portion of the cell phone floor is etched for the antenna design area and the other sides are attached to the metal bezel. There are six antenna modules on the two long sides of the cell phone floor. Each module consists of one-half wavelength slots etched into the cell phone floor (i.e., long slots) and one-quarter wavelength slots etched into the metal bezel (i.e., short slots). Each slot array contains four slot units. The cell phone floor is printed on a 1 mm thick FR4 dielectric substrate (ε_r=4.4, tanδ=0.02), and the metal bezel has a 0.254 mm thick TLY-5 substrate (ε_r=2.2,

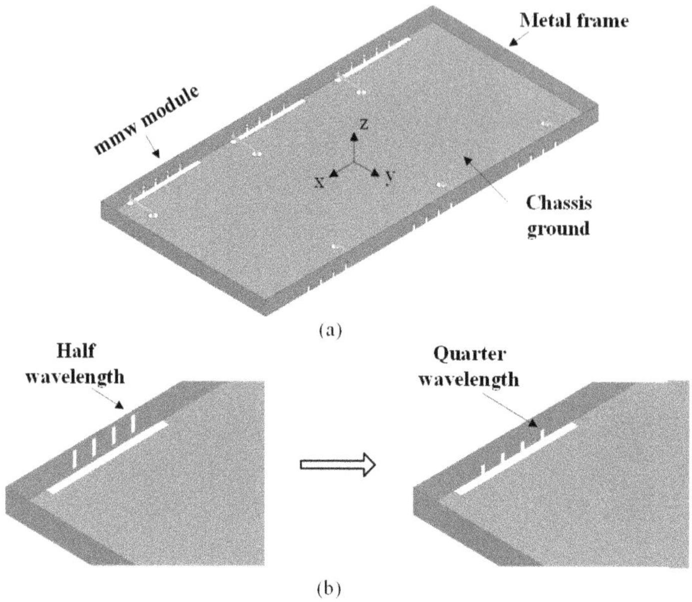

FIGURE 6.80 Joint design process of Sub-6 GHz antenna and millimeter-wave antenna in this design.

tanδ=0.001) on the inside of the metal bezel, and both materials are set to be transparent for convenience. The cell phone floor measures 140 mm×70 mm, the height of the metal bezel is 6 mm, and the dimensions of the long slots are 28 mm×2 mm. each slit is fed by a 50 Ω microstrip line, with the long slots being direct-connected feeds and the short slots being coupled feeds.

The design process of the common caliber antenna is divided into two steps. Firstly, two types of slot antennas are designed separately. The slot (short slot) antenna on the metal bezel is responsible for the 28 GHz band and operates in the one-half wavelength mode with a length of about 5 mm per slot, while the slot (long slot) antenna on the floor of the cell phone is responsible for the Sub-6 GHz band and also operates in the one-half wavelength mode with a length of about 40 mm per slot, and then the joint design of the two slots is considered, so that the non-metallic etching portion on the metal bezel can be reduced. The area of the non-metallic etched portion on the metal bezel is then considered to be reduced by combining the two slot structures. This is done by moving the short slot antenna on the metal bezel vertically downwards and connecting it directly to the long slot antenna on the cell phone floor. In this case, the short-slot antenna operates in quarter-wavelength mode and the size of the millimeter-wave array is reduced by 50%, which can effectively reduce the masking effect caused by human hands touching the metal bezel.

The operation of the Sub-6 GHz and millimeter-wave bands is next analyzed by current distribution. The current distributions in the 3.5 and 28 GHz bands under different port excitations are shown in Figure 6.81. In Figure 6.81a, when port P_1 of the Sub-6 GHz antenna array is excited, the one-half wavelength pattern can be clearly observed along the long slot, and the slot antennas on the metal bezel have currents as well, which helps

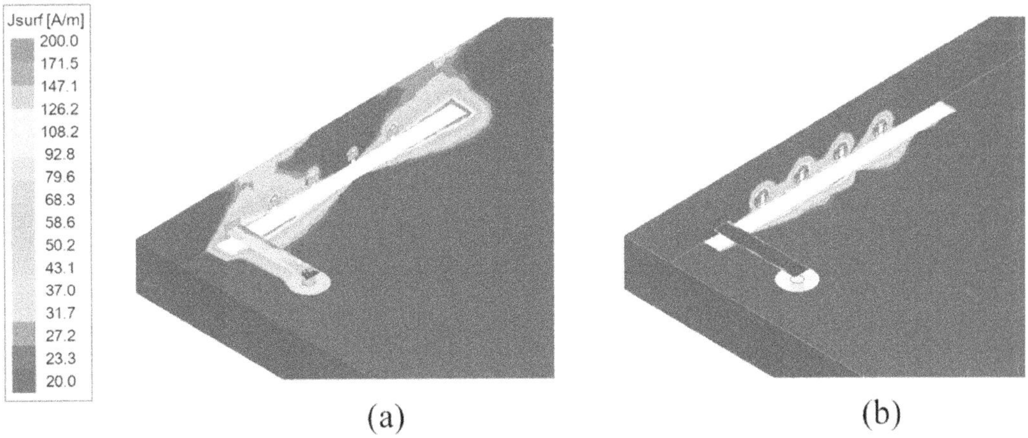

FIGURE 6.81 Current distribution in the 3.5 and 28 GHz bands under different port excitations (a) current distribution in the 3.5 GHz band when port P_1 is excited (b) current distribution in the 28 GHz band when ports S_1–S_4 are excited with equal amplitude and in-phase excitation.

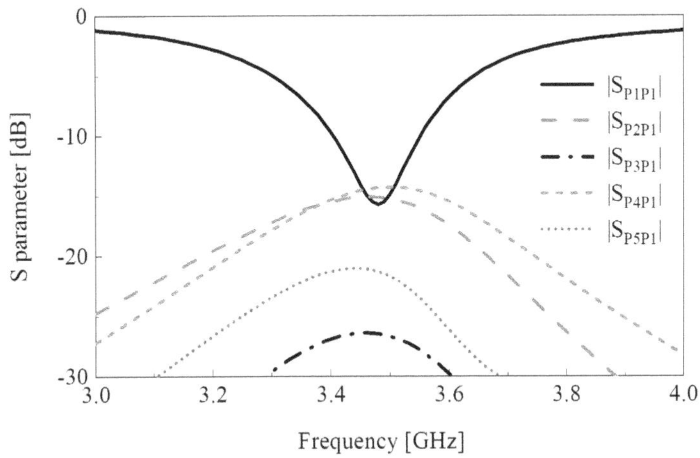

FIGURE 6.82 Simulated S-parameters of the MIMO antenna in the 3.5 GHz band. Note: The subscript P in the legend denotes the low-band antenna.

to extend the total length of the slot. In Figure 6.81b, when the four ports S_1–S_4 of the millimeter-wave antenna array are excited with equal amplitude and in-phase, the currents are concentrated on the millimeter-wave slot array, and the slot array operates in the one-quarter wavelength mode.

In the 3.5 GHz band, six long slots form a MIMO multi-antenna system. The simulated S-parameters of the MIMO antenna in the 3.5 GHz band are shown in Figure 6.82. The −6 dB reflection coefficient bandwidth covers 3.33–3.62 GHz. In each antenna module, the focus is on the port isolation between the long and short slots since they are directly connected. As shown in Figure 6.82, the mutual coupling between any two ports in the long and short slots is less than −16 dB. Therefore, the proposed design achieves MIMO performance in the 3.5 GHz band.

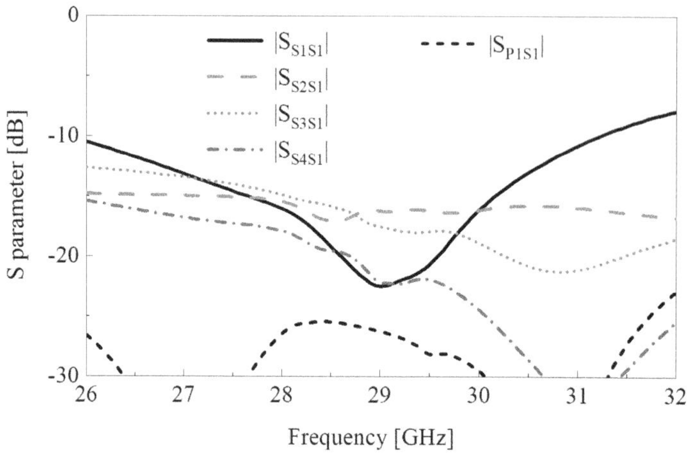

FIGURE 6.83 Simulated S-parameters in the millimeter-wave band. Note: The subscript *S* in the legend represents the high-band antenna.

The simulated S-parameters for the millimeter-wave band are shown in Figure 6.83. The impedance bandwidth is wide, covering 25.8–31.2 GHz. The mutual coupling between any two ports is less than −12 dB, and it can be reduced by increasing the slot cell spacing. In addition, the coupling between the millimeter-wave antenna array and the Sub-6 GHz antenna array is low. Therefore, this design can operate independently in the millimeter-wave band.

Four slot cells were used to realize the beam sweep. The slot cell spacing is 0.44 λ_0 (λ_0 is the free-space wavelength) in the 28 GHz band. Wide angle beam scanning can be realized by changing the feed phase gradient of the array. Figure 6.84a shows the three-dimensional radiation direction of the short-slot array at four typical phase gradients in the 28 GHz band. The main beam is located in the lower half space of the cell phone floor, and the peak gain of the array is about 10 dBi when all slot units are fed in the same phase. When the phase gradient is 150°, the scanning angle of the array can be up to 66°, and the change of gain is less than 3 dBi. Higher gain can also be realized by adding more slot units.

Spatial coverage is very important for the design of mobile terminal antennas. In order to obtain the maximum signal strength, the radiation pattern of the short slot array is synthesized. When the phase gradient is varied from −170° to +170° in steps of 10°, the array will have 35 beams in the entire space. The maximum electric field intensity in the field distribution is selected from the 35 beams at any point in space. The signal strength distribution in the 28 GHz band in the array synthesis mode is shown in Figure 6.84b. When the pitch angle θ is varied in the range of 90°–180°, the array radiates strongly because the main beam is located in the lower half space.

According to the synthesized array orientation diagram, the short-slot array can provide 50% spatial coverage efficiency with a gain greater than 3.8 dBi. Moreover, higher spatial coverage efficiency can be achieved by deploying multiple antenna modules.

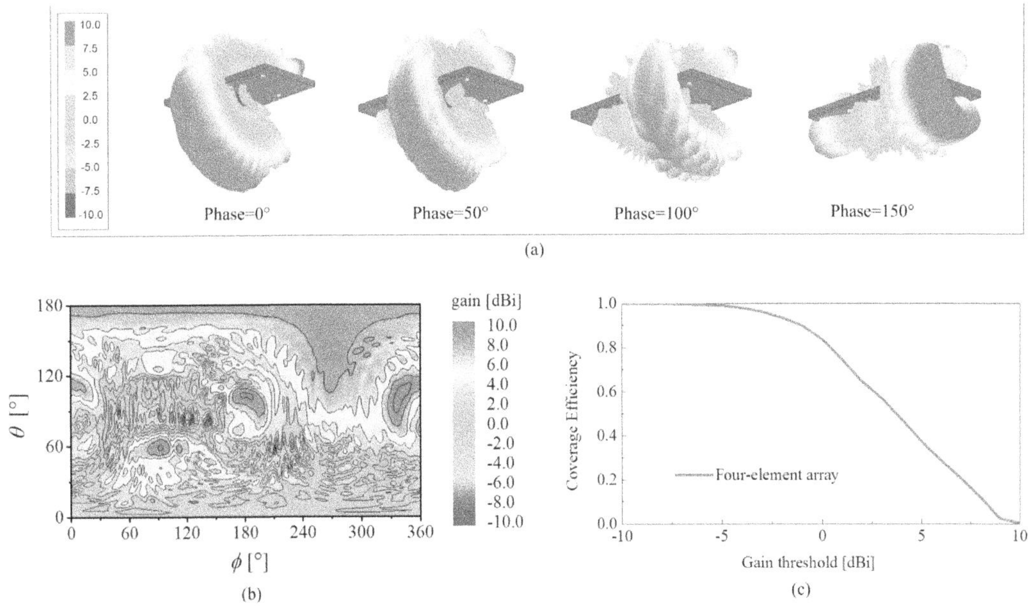

FIGURE 6.84 Beam scan of a short slot array in the 28 GHz band (a) 3D radiation phase map (b) signal intensity distribution in the 28 GHz band in array synthesis mode (c) spatial coverage efficiency with different gain thresholds.

REFERENCES

1. Liu D, Hong W, Rappaport TS, et al. What will 5G antennas and propagation be?. *IEEE Transactions on Antennas and Propagation*, 2017, 65(12): 6205–6212.
2. Hong W, Jiang ZH, Yu C, et al. The role of millimeter-wave technologies in 5G/6G wireless communications. *IEEE Journal of Microwaves*, 2021, 1(1): 101–122.
3. Hong W. Solving the 5G mobile antenna puzzle: assessing future directions for the 5G mobile antenna paradigm shift. *IEEE Microwave Magazine*, 2017, 18(7): 86–102.
4. Liu D, Gu X, Baks CW, et al. Antenna-in-package design considerations for Ka-band 5G communication applications. *IEEE Transactions on Antennas and Propagation*, 2017, 65(12): 6372–6379.
5. Syrytsin I, Zhang S, Pedersen GF, et al. User-shadowing suppression for 5G mm-wave mobile terminal antennas. *IEEE Transactions on Antennas and Propagation*, 2019, 67(6): 4162–4172.
6. Hong W, Baek K, Ko S. Millimeter-wave 5G antennas for smartphones: overview and experimental demonstration. *IEEE Transactions on Antennas and Propagation*, 2017, 65(12): 6250–6261.
7. Chou HT, Chou SJ, Deng JDS, et al. LTCC-based dual-polarized aip module by multilayered cross-dipole antennas for 5G mobile terminal applications at 28 GHz band. *IEEE Transactions on Components, Packaging and Manufacturing Technology*, 2023, 13(10): 1663–1672.
8. He Y, Lv S, Zhao L, et al. A compact dual-band and dual-polarized millimeter-wave beam scanning antenna array for 5G mobile terminals. *IEEE Access*, 2021, 9: 109042–109052.
9. Luo Y, Xu J, Chen Y, et al. A zero-mode induced mmWave patch antenna with low-profile, wide-bandwidth and large-angle scanning for 5G mobile terminals. *IEEE Access*, 2019, 7: 177607–177615.

10. Zhang S, Chen X, Syrytsin I, et al. A planar switchable 3D-coverage phased array antenna and its user effects for 28-GHz mobile terminal applications. *IEEE Transactions on Antennas and Propagation*, 2017, 65(12): 6413–6421.
11. Yoo JU, Son HW. 5G mm-wave patch antenna array with proximity sensing function for detecting user's hand grip on mobile terminals. *Electronics Letters*, 2023, 59(7): e12769.
12. Chen X, Abdullah M, Li Q, et al. Characterizations of mutual coupling effects on switch-based phased array antennas for 5G millimeter-wave mobile communications. *IEEE Access*, 2019, 7: 31376–31384.
13. Xia X, Yu C, Wu F, et al. Millimeter-wave phased array antenna integrated with the industry design in 5G/B5G smartphones. *IEEE Transactions on Antennas and Propagation*, 2023, 71(2): 1883–1888.
14. Xia X, Wu F, Yu C, et al. Millimeter-wave ±45° dual linearly polarized end-fire phased array antenna for 5G/B5G mobile terminals. *IEEE Transactions on Antennas and Propagation*, 2022, 70(11): 10391–10404.
15. Moreno RM, Ala-Laurinaho J, Khripkov A, et al. Dual-polarized mm-wave endfire antenna for mobile devices. *IEEE Transactions on Antennas and Propagation*, 2020, 68(8): 5924–5934.
16. Wu S, Zhao A, Ren Z. Dual-polarized ring-slot 5G millimeter-wave antenna and array based on metal frame for mobile phone applications. (2019-06-20) [2024-07-01].
17. Moreno RM, Kurvinen J, Ala-Laurinaho J, et al. Dual-polarized mm-wave end-fire chain-slot antenna for mobile devices. *IEEE Transactions on Antennas and Propagation*, 2020, 69(1): 25–34.
18. Li H, Cheng Y, Mei L, et al. Dual-polarized frame-integrated slot arrays for 5G mobile handsets. *IEEE Antennas and Wireless Propagation Letters*, 2020, 19(11): 1953–1957.
19. Hsu YW, Huang TC, Lin HS, et al. Dual-polarized quasi Yagi-Uda antennas with endfire radiation for millimeter-wave MIMO terminals. *IEEE Transactions on Antennas and Propagation*, 2017, 65(12): 6282–6289.
20. Li H, Li Y, Chang L, et al. A wideband dual-polarized endfire antenna array with overlapped apertures and small clearance for 5G millimeter-wave applications. *IEEE Transactions on Antennas and Propagation*, 2020, 69(2): 815–824.
21. Khajeim MF, Moradi G, Shirazi RS, et al. Broadband dual-polarized antenna array with endfire radiation for 5G mobile phone applications. *IEEE Antennas and Wireless Propagation Letters*, 2021, 20(12): 2427–2431.
22. Zhang J, Zhao K, Wang L, et al. Wideband low-profile dual-polarized phased array with endfire radiation patterns for 5G mobile applications. *IEEE Transactions on Vehicular Technology*, 2021, 70(9): 8431–8440.
23. Sun L, Li Y, Zhang Z. Wideband dual-polarized endfire antenna based on compact open-ended cavity for 5G mm-wave mobile phones. *IEEE Transactions on Antennas and Propagation*, 2021, 70(3): 1632–1642.
24. Parchin NO, Zhang J, Abd-Alhameed RA, et al. A planar dual-polarized phased array with broad bandwidth and quasi-endfire radiation for 5G mobile handsets. *IEEE Transactions on Antennas and Propagation*, 2021, 69(10): 6410–6419.
25. Taheri MMS, Abdipour A, Zhang S, et al. Integrated millimeter-wave wideband end-fire 5G beam steerable array and low-frequency 4G LTE antenna in mobile terminals. *IEEE Transactions on Vehicular Technology*, 2019, 68(4): 4042–4046.
26. Ding XH, Zhang QH, Yang WW, et al. A dual-band antenna for LTE/mmWave mobile terminal applications. *IEEE Transactions on Antennas and Propagation*, 2023, 71(3): 2826–2831.
27. Ikram M, Nguyen-Trong N, Abbosh A. Hybrid antenna using open-ended slot for integrated 4G/5G mobile application. *IEEE Antennas and Wireless Propagation Letters*, 2020, 19(4): 710–714.

28. Ikram M, Abbas E, Nguyen-Trong N, et al. Integrated frequency-reconfigurable slot antenna and connected slot antenna array for 4G and 5G mobile handsets. *IEEE Transactions on Antennas and Propagation*, 2019, 67(99): 7225–7233.
29. Wang Y, Xu F. Shared-aperture 4G LTE and 5G mm-wave antenna in mobile phones with enhanced mm-wave radiation in the display direction. *IEEE Transactions on Antennas and Propagation*, 2023, 71(6): 4772–4782.
30 Yu Z. *Millimeter-Wave Dual-Polarized End-Fire Phased Array Antenna Array for 5G Mobile Terminals*. Beijing: Beijing Institute of Technology, 2022.
31. Li A, Luk KM, Li Y. A dual linearly polarized end-fire antenna array for the 5G applications. *IEEE Access*, 2018, 6: 78276–78285.
32. Zhang J, Zhao K, Wang L, et al. Dual-polarized phased array with end-fire radiation for 5G handset applications. *IEEE Transactions on Antennas and Propagation*, 2020, 68(4): 3277–3282.
33. Lu R, Yu C, Zhu Y, et al. Compact millimeter-wave endfire dual-polarized antenna array for low-cost multibeam applications. *IEEE Antennas and Wireless Propagation Letters*, 2020, 19(12): 2526–2530.
34. Li A, Luk KM. Single-layer wideband end-fire dual-polarized antenna array for device-to-device communication in 5G wireless systems. *IEEE Transactions on Vehicular Technology*, 2020, 69(5): 5142–5150.
35. Li H, Cheng Y, Mei L, et al. Frame integrated wideband dual-polarized arrays for mm-wave/sub 6-GHz mobile handsets and its user effects. *IEEE Transactions on Vehicular Technology*, 2020, 69(12): 14330–14340.
36. Moreno RM, Kurvinen J, Ala-Laurinaho J, et al. Dual-polarized mm-wave end-fire chain-slot antenna for mobile devices. *IEEE Transactions on Antennas and Propagation*, 2021, 69(1): 25–34.
37. Zhu Y, Deng C. Millimeter-wave dual-polarized multibeam endfire antenna array with a small ground clearance. *IEEE Transactions on Antennas and Propagation*, 2022, 70(1): 756–761.
38. Zhu Y, Deng C. Wideband dual-polarized end-fire phased array antenna with small ground clearance for 5G mmwave mobile terminals. *IEEE Transactions on Antennas and Propagation*, 2023, 73(6): 5469–5474.
39. Zhu Y, Deng C. Single-layer dual-polarized end-fire phased array antenna for 5G mm-wave mobile terminals. *IEEE Antennas and Wireless Propagation Letters*, 2024, 23(6): 1939–1943.
40. Zhu Y, Zhao Z, Deng C. Millimeter-wave dual-polarized frame-integrated patch antenna array for 5G mobile handsets. *2020 IEEE 3rd International Conference on Electronic Information and Communication Technology (ICEICT)*. Piscatway, USA: IEEE, 2020: 613–615.
41. Zhu Y, Deng C. Millimeter-wave dual-polarized dielectric rod antenna array with planar slot excitation for 5G mobile terminals. *International Conference on Microwave and Millimeter Wave Technology (ICMMT)*. Piscatway, USA: IEEE, 2022: 1–3.
42. Deng C, Liu D, Yektakhah B, et al. Series-fed beam-steerable millimeter-wave antenna design with wide spatial coverage for 5G mobile terminals. *IEEE Transactions on Antennas and Propagation*, 2020, 68(5): 3366–3376.
43. Ren T, Deng C. Co-design of sub-6 GHz and mm-wave array antennas with metal frame for 5G mobile terminals. *Microwave and Optical Technology Letters*, 2022, 64(3): 589–595.

Glossary

5G移动终端	5G Mobile Terminals
多天线	Multi-antenna
第三代合作伙伴计划	3rd Generation Partnership Project
高级移动电话系统	Advanced Mobile Phone System
封装天线	Antenna-in-Package
人工磁导体	Artificial Magnetic Conductor
码分多址	Code-Division Multiple Access
缺陷地结构	Defected Ground Structure
介质谐振天线	Dielectric Resonator Antenna
有效全向辐射功率	Effective Isotropic Radiated Power
有效全向灵敏度	Effective Isotropic Sensitivity
包络相关系数	Envelope Correlation Coefficient
第一零点波束宽度	First Null Beam Width
频率带宽	Frequency Bandwidth
频分双工	Frequency-Division Duplex
频分双工长期演进技术	Frequency-Division Duplex Long Term Evolution
全球定位系统	Global Positioning System
全球移动通信系统	Global System for Mobile Communications
接地共面波导	Grounded Coplanar Waveguide
半功率波束宽度	Half-Power Beam Width
高密度聚乙烯	High Density Polyethylene
临时标准-95	Interim Standard 95
国际电信联盟	International Telecommunication Union
互联网协议	Internet Protocol
倒F天线	Inverted-F Antenna
长期演进技术	Long Term Evolution
低温共烧陶瓷	Low Temperature Co-fired Ceramic
多输入多输出	Multiple-Input Multiple-Output
近场通信	Near Field Communication
通过空中激活	Over The Air
完美电导体	Perfect Electric Conductor
完美磁导体	Perfect Magnetic Conductor
物理媒体连接	Physical Medium Attachment
平面倒F天线	Planar Inverted-F Antenna

印制电路板	Printed-Circuit Board
比吸收率	Specific Absorption Rate
基片集成波导	Substrate Integrated Waveguide
补充上行链路	Supplementary Uplink
时分同步码分多址	Time Division-Synchronous Code Division Multiple Access
时分双工	Time-Division Duplex
时分长期演进技术	Time-Division Long Term Evolution
全接入通信系统	Total Access Communication System
总全向灵敏度	Total Isotropic Sensitivity
总辐射功率	Total Radiated Power
电压驻波比	Voltage Standing Wave Ratio
宽带码分多址	Wideband CDMA
无线保真	Wireless Fidelity

For Product Safety Concerns and Information please contact our EU representative GPSR@taylorandfrancis.com
Taylor & Francis Verlag GmbH, Kaufingerstraße 24, 80331 München, Germany

www.ingramcontent.com/pod-product-compliance
Ingram Content Group UK Ltd.
Pitfield, Milton Keynes, MK11 3LW, UK
UKHW052221190926
475702UK00006B/1134

* 9 7 8 1 0 4 1 2 5 7 9 7 4 *